Encyclopaedia of Mathematical Sciences

Volume 8

Editor-in-Chief: R. V. Gamkrelidze

G. M. Khenkin A. G. Vitushkin (Eds.)

Several
Complex Variables II

Function Theory in Classical Domains
Complex Potential Theory

With 19 Figures

Springer-Verlag Berlin Heidelberg GmbH

Consulting Editors of the Series:
A. A. Agrachev, A. A. Gonchar, E. F. Mishchenko,
N. M. Ostianu, V. P. Sakharova, A. B. Zhishchenko

Title of the Russian edition:
Itogi nauki i tekhniki, Sovremennye problemy matematiki,
Fundamental'nye napravleniya, Vol. 8, Kompleksnyj analiz - mnogie peremennye 2
Publisher VINITI, Moscow 1985

Mathematics Subject Classification (1991):
32-02, 32A07, 32A27, 32A35, 32A40, 32F05

ISBN 978-3-540-18175-0

Library of Congress Cataloging-in-Publication Data
Kompleksnyĭ analiz–mnogie peremennye 2. English
Several complex variables II: function theory in classical domains: complex potential theory /
G. M. Khenkin, A. G. Vitushkin (eds.)
p. cm. – (Encyclopaedia of mathematical sciences; v. 8)
Includes bibliographical references and indexes.
ISBN 978-3-540-18175-0   ISBN 978-3-642-57882-3 (eBook)
DOI 10.1007/978-3-642-57882-3
1. Functions of several complex variables. I. Khenkin, G. M. II. Vitushkin, A. G. (Anatoliĭ Georgievich)
III. Title. IV. Title: Several complex variables 2. V. Series.
QA331.K7382513   1994   515'.94–dc20   92-45735

Typesetting: Asco Trade Typesetting Ltd., Hong Kong
41/3140 - 5 4 3 2 1 0 - Printed on acid-free paper

List of Editors, Authors and Translators

Editor-in-Chief

R.V. Gamkrelidze, Russian Academy of Sciences, Steklov Mathematical Institute, ul. Vavilova 42, 117966 Moscow, Institute for Scientific Information (VINITI), ul. Usievicha 20a, 125219 Moscow, Russia

Consulting Editors

G. M. Khenkin, Central Economic and Mathematical Institute of the Russian Academy of Sciences, ul. Krasikova 32, 117418 Moscow, Russia
A. G. Vitushkin, Steklov Mathematical Institute, ul. Vavilova 42, 117966 Moscow, Russia

Authors

L. A. Aizenberg, Akademgorodok, Institute of Physics, 660036 Krasnoyarsk 36, Russia
A. B. Aleksandrov, Petrodvorets, St. Petersburg State University, 198904 St. Petersburg, Russia
A. Sadullaev, Vuzgorodok, Tashkent State University, 700095 Tashkent, Usbekistan
A. G. Sergeev, Steklov Mathematical Institute, ul. Vavilova 42, 117966 Moscow, Russia
A. K. Tsikh, Akademgorodok, Institute of Physics, 660036 Krasnoyarsk 36, Russia
V. S. Vladimirov, Steklov Mathematical Institute, ul. Vavilova 42, 117966 Moscow, Russia
A. P. Yuzhakov, Akademgorodok, Institute of Physics, 660036 Krasnoyarsk 36, Russia

Translators

P. M. Gauthier, Département de Mathématiques et de Statistique, Université de Montréal, CP 6128-A, Montréal QC H3C 3J7, Canada
J. R. King, Department of Mathematics, GN-50, Seattle, WA 98195, USA

Contents

I. Multidimensional Residues and Applications

L.A. Aizenberg, A.K. Tsikh, A.P. Yuzhakov

Translated from the Russian
by J.R. King

Contents

Chapter 1
Methods for Computing Multidimensional Residues

A.P. Yuzhakov

Introduction

One of the problems in the theory of multidimensional residues is the problem of studying and computing integrals of the form

$$\int_\gamma \omega, \tag{1}$$

where ω is a closed differential form of degree p on a complex analytic manifold X with a singularity on an analytic set $S \subset X$, and where γ is a compact p-dimensional cycle in $X \backslash S$. A special case of this problem is computing the integral (1) when ω is a holomorphic (meromorphic) form of degree $p = n = \dim_\mathbb{C} X$; in local coordinates the form can be written as $\omega = f(z)\,dz = f(z_1, \ldots, z_n)\,dz_1 \wedge \cdots \wedge dz_n$, where f is a holomorphic (meromorphic) function. According to the Stokes formula, the integral (1) depends only on the homology class[1] $[\gamma] \in H_p(X \backslash S)$ and the De Rham cohomology class $[\omega] \in H^p(X \backslash S)$. Thus in integral (1) the cycle γ can be replaced by a cycle γ_1 homologous to it ($\gamma_1 \sim \gamma$) in $X \backslash S$ and the form ω can be replaced by a cohomologous form $\omega_1 (\omega_1 \sim \omega)$ which may perhaps be simpler; for example, it could have poles of first order on S (see § 1, Subsection 4). If $\{\gamma_j\}$ is a basis for the p-dimensional homology of the manifold $X \backslash S$, then by Stokes formula for any compact cycle $\gamma \in Z_p(X \backslash S)$ the integral (1) is equal to

$$\int_\gamma \omega = \sum_j k_j \int_{\gamma_j} \omega, \tag{2}$$

where the k_j are the coefficients of the cycle γ as a combination of the basis elements $\{\gamma_j\}$, $\gamma \sim \sum_j k_j \gamma_j$. Formula (2) shows that the problem of computing integral (1) can be reduced to
  1) studying the homology group $H_p(X \backslash S)$ (finding its dimension and a basis);
  2) determining the coefficients of the cycle γ with respect to a basis;
  3) computing the integrals over the cycles in the basis.
Solving problems 1) and 2) is a difficult topological problem in the multidimensional case and requires the machinery of algebraic topology. In some

---

[1] In this chapter we will denote by H_p the group of compact singular homology; this group was denoted by H_p^c in the contribution of Dolbeault (Dolbeault, 1985) in Volume 7 of the Encyclopaedia of Mathematical Sciences.

cases, to solve this it helps to apply the dualities of Alexander-Pontryagin and De Rham (§ 2). Simple and multiple Leray coboundaries (Subsection 1.1) give a construction of standard cycles in $X \setminus S$. The general structure of the homology group $H_p(X \setminus S)$ is described in "good cases" by the decomposition theorem of Froissard (Subsection 1.3). Integrals on coboundary cycles can be reduced to integrals of lower degree by the simple and multiple Leray residue formulas (Subsection 1.2). The computation of an important class of residues, the Grothendieck residues, and a special case of them, the logarithmic residue, is considered in Section 2 and Section 3 of this article; § 3 is devoted to the application of residues to the study of integrals depending on parameters and to combinatorial analysis.

§ 1. Leray Theory. Froissart Decomposition Theorem

Here we will pause to study in more detail the computational side of the Leray theory of residues expounded in (Dolbeault (1985)). To start with, we consider the case of codimension 1.

1.1. Leray Coboundary. We give a constructive description of the coboundary homomorphism δ which was introduced in (Dolbeault (1985), Sect. 0.3). In the one-dimensional case the simplest cycle (contour) of integration is a circle of sufficiently small radius around an isolated singular point. Leray (1959) constructed the analog of this for complex analytic manifolds, the coboundary homomorphism δ. The construction of δ^{-1} was first considered by Poincaré (1887).

Let X be a complex analytic manifold of complex dimension n. Let S be a complex-analytic submanifold of X of codimension 1. We consider a tubular neighborhood V of the submanifold S, which is a locally-trivial fiber bundle with base S and fiber V_a, $a \in S$, homeomorphic to the disk. In order to construct such a fiber bundle we choose a Riemannian metric on X and take as V_a the union of geodesic segments of length $\rho(a)$, beginning at a and orthogonal to S, where $\rho(a)$ is sufficiently small. We assume that the function $\rho(a)$ is smooth; this implies the smoothness of ∂V. To each $(p - 1)$-dimensional element of a chain (a simplex, a rectangle) σ_{p-1} in S we associate a p-dimensional chain in $X \setminus S$. The chain is $\delta \sigma_{p-1} = \bigcup_{a \in |\sigma_{p-1}|} \delta a$ where $\delta a = \partial V_a$; it is homeomorphic to $\partial V_a \times \sigma_{p-1}$ with the natural orientation. Thus a homomorphism of homology groups is defined,

$$\delta : H_{p-1}(S) \to H_p(X \setminus S),$$

since $\partial \delta = - \delta \partial$. Then the *Leray homology exact sequence* is defined:

$$\cdots \to H_{p+1}(X) \xrightarrow{\bar{\omega}} H_{p-1}(S) \xrightarrow{\delta} H_p(X \setminus S) \xrightarrow{i} H_p(X) \to \cdots \qquad (3)$$

where i is the homomorphism induced by the inclusion $X \setminus S \subset X$ and $\bar{\omega}$ in induced by the intersection of chains in X, transversal to S, with the submanifold S.

If a family of $S_1, \ldots, S_m$ of submanifolds of codimension 1 is in general position, the multiple Leray coboundary is defined:

$$\delta^m : H_{p-m}(S_1 \cap \cdots \cap S_m) \to H_p(X \backslash (S_1 \cup \cdots \cup S_m)),$$

which is anticommutative with respect to the order of $S_1, \ldots, S_m$ (for the cohomological multiple coboundary, see (Dolbeault (1985), Sect. 03).

1.2. Form-Residue, Class-Residue, Leray Residue Formula. As was pointed out in (Dolbeault (1985), Sect. 03), if ϕ is a closed regular differential form in $X \backslash S$ with a pole of first order on S, then in some neighborhood of any point $a \in S$ the form ϕ can be represented as

$$\phi = \frac{ds}{s} \wedge \psi + \theta, \tag{4}$$

where $s = s_a(z)$ is the defining function of the manifold S in U_a and ψ, θ are forms which are regular on U_a. Here the form $\psi|_S$ is globally defined, is closed, and is uniquely determined by the form ϕ. This restriction $\psi|_S$ is called the *form-residue* of the form ϕ and is denoted by res[ϕ]. We remark that if ϕ is holomorphic on $X \backslash S$, then the form residue res[ϕ] is holomorphic on S.

Example 1. Let $X = \mathbb{C}^n$, with $S = \{z \in \mathbb{C}^n : s(z) = 0\}$ and $\phi = f(z) \, dz_1 \wedge \cdots \wedge dz_n / s(z)$. Since $\phi = (-1)^{j-1} ds \wedge dz_{[j]} / s \cdot s'_{z_j}$, where $dz_{[j]} = dz_1 \wedge \cdots \wedge dz_{j-1} \wedge dz_{j+1} \wedge \cdots \wedge dz_n$, then res[$\phi$] $= (-1)^{j-1} f(z) \, dz_{[j]} / s'_{z_j}|_S$ at the points where $s'_{z_j} \neq 0$.

Remark. The map

$$f(z) \, dz / s(z) \to (-1)^{j-1} f(z) \, dz_{[j]} / s'_{z_j}|_S$$

is called the Poincaré residue map and is denoted by P.R. If we denote by Ω^n_X, $\Omega^n_X(S)$, Ω^{n-1}_S, the sheaves of germs, respectively, of holomorphic n-forms on X, meromorphic n-forms having only simple poles on S, and holomorphic $(n-1)$-forms on S, then there is an exact sequence of sheaves

$$0 \to \Omega^n_X \to \Omega^n_X(S) \to \Omega^{n-1}_S \to 0$$

which defines an exact sequence on cohomology

$$H^0(X, \Omega^n_X(S)) \overset{\text{P.R.}}{\to} H^0(S, \Omega^{n-1}_S) \overset{\delta}{\to} H^1(X, \Omega^n_X).$$

Therefore, the Poincaré residue map is surjective on global sections if $H^1(X, \Omega^n_X) = 0$. In particular, this is true for projective space $X = \mathbb{CP}^n$, $n > 1$. Thus for $n > 1$ every holomorphic form of degree $n - 1$ on the submanifold S is the Poincaré residue of a meromorphic n-form on $\mathbb{CP}^n$.

By the theorem of (Dolbeault (1985), Sect. 0.3), for every closed regular differentiable form ϕ on $X \backslash S$, there is a form $\tilde\phi$ cohomologous to it which has a pole of first order on S. In this case the cohomology class of the form res[$\tilde\phi$] depends only on the cohomology class of the form ϕ.

The cohomology class of the form res$[\tilde{\phi}]$ is called the *class residue* of the form ϕ and is denoted by Res$[\phi]$. Since the operator res is linear, Res : $H^p(X\backslash S) \to H^{p-1}(S)$ is a homomorphism.

We observe that the form residue and the class residue also exist in the case when S is an analytic subset. However in this case the form residue has singularities on the set S^* of singular points of S. If the singular set S^* of the set S is resolved to a divisor with normal crossings, then the form residue only has simple poles on the resolution of S^* (Gordan, 1974).

The abstract residue formula (0.3.2) of (Dolbeault (1985)) is written thus:

Theorem 1.1 (Leray residue formula). *For an arbitrary closed form ϕ of degree p on $X\backslash S$ and a cycle $\sigma \in Z_{p-1}(S)$ there is a formula*

$$\int_{\delta\sigma} \phi = 2\pi i \int_{\sigma} \text{Res}[\phi]. \tag{5}$$

If the form $\phi \in Z^p(X\backslash(S_1 \cup \cdots \cup S_m))$ has a pole of first order on $S_1, \ldots, S_m$, then by applying formula (4) one can define iterated form residues res$^m[\phi] \in Z^{p-m}(S_1 \cap \cdots \cap S_m)$ and a homomorphism

$$\text{Res}^m : H^p(X\backslash(S_1 \cup \cdots \cup S_m)) \to H^{p-m}(S_1 \cap \cdots \cap S_m),$$

as the composition of homomorphisms

$$H^p(X\backslash(S_1 \cup \cdots \cup S_m)) \overset{\text{Res}}{\to} H^{p-1}(S_1\backslash(S_2 \cup \cdots \cup S_m)) \to \cdots$$
$$\to H^{p-m+1}(S_1 \cap \cdots \cap S_{m-1}\backslash S_m) \overset{\text{Res}}{\to} H^{p-m}(S_1 \cap \cdots \cap S_m).$$

Iterating Theorem 1.1 we obtain the compound Leray residue.

Theorem 1.2 (Leray (1959)). *For an arbitrary form $\phi \in Z^p(X\backslash(S_1 \cup \cdots \cup S_m))$ and any cycle $\sigma \in Z_{p-m}(S_1 \cap \cdots \cap S_m)$, a compound Leray formula holds:*

$$\int_{\delta^m\sigma} \phi = (2\pi i)^m \int_{\sigma} \text{Res}^m[\phi]. \tag{6}$$

The Leray formulas (5) and (6) allow one to lower the degree of the multiple integral (1) when the cycle of integration belongs to a coboundary class: $\gamma \in \delta H_{p-1}(S)$, $\gamma \in \delta^m H_{p-m}(S_1 \cap \cdots \cap S_m)$. Since, for a form ϕ having a first-order pole, the form-residue res$[\phi]$ (res$^m[\phi]$) is found constructively, the problem arises of how to lower the order of poles of *semimeromorphic forms* $\phi \in Z^p(X\backslash S)$ ($\phi \in Z^p(X\backslash(S_1 \cap \cdots \cap S_m))$); in other words, how can one find a form ϕ_1 cohomologous to ϕ which has a first-order pole.

1.3. Tests for Leray Coboundaries. Froissart Decomposition Theorem. From the Leray exact sequence (3) it follows that a cycle $\gamma \in Z_p(X\backslash S)$ is a coboundary ($\gamma \sim \delta\sigma$ for some $\sigma \in Z_{p-1}(S)$) if and only if $\gamma \sim 0$ in X. If $H_p(X) = 0$, then $H_p(X\backslash S) = \delta H_{p-1}(S)$, i.e., every cycle in $X\backslash S$ is a coboundary. If $H_{p+1}(X) = 0$, then δ is a monomorphism.

Proposition 1.3 (Griffiths (1969)). *If S is an algebraic manifold in complex projective space $\mathbb{C}\mathbb{P}^n$, then $\delta : H_{n-1}(S) \to H_n(\mathbb{C}\mathbb{P}^n \backslash S)$ is always surjective and is injective for even n.*

Proposition 1.4. *For a cycle $\gamma \in Z_p(X \backslash (S_1 \cup \cdots \cup S_m))$ to be a compound Leray coboundary ($[\gamma] \in \delta^m H_{p-m}(S_1 \cap \cdots \cap S_m)$), it is necessary (and sufficient in the case when X is a Stein manifold) that*

$$\gamma \sim 0 \text{ in } X \backslash (S_1 \cup \cdots \cup S_{j-1} \cup S_{j+1} \cup \cdots \cup S_m), j = 1, \ldots, m.$$

Under some assumptions about X and $S_1, \ldots, S_m$, the structure of the homology group $H_p(X \backslash (S_1 \cup \cdots \cup S_m))$ is described by the following theorem.

Theorem 1.5 (Froissart decomposition (Fotiadi et al. (1965)). *Let $S_0, S_1, \ldots, S_m$ and $\Sigma_1, \ldots, \Sigma_k$ be two families of submanifolds of codimension 1 in the complex projective space $\mathbb{C}\mathbb{P}^n$ such that the families are in general position. $S_0 = \mathbb{C}\mathbb{P}_\infty^{n-1} = \mathbb{C}\mathbb{P}^n \backslash \mathbb{C}^n$ is the hyperplane at infinity. Let $Y = \Sigma_1 \cap \cdots \cap \Sigma_k$. $X = \mathbb{C}^n \cap Y = Y \backslash S_0$. Then the cohomology group*

$$H_p\left(X \backslash \bigcup_{j=1}^m S_j \right) \simeq \bigoplus_{h \subset \{1, \ldots, m\}} \delta^{|h|} H_{p-|h|} \left(\bigcap_{j \in h} (S_j \cap X) \right)$$

$$= H_p(X) \oplus \sum_{j=1}^m \delta H_{p-1}(S_j \cap X) \oplus \sum_{1 \le j \le q \le m} \delta H_{p-2}(X \cap S_j \cap S_q)$$

$$\oplus \cdots$$

where $|h|$ is the number of elements in the set h.

Example 2. Let $X = \mathbb{C}^n$, and let $S_j = \{z \in \mathbb{C}^n : L_j(z) = \sum_{k=1}^n a_{jk} z_k + b_j = 0\}$, $j = 1, \ldots, m$ be analytic hyperplanes in general position (if $S_{j_1} \cap \cdots \cap S_{j_k} \ne \varnothing$, then $L_{j_1}, \ldots, L_{j_k}$ are linearly independent). Since $H_n(\mathbb{C}^n) = 0$, $H_{n-r}(S_{j_1} \cap \cdots \cap S_{j_r}) = 0$ for $r < n$, and $H_0(S_{j_1} \cap \cdots \cap S_{j_n}) \simeq \mathbb{Z}$ for $S_{j_1} \cap \cdots \cap S_{j_n} \ne \varnothing$, then there is a basis of the group $H_n(\mathbb{C}^n \backslash (S_1 \cup \cdots \cup S_m))$ consisting of cycles of the form $\delta^n(S_{j_1} \cap \cdots \cap S_{j_n}) = \{z : |L_\nu(z)| = \varepsilon, \nu = j_1, \ldots, j_n\}$. The residue of the form $\phi = h \, dz / L_1^{r_1} \ldots L_m^{r_m}$, where $h \in A(\mathbb{C}^m)$, relative to a basis cycle $\delta^n(S_{j_1} \cap \cdots \cap S_{j_n})$ is the Grothendieck residue (see Chapter 3) at the point $S_{j_1} \cap \cdots \cap S_{j_n}$. After a linear change of variables, it is computed as the derivative of a multiple Cauchy integral.

1.4. Cohomological Lowering of Pole Order. In Subsection 1.2 only the existence of the class residue of a form $\phi \in Z^p(X \backslash S)$ was discussed, but no algorithm was demonstrated for computing it, that is, for finding a form $\phi_1 \sim \phi$ having a pole of first order. The same is true for the compound class residue of a form $\phi \in Z^p(X \backslash (S_1 \cup \cdots \cup S_m))$. In some cases the problem of the cohomological reduction of a semimeromorphic form (see (Dolbeault (1985)) Subsection 3.5) to a form having a pole of order 1 can be solved constructively. For example,

Proposition 1.6 (Pham (1967)). *Let $S = \{z : s(z) = 0\}$, where s is a holomorphic function in a neighborhood $V(S)$ of the manifold S, with grad $s|_S \ne 0$, then any*

form $\phi \in Z^p(X \setminus S)$ *having a pole of order k on S can be represented in the form*

$$\phi = \frac{ds}{s^k} \wedge \psi + \frac{\theta}{s^{k-1}} = d\left(\frac{\psi}{(k-1)s^{k-1}}\right) + \frac{\theta_1}{s^{k-1}} \sim \frac{\theta_1}{s^{k-1}}, \tag{7}$$

where θ *and* ψ *are regular forms in* $V(S)$ *and* $\theta_1 = d\psi/(k-1) + \theta$.

Theorem 1.7 (Leray (1959)). *Let the submanifolds* S_j *in a neighborhood* V *of the set* $S_1 \cap \cdots \cap S_m$ *be defined by the equations* $s_j(z) = 0$, *where the* s_j *are functions holomorphic in* $V, j = 1, \ldots, m$. *If the form* $\phi \in Z^p(X \setminus (S_1 \cup \cdots \cup S_m))$ *is represented in* V *as*

$$\phi = ds_1 \wedge \cdots \wedge ds_m \wedge \omega/s_1^{r_1+1} \ldots s_m^{r_m+1},$$

then

$$\mathrm{Res}^m[\phi] \ni \frac{1}{r_1! \ldots r_m!} \frac{\partial^{r_1 + \cdots + r_m}\omega}{\partial s_1^{r_1} \ldots \partial s_m^{r_m}}\bigg|_{S_1 \cap \cdots \cap S_m},$$

where the $\partial^{r_1 + \cdots + r_m}/\partial s_1^{r_1} \ldots \partial s_m^{r_m}$ *are found recursively from the equation*

$$d\omega = ds_1 \wedge \omega_1 + \cdots + ds_m \wedge \omega_m, \quad \omega_j \stackrel{\mathrm{def}}{=} \partial\omega/\partial s_j, \tag{8}$$

which is a consequence of the condition $d\omega \wedge ds_1 \wedge \cdots \wedge ds_m = 0$ $(d\phi = 0)$.

We observe that in order to find the forms ψ and θ in (7) and ω_j in (8) it is necessary to employ a partition of unity. The apparatus of partial derivatives for exterior differntial forms was developed by Norguet (1959).

Theorem 1.8 (Leinartas, Yuzhakov (cf. Aizenberg-Yuzhakov (1979))). *Let* Q_1, $\ldots, Q_m$ *be irreducible polynomials in* $\mathbb{C}^n$ *and let the* $S_j = \{z \in \mathbb{C}^n : Q_j(z) = 0\}, j = 1, \ldots, m$ *be manifolds in general position. Then the form*

$$\omega = P \, dz_1 \wedge \cdots \wedge dz_n/Q_1^{r_1} \ldots Q_m^{r_m}$$

is cohomologous in $\mathbb{C}^n \setminus (S_1 \cup \cdots \cup S_m)$ *to a form of type*

$$\omega^* = \sum_J P_J \, dz_1 \wedge \cdots \wedge dz_n/Q_{j_1} \ldots Q_{j_k},$$

where $J = \{j_1, \ldots, j_k\}, k \leq n$, *and the* P_J *are polynomials.*

Remark. The form ω^* is found constructively using elimination theory (the Hilbert Nullstellensatz).

Theorem 1.9 (Griffiths (1969)). *Let* Q *be an irreducible homogeneous polynomial and let* $S = \{\zeta \in \mathbb{CP}^n : Q(\zeta) = 0\}$ *be a manifold. Then any closed rational n-form* ω *in* $\mathbb{CP}^n$ *with poles on* S,

$$\omega = P(\zeta)\Omega(\zeta)/Q^m(\zeta), \tag{9}$$

where $\Omega(\zeta) = \sum_{j=1}^n (-1)^j \zeta_j \, d\zeta_0 \wedge \cdots [j] \cdots \wedge d\zeta_n$, *can be replaced by another form of type* (9) *which is cohomologous to it but which has* $m \leq n - E(n/q)$, *where* $q = \deg Q$ *(the symbol* $[j]$ *signifies that the term* $d\zeta_j$ *is omitted).*

We remark that in $\mathbb{CP}^n$ it is not always possible to lower the order of the pole to one. This follows from the fact that in $H^n(\mathbb{CP}^n \backslash S)$ there can exist classes which contain no rational forms with first order poles. For example, let

$$S = \{\zeta \in \mathbb{CP}^2 : \zeta_0^3 + \zeta_1^3 + \zeta_2^3 = 0\}.$$

Since S is a curve of genus 1, then

$$\dim H^2(\mathbb{CP}^2 \backslash S) = \dim H_2(\mathbb{CP}^2 \backslash S) = 2.$$

Since any rational form ω with a pole of first order on S can be written as $const \cdot \Omega/(\zeta_0^3 + \zeta_1^3 + \zeta_2^3)$, the rational forms with first order poles cannot generate the entire group $H_2(\mathbb{CP}^2 \backslash S)$.

In the general case, if S is an algebraic submanifold of $\mathbb{CP}^n$, with $q(z) = q(z_1, \ldots, z_n) = 0$ being its equation in affine coordinates, then any differential form of degree n in $\mathbb{CP}^n \backslash S$ with a pole of order one on S has the form $\omega = p(z) \, dz/q(z)$ in the coordinates $z = (z_1, \ldots, z_n)$, where $\deg p \leq \deg q - n - 1$. The latter condition means that ω does not have a pole on the hyperplane at infinity. The class residues of such forms are represented by Poincaré residues: P.R.$[\omega] = [(p \, dz_{[j]}/(\partial q/\partial z_j)|_S]$. According to the remark in 1.2, this coincides with the cohomology of holomorphic forms of degree $n - 1$ on S. In general, on a compact manifold S this part of the cohomology comprises a proper subspace of $H^{n-1}(S)$.

Let $A_k^p(S)$ be the set of rational p-forms with poles of order k on a submanifold $S \subset \mathbb{CP}^n$, $\mathcal{H}_k = A_k^n(S)/dA_{k-1}^{n-1}(S)$. Then the image of the mapping P.R.: $\mathcal{H}_1 \to H^{n-1}(S)$ lies in what is called the primitive subgroup $H_{\text{prim}}^{n-1}(S)$ (see Griffiths (1969)). Moreover there exists a natural map $R_k: \mathcal{H}_k \to F^{(n-1,k-1)}(S)$, mapping $\mathcal{H}_k$ to the Hodge filtration

$$F^{(n-1,0)}(S) \subset F^{(n-1,1)}(S) \subset \cdots \subset F^{(n-1,n-1)}(S) = H^{n-1}(S)$$

of the group $H^{n-1}(S)$, where $R_1 = $ P.R. and the image $R_k(\mathcal{H}_k)$ is the primitive subgroup $F_{\text{prim}}^{(n-k,k-1)}(S)$.

1.5. Generalization of the Leray Theory to the Case of Submanifolds of Codimension $q > 1$ (Norguet (1971)).

Let X be a complex analytic manifold, $\dim_{\mathbb{C}} X = n$, and let S be a complex submanifold of codimension q (in some neighborhood U_a of any point $a \in S$ the set $S \cap U_a = \{z \in U_a : s_1(z) = \cdots = s_q(z) = 0\}$, where the s_j are holomorphic functions in U_a and the vectors grad s_j, $j = 1, \ldots, q$ are linearly independent). There is an exact homology sequence analogous to (3), where the homomorphism δ is induced from a fiber bundle with base S and fiber δ_a homeomorphic to the $(2q - 1)$-dimensional sphere. A differential form ϕ, regular in $X \backslash S$, is called a *simple form* (cf. [Dolbeault (1985), 4.1]) if there exists a form ψ, regular on X, such that for any point $a \in S$, in some neighborhood U_a of the point, the form ϕ can be represented as

$$\phi|_{U_a} = K_a \wedge \psi|_{U_a} + \theta_a,$$

where θ_a is a form regular in U_a and

$$K_a = \left(\sum_{j=1}^{q} s_j \bar{s}_j \right)^{-q} \sum_{v=1}^{q} (-1)^{v-1} \bar{s}_v \, d\bar{s}_1 \wedge \cdots [v] \cdots \wedge d\bar{s}_q \wedge ds_1 \wedge \cdots \wedge ds_q.$$

Theorem 1.10 (Norguet). *Every form $\phi \in Z^p(X \backslash S)$ is cohomologous to a simple form $\phi_1 \in Z^p(X \backslash S)$. Thus for any cycle $\sigma \in Z_{p-2q+1}(S)$, the residue formula holds*:

$$\int_{\delta\sigma} \phi = \frac{(2\pi i)^q}{(q-1)!} \int_\sigma \phi|_S.$$

§2. Application of Alexander-Pontryagin Duality and De Rham Duality

2.1. Application of Alexander-Pontryagin Duality.[2] To find the dimension and a basis for the group $H_p(X \backslash S)$, it is sometimes useful to apply the topological Alexander-Pontryagin duality theory which establishes an isomorphism of homology groups:

$$H_p(S^n \backslash T) \simeq H_r(T)$$

where S^n is a manifold homeomorphic to the n-dimensional sphere and T is a compact subset of S^n, where $p + r = n - 1$. There exist dual bases $\{\gamma_j\}$, $\{\sigma_j\}$ such that $\mathfrak{o}(\sigma_j, \gamma_k) = \delta_{jk}$, where $\mathfrak{o}(\sigma_j, \gamma_k)$ is the linking coefficient of the cycles σ_j and γ_k.

Let the form ω be regular in the domain $D = \mathbb{C}^n \backslash T$ (T is the singular set of the form ω). We compactify the space $\mathbb{C}^n$ to form the sphere $S^{2n} = \mathring{\mathbb{C}}^n = \mathbb{C}^n \cup \{\infty\}$ by attaching a single point $\{\infty\}$ at infinity. Since T is closed in $\mathbb{C}^n$, $\mathring{T} = T \cup \{\infty\}$ is compact, and $D = \mathring{\mathbb{C}}^n \backslash \mathring{T}$. Then by Alexander-Pontryagin duality, $H_p(D) \simeq H_{2n-p-1}(\mathring{T})$. Thus to find the dimension and a basis $\{\gamma_j\}$ of the group $H_p(D)$, it is necessary and sufficient to find the dimension q and a basis σ_j of the $(2n - p - 1)$-dimensional homology group of the singular set $\mathring{T}$. Then the coefficients k_j of an expansion of an arbitrary cycle $\gamma \in Z_p(D)$ with respect to the basis $\{\gamma_j\}_{j=1}^q$, the basis dual to $\{\sigma_j\}_{j=1}^q$, are the linking coefficients of the cycle γ with the cycles of the dual basis:

$$\mathfrak{o}(\sigma_j, \gamma) = \mathfrak{o}\left(\sigma_j, \sum_{v=1}^{q} k_v \gamma_v \right) = \sum_{v=1}^{q} k_v \delta_{vj} = k_j.$$

To find the integrals with repect to the basis cycles, $\int_{\gamma_j} \omega$, it is sufficient to take q homologically independent p-cycles in D, $\Gamma_1, \ldots, \Gamma_q$. Then the integrals over

---

[2] The first to apply Alexander-Pontryagin duality to complex analysis was Martinelli (1953), who used it to deduce a generalization of the Cauchy integral formula to $\mathbb{C}^n$ for multiple integrals of degree $n + l, 0 \leq l \leq n - 1$.

the basis cycles are found from this system of linear equations,

$$\int_{\Gamma_j} \omega = \sum_{v=1}^{q} k_{jv} \int_{\gamma_v} \omega, \quad j = 1, \ldots, q,$$

where $k_{jv} = \mathfrak{o}(\sigma_v, \Gamma_j)$. We observe that $\det \|k_{jv}\| \neq 0$ is the condition of homo-logical independence of the cycles $\Gamma_1, \ldots, \Gamma_q$.

Let $p = n$ and let $\omega = f(z) \, dz$ be a holomorphic form in D. From the preceding this then follows:

Theorem 2.1 (On residues). *Let the function f be holomorphic in the domain $D \subset \mathbb{C}^n$, let $T = \mathbb{C}^n \backslash D$, and let $\mathring{T} = T \cup \{\infty\}$ be a subpolyhedron in the spherical compactification of $\mathbb{C}^n$, $\mathring{\mathbb{C}}^n \cup \{\infty\}$. If $\{\sigma_j\}_{j=1}^q$ is a basis for the $(n-1)$-dimensional homology of the singular set $\mathring{T}$, and if $\{\gamma_j\}_{j=1}^q$ is the basis dual to this in the n-dimensional homology of D, then for any cycle $\gamma \in Z_n(D)$ this equation holds:*

$$\int_{\gamma} f(z) \, dz = (2\pi i)^n \sum_{j=1}^{q} k_j R_j,$$

where $k_j = \mathfrak{o}(\sigma_j, \gamma)$ and $R_j = (2\pi i)^{-n} \int_{\gamma_j} f(z) \, dz$.

The application of Alexander-Pontryagin duality is especially effective in the case when $n = 2$ and T is an analytic set. In this case the study of the two-dimensional homology group of a domain in real four-dimensional space reduces to the study of the one-dimensional homology group of a surface (a complex curve). As an example we consider the residue of a rational function of two variables.

2.2. Residues of Rational Functions of Two Variables. We apply the approach described above to integrals of the form

$$\int\int_{\gamma} \frac{P(w, z)}{Q(w, z)} \, dw \wedge dz, \tag{10}$$

where P and Q are polynomials (or we may assume that P is an entire function), the cycle $\gamma \in Z_2(\mathbb{C}^2 \backslash T)$ and $T = \{Q(w, z) = 0\}$. Let $Q = Q_1^{r_1} \ldots Q_m^{r_m}$, where $Q_1^{r_1}, \ldots, Q_m^{r_m}$ are irreducible polynomials. Then $T = \bigcup_{j=1}^m T_j$, where $T_j = \{Q_j(w, z) = 0\}$. Using the Euler-Poincaré formula, it is not difficult to compute the dimension of the group $H_1(\mathring{T})$ and consequently that of its dual group $H_2(\mathbb{C}^2 \backslash T)$.

Theorem 2.2 (Yuzhakov, see Aizenberg-Yuzhakov (1979)). *The dimension of the homology groups $H_1(\mathring{T}) \simeq H_2(\mathbb{C}^2 \backslash T)$ is defined by the formula*

$$q = \sum_{j=1}^{m} 2\rho_j + \sum_{i=1}^{s} (q_i - 1) + l - m,$$

where ρ_j is the genus of the surface (the genus of the Riemann surface defined by the equation $Q_j(w, z) = 0$) and q_i is the number of irreducible elements of the subset T intersecting at the point A_i; s is the number of such points of self-intersection, and l is the number of elements at infinity of the set T (the number of connected cmponents of the set $T \cap \{|w|^2 + |z|^2 > R^2\}$ for R sufficiently large).

We construct the dual bases of $H_1(\overset{\circ}{T})$ and $H_2(\mathbb{C}^2 \backslash T)$. For this, on each irreducible component T_j we take $2\rho_j$ canonical cycles $\sigma_{js}^1, s = 1, \ldots, 2\rho_j$, (a basis of the one-dimensional homology of the corresponding Riemann surface) so that the cycles $\sigma_{j,2k-1}^1$ and $\sigma_{j,2k}^1$ intersect each other only at one point and do not intersect the other σ_{js}^1. Moreover we assume that the curves σ_{js}^1 do not pass through any self-intersection point A_i. We take $\gamma_{js}^1 = \delta \omega_{js}^1$, where ω_{js}^1 is obtained from $\sigma_{jr}, r = s - (-1)^s$, by a small perturbation, and δ is the Leray coboundary. Further, suppose that through the point A_i pass q_i elements $S_r, r = 1, \ldots, q_i$, of the set T. On each of these, except for S_{q_i} we construct a simple closed curve ω_{ir}^2 surrounding the point A_i. We set $\gamma_{ir}^2 = \delta \omega_{ir}^2$. As the dual cycle σ_{ir}^2 we take a simple closed curve which begins as a curve on S_r with initial point A_i and which ends as a curve which returns to A_i on S_{q_i}. We obtain $\sum_{i=1}^s (q_i - 1)$ pairs of cycles. Analogously we construct $l - m = \sum_{j=1}^m (l_j - 1)$ pairs of cycles $\gamma_{jv}^3 = \delta \omega_{jv}^3$ and $\sigma_{jv}^3, v = 1, \ldots, l_j - 1, j = 1, \ldots, m$, in a neighborhood of the point $\{\infty\}$.

Theorem 2.3. *The cycles $\{\gamma_{js}^r\}$ and $\{\sigma_{js}^r\}, r = 1, 2, 3$, form dual bases of the homology groups $H_2(\mathbb{C}^2 \backslash T)$ and $H_1(\overset{\circ}{T})$.*

We may assume that $\partial Q_j / \partial w \neq 0$ at the points of the curves ω_j^r. Then the integral over a basis cycle is equal to

$$\iint_{\gamma_{js}^r} \frac{P(w, z)}{Q(w, z)} \, dw \wedge dz = \int_{\omega_{js}^r} \operatorname{Res}[P \, dw \wedge dz/Q], \tag{11}$$

where

$$\operatorname{Res}[P \, dw \wedge dz/Q] = \operatorname{Res}[F \, dQ_v \wedge dz/Q] \ni [1/(r_v - 1)!]$$
$$\times \, \partial^{r_v - 1} F/\partial Q_v^{r_v - 1} \, dz|_{Q_v = 0},$$
$$F = P Q_v^{r_v}/Q \cdot (\partial Q_v / \partial w),$$

if $|\omega_{js}^r| \subset T_v$. Since $\partial^{r_v - 1} F/\partial Q_v^{r_v - 1}$ is a rational function, we have

Theorem 2.4 (Poincaré (1887)). *The integral of a rational function of two variables over an arbitrary cycle $\gamma \in Z_2(\mathbb{C}^2 \backslash T)$ can be expressed in terms of the periods of abelian integrals on the Riemann surfaces defined by the equations $Q_j(w, z) = 0, j = 1, \ldots, m$.*

It is clear that for $r = 2, 3$ the computation of the integral (11) reduces to the computation of the residue of an algebraic function relative to its pole.

Example 3. If p and q are relatively prime integers, the integral $\iint_\gamma P(w, z) \, dw \wedge dz/(z^p - w^q) = 0$ for any cycle $\gamma \in Z_2(\mathbb{C}^2 \backslash T)$, where $T = \{z^p - w^q = 0\}$, since $\overset{\circ}{T}$ is homeomorphic to the two-dimensional sphere and $H_2(\mathbb{C}^2 \backslash T) \simeq H_1(\overset{\circ}{T})$.

Example 4. Let $Q(w, z) = wz - 1$. Then $\overset{\circ}{T}$ is homeomorphic to the Riemann sphere with two points identified ($z = \infty$ and $z = 0$, the pole of the function $w = 1/z$). Thus $H_1(\overset{\circ}{T}) \simeq \mathbb{Z}$. The cycles $\gamma_1 = \{|w| = |z| = 2\}$ and $\sigma_1 = \{w = 1/t, z = t, 0 \leq t \leq \infty\}$ form dual bases of the homology groups $H_2(\mathbb{C}^2 \backslash T)$ and

$H_1(\overset{\circ}{T})$. For any cycle $\gamma \in Z_2(\mathbb{C}^2 \setminus T)$ the integral

$$\iint_\gamma P(w, z)\, dw \wedge dz/(wz - 1) = (2\pi i)^2 k \cdot R,$$

where $k = o(\gamma, \sigma_1)$,

$$R = (2\pi i)^{-2} \iint_{\gamma_1} P\, dw \wedge dz/(wz - 1)$$

$$= \sum_{m=0}^{\infty} (2\pi i)^{-2} \iint_{\gamma_1} P\, dw \wedge dz/(wz)^{m+1} = \sum_{m=0}^{\infty} \frac{1}{(m!)^2} \frac{\partial^{2m} P(0, 0)}{\partial w^m \partial z^m}.$$

2.3. Application of De Rham Duality. In some cases, for the study of residues it is useful to apply De Rham's Theorem, which establishes a duality between the homology groups and the cohomology of exterior differential forms. The theorem can be stated thus: Let X be a differentiable manifold. For any homomorphism $\lambda : H_p(X) \to \mathbb{C}$, there exists a unique element $h = [\omega] \in H^p(X)$ of the De Rham cohomology group such that $\lambda(g) = \int_g h = \int_\gamma \omega$ for any $g = [\gamma] \in H_p(X)$. From De Rham's Theorem we get the following proposition, which is useful in applications.

Proposition 2.5. *If for cycles $\gamma_j \in Z_p(X)$, $j = 1, \ldots, q$, and forms $\omega_j \in Z^p(X)$, $j = 1, \ldots, q$ the determinant $\det \|a_{jk}\| \neq 0$, where $a_{jk} = \int_{\gamma_j} \omega_k$; and if any form $\omega \in Z^p(X)$ can be represented in the form $\omega \sim \sum_{j=1}^{q} c_j \omega_j$, where the c_j are complex numbers, then $\dim H^p(X) = \dim H_p(X) = q$ and the $\{\gamma_j\}_{j=1}^{q}$ and $\{\omega_j\}_{j=1}^{q}$ are bases of the p-dimensional homology and cohomology of the manifold X.*

If $a_{jk} = \delta_{jk}$, then these bases are called *dual* in the sense of De Rham.

Theorem 2.6. *Let $\{\gamma_j\}$ and $\{\omega_j\}$ be bases of the p-dimensional homology and cohomology of the manifold X, dual in the sense of De Rham. Then for any cycle $\gamma \in Z_p(X)$ and any cocycle $\omega \in Z^p(X)$, the integral*

$$\int_\gamma \omega = \sum_{j=1}^{q} k_j R_j,$$

where the $k_j = \int_\gamma \omega_j$ are the coefficients of the expansion $\gamma \sim \sum_{j=1}^{q} k_j \gamma_j$ of the cycle γ with respect to the basis $\{\gamma_j\}$ and where the $R_j = \int_{\gamma_j} \omega$ are the coefficients of the expansion $\omega \sim \sum_{j=1}^{q} R_j \omega_j$ of the De Rham class of the form ω with respect to the basis $\{\omega_j\}$.

In the case of complex analytic manifolds (for example, domains in $\mathbb{C}^n$) the next theorems are also useful. (See, e.g., the paper of Onishchik in Volume 10, I of this series.)

Theorem 2.7 (Serre). *If X is a Stein manifold (for example a domain of holomorphy in $\mathbb{C}^n$), then any cohomology class $h \in H^p(X)$ contains a holomorphic form $\omega \in h$.*

Theorem 2.8 (Grothendieck). *If either X is a Stein manifold and S is a submanifold of X of codimension 1 or if X is an algebraic manifold in $\mathbb{CP}^n$ and S is a positive divisor on X, then any cohomology class $h \in H^p(X \setminus S)$ contains a holomorphic form $\omega \in h$ with a pole on S or, respectively, a meromorphic form $\omega \in h$ whose only pole is on S. In particular, if $X = \mathbb{CP}^n$ or $\mathbb{C}^n$ and S is the zero set of a polynomial, then any class $h \in H^p(X \setminus S)$ contains a rational form.*

Example 4. Let $Q(z) = z_1^2 + \cdots + z_n^2 - 1$ and let $T = \{z \in \mathbb{C}^n : Q(z) = 0\}$ and $D = \mathbb{C}^n \setminus T$. By Theorem 2.8 any closed form $\omega \in Z^n(D)$ is cohomologous in D to a rational form $P\, dz/Q^r$, where P is a polynomial. It is sufficient to consider the case when P is a homogeneous polynomial of degree $p \geq 0$. Computing

$$d\left(P \sum_{j=1}^{n} (-1)^j z_j\, dz_{[j]}/Q^r\right),$$

where $dz_{[i]} = dz_1 \wedge \cdots [j] \cdots \wedge dz_n$, and also computing $d[(z^\alpha/z_j)(-1)^{j-1} \times dz_{[j]}/(\alpha_j - 1)Q]$, $\alpha = (\alpha_1, \ldots, \alpha_n)$, $\alpha_j \geq 2$, we obtain the recurrence formulas

$$P\, dz/Q^{r+1} \sim [(n + p - 2r)/2r]P\, dz/Q^r, \tag{12}$$

$$z^\alpha\, dz/Q \sim (\alpha_j - 1)(z^\alpha/z_j^2)\, dz/(n + |\alpha| + 2)Q, \tag{13}$$

from which it follows that $P\, dz/Q^r \sim c\phi$, where $\phi = dz/Q$, since $z^\alpha\, dz/Q \sim 0$ if $\alpha_j = 1$ for any j. The constant c is found from formulas (12) and (13). We take the cycle

$$\gamma = \{z : x_1^2 + \cdots + x_n^2 = 1, y_1 = \cdots = y_n = 0\},$$

where $x_j = \operatorname{Re} z_j$ and $y_j = \operatorname{Im} z_j$. Then

$$\int_{\delta\gamma} \phi = 2\pi i \int_{\gamma} \operatorname{res}[\phi] = \pi i \int_{\gamma} dz_1 \wedge \cdots \wedge dz_{n-1}/z_n = \pi i \Sigma_{n-1},$$

where $\Sigma_{n-1} = 2\pi^{n/2} \Gamma(n/2)$ is the volume of the unit sphere in $\mathbb{R}^n$. Thus according to Proposition 2.5, $\dim H^n(D) = \dim H_n(D) = 1$ and $\delta\gamma$ and $\phi/\pi i \Sigma_{n-1}$ are bases of the homology and cohomology of D which are De Rham dual. For any cycle $\Gamma \subset Z_n(D)$ the integral $\int_\Gamma P\, dz/Q^r = m \cdot c \cdot \pi i \Sigma_{n-1}$, where $\Gamma \sim m \cdot \delta\gamma$.

As another example, we consider local residues in $\mathbb{C}^n$. Let $\phi_1, \ldots, \phi_N$ be holomorphic functions in a neighborhood U_α of a point $a \in \mathbb{C}^n$ and $\Delta_\alpha = \partial(\phi_{\alpha_1}, \ldots, \phi_{\alpha_n})/\partial(z_1, \ldots, z_n) \neq 0$ for any $\alpha = (\alpha_1, \ldots, \alpha_n) \in \{1, \ldots, N\}$. We define $T_j = \{z \in U_\alpha : \phi_j(z) = 0\}$. From the assertions of 2.5–2.8 and from the separation of singularities of holomorphic functions this follows:

Proposition 2.9. *The dimension of the n-dimensional homology and cohomology groups of the domain $U_\alpha \setminus (\bigcup_{j=1}^N T_j)$ is equal to $\binom{N-1}{n-1}$, and their De Rham-dual bases consist of the cycles*

$$\gamma_\alpha = \{z \in U_\alpha : |\phi_{\alpha_1}(z)| = \cdots = |\phi_{\alpha_{n-1}}(z)| = \delta, |\phi_N(z)| = \varepsilon\}$$

and forms

$$\omega_\alpha = (2\pi i)^{-n} d\phi_{\alpha_1} \wedge \cdots \wedge d\phi_{\alpha_{n-1}} \wedge d\phi_N / \phi_{\alpha_1} \cdots \phi_{\alpha_{n-1}} \phi_N,$$

where ε and δ/ε are sufficiently small positive numbers and $\alpha = (\alpha_1, \ldots, \alpha_n)$, $\alpha_n = N$ and $\{\alpha_1, \ldots, \alpha_{n-1}\} \in \{1, \ldots, N-1\}$.

Let the holomorphic form $\omega = f\, dz = f\, dz_1 \wedge \cdots \wedge dz_n$, where $f \in A(U_\alpha \setminus \bigcup_{j=1}^N T_j)$. The residue of this form with respect to the basis cycle γ_α can be computed in this way. We perform a change of coordinates $\zeta_j = L_{\alpha_j}(z), j = 1, \ldots, n$, where $L_j(z)$ is the linear part of the function f_j at the point a. For sufficiently small ε and δ/ε the cycle $\gamma_\alpha \sim \bar\gamma_\alpha = \{|\zeta_1| = \cdots = |\zeta_{n-1}| = \delta, |\zeta_n| = \varepsilon\}$. Then the residue

$$R_\alpha = (2\pi i)^{-n} \int_{\gamma_\alpha} \omega = (2\pi i)^{-n} \int_{\bar\gamma_\alpha} \tilde f(\zeta)\, d\zeta,$$

where $\tilde f(\zeta) = f(L_\alpha^{-1}(\zeta))$. The function $\tilde f$ is holomorphic in a neighborhood of the cycle $\bar\gamma_\alpha$ and can be expanded there into a Laurent series $\tilde f(\zeta) = \sum_{|\beta|=-\infty}^\infty c_\beta \zeta^\beta$. Thus the residue $R_\alpha = c_{-I} = c_{-1,\ldots,-1}$. In the case of a meromorphic function $f = g/\phi_1^{r_1} \cdots \phi_N^{r_N}$ the coefficient c_{-I} can be found directly. Here $\phi_{\alpha_j} = \zeta_j + g_{\alpha_j}(\zeta)$, where $g_{\alpha_j}(\zeta) = \sum_{|\beta|>1} c_{\alpha_j\beta} \zeta^\beta$, $j = 1, \ldots, n$, $\phi_\nu = c_{\nu n}\zeta_n + g_\nu(\zeta)$, where $g_\nu(\zeta) = \sum_{k=1}^{n-1} c_{\nu k}\zeta_k + \sum_{|\beta|>1} c_{\nu\beta}\zeta^\beta$, $\nu \neq \alpha_1, \ldots, \alpha_n$, $c_{\nu n} \neq 0$.

For ε and ε/δ sufficiently small, these inequalities hold on $\bar\gamma_\alpha$: $|g_\nu(\zeta)| < |c_{\nu n}\zeta_n|$, $\nu \neq \alpha_1, \ldots, \alpha_n$, $|g_{\alpha_j}(\zeta)| < |\zeta_j|$, $j = 1, \ldots, n$. Expanding the fractions $1/\phi_{\alpha_j} = 1/\zeta_j[1 + g_{\alpha_j}(\zeta)/\zeta_j]$, $1/\phi_\nu = 1/c_{\nu n}\zeta_n[1 + g_\nu(\zeta)/c_{\nu n}\zeta_n]$ into series of geometric progressions and cross-multiplying the resulting series, we obtain the desired Laurent series and its coefficient c_{-I}.

§3. Homological Methods for Studying Integrals that Depend upon Parameters. Application to Combinatorial Analysis

Leray (1959) applied the theory that he had developed about residues on a complex analytic manifold, along with the topological Picard-Lefschetz Theorem, to the study of integrals depending on parameters. The integrals arise in the solution of the Cauchy problem for partial differential equations in the complex domain.[3] Beyond this, homological methods were developed in connection with the study of singularities and of the character of the branching of the Feynman integrals that arise in theoretical physics (see Fotiadi et al. (1965), Pham (1967), Golubeva (1976)). Various multidimensional analogs of the Hadamard product also lead to the study of holomorphic functions defined by integrals over cycles in $\mathbb{C}^n$. Consider how the integral of a closed form over a cycle contained in

---

[3] Earlier topological methods were used to study integrals by Picard, Lefshetz, I.G. Petrovskij, and A.A. Borovikov.

the level surface of a holomorphic function in $\mathbb{C}^n$ depends on a parameter, the constant defining the level surface. This defines a function of the parameter; the branching of this function has been studied in detail in (Arnol'd et al. (1984)).

3.1. Analytic Continuation of Integrals Depending on Parameters. Isotopy Theorem.

Let X and T be complex manifolds of dimension n and q and let $\omega_t(z) = \omega(t, z)$ be a closed differential form of degree p in $z \in X \backslash S_t$ which depends holomorphically on the parameter $t \in T$. Let S_t be an analytic set in X (the singular set of the form $\omega_t(z)$) which also depends holomorphically on t. We will consider the problem of analytic continuation of the function defined by the integral

$$I(t) = \int_\Gamma \omega(z, t), \tag{14}$$

where Γ is a compact cycle in $X \backslash S_{t_0}$. By the compactness of Γ the integral (14) is a holomorphic function in a neighborhood of the point t_0.

Theorem 3.1 (Analytic continuation of integrals (Fotiadi et al. (1965), Pham (1967)). *If the projection*

$$\pi : (X \times T, \mathcal{T}) \to T, \tag{15}$$

where $\mathcal{T} = \{(z, t) \in X \times T : z \in S_t\}$, *is a locally trivial fibration of the pair* $(X \times T, \mathcal{T})$ *with fiber* (X, S_{t_0}), *then for any cycle* $\Gamma \in Z_p(X \backslash S_{t_0})$, *the integral (14) can be continued holomorphically along any path* λ *in* T.

In fact for a path $\lambda : [0, 1] \to T$, $\lambda(0) = t_0$, using local trivializations one can construct a covering isotopy, a homeomorphism of the pair

$$g_\tau : (X, S_{t_0}) \to (X, S_{\lambda(t)}), \quad 0 \le \tau \le 1, \tag{16}$$

continuously depending on τ which establishes a continuous deformation of the cycle Γ in $X \times T \backslash \mathcal{T}$, denoted by $\Gamma_\tau = g_\tau(\Gamma) \in Z_p(X \backslash S_{\lambda(\tau)})$. The holomorphic elements $\int_{\Gamma_\tau} \omega(z, t)$ define an analytic continuation of the integral (14) along the path λ, since for nearby values τ_1, τ_2 and for t close to $\lambda(\tau_1)$ and $\lambda(\tau_2)$, the cycles Γ_{τ_1} and Γ_{τ_2} are homologous in $X \backslash S_t$. Consequently

$$\int_{\Gamma_{\tau_1}} \omega(z, t) = \int_{\Gamma_{\tau_2}} \omega(z, t).$$

In applications an important case is when S_t is a family of algebraic sets depending algebraically on t. In this case the projection (15) is a locally trivial fibration over $T \backslash \{\text{algebraic set}\}$. A sufficient condition for the local triviality of the fibration (15) is given by Thom's theorem on covering isotopies.

Theorem 3.2 (Thom (Fotiadi et al. (1965), Pham (1967))[4]. *Let $\pi : Y \to T$ be a proper differentiable mapping of a stratified[5] set Y to a connected differentiable manifold T such that the restriction of π to any stratum has rank equal to the dimension of T. Then π is a locally trivial stratified map.*

In the case when $Y = X \times T$, and

$$S = \mathcal{T} = \{(z, t) : t \in T, z \in S_t\},$$

the mapping π is proper only when X is compact.

Corollary 3.3 (Fotiadi et al. (1965)). *Let T and X be differentiable C^∞-manifolds, with X compact. Let $S_t = \bigcup_j S_j(t)$, where $\{S_j(t)\}$ is a finite family of C^∞-manifolds in X depending C^∞ on t and in general position for any $t \in T$. Then the projection of the pair, $\pi : (X \times T, \mathcal{T}) \to T$, where $\mathcal{T} = \{(z, t) : t \in T, z \in S_t\}$ is a locally trivial stratified fibration.*

Theorems 3.1 and 3.2 and Corollary 3.3 give conditions for the analytic continuation of the integral (14) along any path in T for an arbitrary initial cycle $\Gamma \in Z_p(X \setminus S_{t_0})$. However for specially-chosen cycles Γ these conditions can be made more precise. For example, in (14) let $\omega(z, t)$ be a meromorphic differentiable form of degree $n = \dim_{\mathbb{C}} X$ with poles on $S_j(t), j = 1, \ldots, n$, where the $S_j(t)$ are analytic sets of codimension 1 (divisors) depending holomorphically on the parameter $t \in T$. Assume that a is an isolated point of the set $S_1(t_0) \cap \cdots \cap S_n(t_0)$ and that

$$\Gamma = \{z \in U_\alpha : |f_j(z, t_0)| = \varepsilon, j = 1, \ldots, n\},$$

where $f_j(z, t)$ is a defining function of the set $S_j(t)$ in a neighborhood U_α of the point a. In this case the integral (14) is a Grothendieck residue (see Chapter 3) depending on the parameter t. In this case the following holds:

Proposition 3.4 (Tsikh). *The analytic element $I(t)$ defined by the Grothendieck residue depending on the parameter t can be continued holomorphically along any path*

$$\gamma = \{t = t(\tau), 0 \leq \tau \leq 1\},$$

for which there exists a lifting

$$\gamma' = \{z = z(\tau), t = t(\tau), 0 \leq \tau \leq 1\},$$

in $X \times T$ such that $z(0) = a$ and the intersection multiplicity of the divisors $S_j(t(\tau)), j = 1, \ldots, n$, at the point $z(\tau)$ does not depend on τ.

---

[4] See also J.N. Mather, "Stratifications and mappings," *Dynamical Systems,* New York and London, 1973, 195–232.
[5] It is assumed that the stratification satisfies Whitney's Conditions A and B (see Pham (1967)).

3.2. Foliation near a Landau Singularity. Picard-Lefschetz Formula. A holomorphic function defined by the integral of (14) can have a singularity only at those points $t' \in T$ for which the cycle Γ cannot be deformed to one disjoint from the singular set $S_{t'}$. In the case of a compact manifold X, according to Theorems 3.1 and 3.2, such points are the points of the *Landau set* $L = \bigcup_A \pi(\{(z, t) \in A : \text{rank } \pi|_A < \dim T\})$, where $\bigcup_A$ is the union over all strata A of the stratified set $\mathcal{T}$. In this case the integral (14) can be continued holomorphically along any path in $T\backslash L$. This continuation is in general a multivalued function. The Landau set L is an analytic set in T (see Pham (1967)). If the manifold X is noncompact, then the singularities of the integral (14) are not exhausted by the Landau set. Therefore, one usually considers a compactification of X (see Fotiadi et al. (1965), Golubeva (1976)). For any closed path (loop) λ in $T\backslash L$, with $\lambda(0) = \lambda(1) = t_0$, the isotopy (monodromy, see Arnold et al. (1984)) (16) defines an automorphism of the homology group (monodromy transformation)

$$\lambda_* : H_p(X\backslash S_{t_0}) \to H_p(X\backslash S_{t_0}). \tag{17}$$

If two paths λ and λ_1 are homotopic, the corresponding monodromies are also homotopic and consequently $\lambda_* = \lambda_{1_*}$. In other words, the monodromy transformation (17) depends only on the class $[\lambda] \in \pi_1(T\backslash L, t_0)$. Thus (16) and (17) generate a homomorphism (a representation) of the fundamental group $\pi_1(T\backslash L, t_0)$ into the automorphisms of the homology group $H_p(X\backslash S_{t_0})$:

$$\pi_1(T\backslash L, t_0) \to \text{Aut } H_p(X\backslash S_{t_0}). \tag{18}$$

The monodromy group, the image of homomorphism (18), completely describes the character of the multivalued function defined by integral (14). Its jump as it is continued around the loop λ is equal to the integral

$$\int_{\lambda_* \Gamma - \Gamma} \omega(z, t_0).$$

We will assume that T is simply connected, i.e., $\pi_1(T) \equiv 0$. We choose a regular point $b \in L$ and choose a coordinate system in a neighborhood U_b of the point b such that $L \cap U_b = \{t_1 = 0\}$. The loop in U_b given by the equations $t_2 = \cdots = t_q = 0, t_1 = \varepsilon e^{i\theta}, 0 \le \theta \le 1$, is called a simple loop in $T\backslash L$.

Proposition 3.5 (Pham (1967)). *If $\pi_1(T) \equiv 0$, then $\pi_1(T\backslash L)$ is generated by simple loops.*

In a special case, a stronger version holds:

Proposition 3.6 (Pham, Zariski). *Let $L = L_1 \cup \cdots \cup L_m$, where*

$$L_j = \{t \in \mathbb{C}^q : P_j(t) = 0\},$$

for an irreducible polynomial P_j. If the compactification of the sets $L_1, \ldots, L_m$ in $\mathbb{CP}^n$ and the hyperplane at infinity $\mathbb{CP}_\infty^{q-1}$ are manifolds in general position at every point except possibly for an algebraic set of codimension ≥ 3, the fundamental group $\pi_1(\mathbb{C}^q\backslash L)$ is a free abelian group with m generators.

Thus, in the case $\pi_1(T) \simeq 0$ the study of the representation (18) reduces to finding the automorphisms (17) for simple loops λ. In the simplest cases of these automorphisms, the branching of the integral (14) around the Landau set, are described by the Picard-Lefschetz formula (see Leray (1959), Fotiadi et al. (1965), Griffiths-King (1973), Arnold et al. (1984)).

We will show the example of the simplest singularity. Let the point $b \in L$ be the projection of a simple critical point (a, b) of some stratum which does not belong to the projection of the closure of the set of critical points of the other strata. Then in a neighborhood of the point (a, b), the set

$$\mathcal{T} = \{(z, t) : t \in T, z \in S_t\}$$

has the following form: $S_t = \bigcup_{j=1}^m S_j(t)$, $m < n$, which, in suitable local coordinates, can be written as

$$S_j(t) = \{s_j(z, t) = z_j = 0\}, \quad j = 1, \ldots, m - 1,$$

$$S_m(t) = \{S_m(z, t) = t_1 - (z_1 + \cdots + z_{m-1} + z_m^2 + \cdots + z_n^2 = 0\}.$$

We define what are called the vanishing cycles:

$$\tilde{e} = \{z : x_1 = \cdots = x_{m-1} = y_1 = \cdots = y_n = 0, x_m^2 + \cdots + x_n^2 = t\}$$

$$\in Z_{n-m}\left(\bigcap_{j=1}^m S_j(t)\right),$$

where $x_j = \operatorname{Re} z_j$, $y_j = \operatorname{Im} z_j$, and $e = \delta_1 \circ \cdots \circ \delta_m \tilde{e} \in Z_n(X \setminus S_t)$, where δ_j is the Leray coboundary with respect to $S_j(t)$ and also the vanishing square

$$\hat{e} = \{y_1 = \cdots = y_m = 0, x_j \geq 0, j = 1, \ldots, m, x_m^2 + \cdots + x_n^2 \leq t\}.$$

Under the given hypotheses this theorem holds:

Theorem 3.7 (Leray (1959), Fotiadi et al. (1965), Pham (1967)). *A circuit around L along a simple loop λ in the neighborhood of a point b induces a homomorphism (17), which for $p < n$ is the identity but for $p = n$ defines the following Picard-Lefschetz formula:*

$$\lambda_* h = h + N \cdot [e], \quad h \in H_p(X \setminus S_t),$$

where $[e]$ is the homology class of the cycle e, $N = (-1)^{(n+1)(n+2)/2}$, and $\langle \hat{e}, h \rangle$ is the intersection number of the chain $\hat{e}$ with any representative cycle in the class h.

Since

$$\langle \hat{e}, e \rangle = \begin{cases} 2 \cdot (-1)^{(n+1)(n+2)/2+1}, & \text{if } n - m = 2k, \\ 0, & \text{if } n - m = 2k + 1, \end{cases}$$

then

$$\lambda_*[e] = \begin{cases} -[e], & \text{if } n - m = 2k, \\ [e], & \text{if } n - m = 2k + 1, \end{cases}$$

from which follows

Corollary 3.8. *If $n - m$ is even, then $\lambda_*^2 h = h$ (after a double trip around L the class h is carried into itself). This means that this multi-valued function has branching like that of a square root. If $n - m$ is odd, then $\lambda_*^k h = h + k \cdot N \cdot [e]$ (each circuit adds a multiple of the cycle e). In this case the multi-valued function is of logarithmic type. If $\langle \hat{e}, h \rangle = 0$, then there is no branching.*

The jump of the integral (14) as one goes around L along the loop λ is equal to

$$\int_e \omega(z, t) = (2\pi i)^{m-1} \int_{\tilde{e}} \mathrm{Res}^{m-1} [\omega].$$

The study of the asymptotics of the integral by the saddle-point method (Arnol'd et al. (1984), Varchenko (1983)) leads to the study of the branching behavior and asymptotics of integrals of the following type. Let $f : (\mathbb{C}^n, 0) \to (\mathbb{C}, 0)$ be the germ of a holomorphic function having an isolated critical point at 0, and let ω be a holomorphic differential form of degree $n - 1$. We denote by $X_t = \{z : f(z) = t\}$, $t \neq 0$, the level surface of a non-critical value of the function f, and by $\sigma_t \in Z_{n-1}(X_t)$ the family of cycles obtained by a continuous deformation of the cycle $\sigma_{t_0} \in Z_{n-1}(X_{t_0})$ as t varies along some curve in $\mathbb{C} \setminus \{0\}$. Let M be the matrix of the monodromy transformation $H_{n-1}(X_{t_0}) \to H_{n-1}(X_{t_0})$ corresponding to the simple loop $\gamma = \{t = \varepsilon \cdot \exp(2\pi i \tau), 0 \leq \tau \leq 1\}$.

Then we have

Theorem 3.9 (Arnol'd et al. (1984)). *The integral*

$$I(t) = \int_{c(t)} \omega$$

is a multivalued holomorphic function which for small values of $t \neq 0$ is represented by the series

$$I(t) = \sum_{\alpha, k} a_{\alpha, k} t^\alpha \cdot (\ln t)^k,$$

where α is a non-negative rational number and the k are integers. All the coefficients $a_{\alpha, k}$ are zero when $k > 0$. Each number α has the property that $\exp(2\pi i \alpha)$ is an eigenvalue of the matrix M. The coefficient $a_{\alpha, k}$ is zero whenever the matrix of the Jordan form of M has no blocks of size $k + 1$ or larger belonging to the eigenvalue $\exp(2\pi i \alpha)$.

In Arnol'd et al. (1984) a more general situation was considered, where in addition f depends holomorphically on a parameter $y \in \mathbb{C}^k$. In this case $I(y, t)$ also depend holomorphically on y.

3.3. Some Examples of Integrals Depending on Parameters

(1) In (14) let $X = \mathbb{C}^n$, and let $\omega(z, t) = P(z, t) \, dz/Q(z, t)$, where P and Q are polynomials in $z \in \mathbb{C}^n$ and $t \in \mathbb{C}^q$; let $S_t = \{z : Q(z, t) = 0\}$. The polynomial Q factors into a product of terms of degree one in z:

$$Q = Q_1^{r_1} \cdots Q_m^{r_m}$$

where

$$Q_j(z, t) = \sum_{v=1}^n a_{jv}(t)z_v + a_{j0}(t), \quad j = i, \ldots, m.$$

Proposition 3.10. *Under the hypotheses above, the analytic function defined by the integral (14) is a ratonal function whose denominator consists of powers of minors of rank $n + 1$, not identically zero, of the matrix $\|a_{jv}(t)\|_{(j=1,\ldots,m; v=0,1\ldots,n)}$ and minors of rank n of the matrix $\|a_{jv}(z)\|_{(j=1,\ldots,m; v=1,\ldots,n)}$.*

(2) The Feynman integral for a one-loop graph with vertices has the form (see Golubeva (1976)):

$$I(p) = \int_\Gamma f(z) \, dz \Big/ \prod_{j=1}^m (p_j z - 1)^{q_j},$$

where

$$z = (z_0, z_1, \ldots, z_n) \in \Sigma = \{z \in \mathbb{C}^{n+1} : z_0^2 + z_1^2 + \cdots z_n^2 = 1\};$$

$$p = (p_1, \ldots, p_m), \quad p_j = (p_{j0}, p_{j1}, \ldots, p_{jn}) \in \mathbb{C}^{n+1},$$

$$p_j z = p_{j0} z_0 + \cdots + p_{jn} z_n;$$

$$P_j = \{z : p_j z = 1\}, \quad \Gamma \in Z_n(\Sigma \backslash \bigcup_{j=1}^m P_j),$$

$$\omega(z) = \sum_{v=1}^n (-1)^v z_v \, dz_0 \wedge \cdots [v] \cdots \wedge dz_n.$$

Theorem 3.11 (Boiling, Golubeva). *The integral $I(p)$ extends holomorphically along any path in $\mathbb{C}^{m(n+1)} \backslash \bigcup_{\beta \in B} L_\beta$, where*

$$L_\beta = \{p : \det \|p_j p_v - 1\|_{j,v \in \beta} = 0\},$$

$$B = \{\beta = \{\beta_1, \ldots, \beta_k\} \in \{1, \ldots, m\}, 1 \leq k \leq n + 1\}.$$

The character of the branching of $J(p)$ around the Landau set is determined by the Picard-Lefschetz formula.

One can find a survey of results and a bibliography of research on Feynman integrals in Golubeva (1976). The branching of several integrals depending on parameters is examined in the works of Varchenko (1983), Pedan[6] and a series of others.

(3) Multidimensional analogs of the Hadamard product. The *Hadamard product* of two power series in n variables

$$f(z) = \sum_{\alpha \geq 0} a_\alpha z^\alpha, \quad g(z) = \sum_{\alpha \geq 0} b_\alpha z^\alpha, \tag{19}$$

---

[6] Yu.V. Pedan, "Investigation of the Riemann surfaces of some multiple integrals that depend on a complex parameter II: the Riemann surface of the element $I_r(t)$," Izv. Vuzov Matem., 1976, No. 12, 66–76.

is defined to be the series

$$h(z) = \sum_{\alpha \geq 0} a_\alpha b_\alpha z^\alpha.$$

If the series (19) converges in a closed polydisk of radius ρ, then

$$h(z) = \int_{\Gamma_\rho} \frac{f(\zeta) \cdot g(\zeta) \, d\zeta \wedge d\eta}{\prod_{j=1}^n (\zeta_j \eta_j - z_j)}, \tag{20}$$

where

$$\Gamma_\rho = \{(\zeta, \eta) \in \mathbb{C}^{2n} : |\zeta_j| = |\eta_\nu| = \rho, \, j, \nu = 1, \ldots, n\}.$$

The integral representation (2) allows one to study the singularity of the Hadamard product $h(z)$ (Odoni, Dyakovich, Haustus and Klarner, A.I. Yanashauskas, E.K. Leinartas, K.V. Safonov et al).

3.4. Application of Residues to Combinatorial Analysis (see Egorychev (1977), Aizenberg-Yuzhakov (1979)). The fundamental idea of using multidimensional residues to compute combinatorial sums and discover generating functions was developed, with many illustrative examples, by G.P. Egorychev (1977). The combinatorial expressions in a sum (series) are represented as the Taylor coefficients in integral form of their generating functions; replacing the products of integrals with multiple integrals, interchanging the integration and summation signs, then combining the terms of the sum under the integral sign, we obtain the integral of a holomorphic form over a cycle. Computing it using residues, we find the desired sum (generating function). This idea is realized in quite general circumstances by the following theorem:

Theorem 3.12 (Egorychev, Yuzhakov). *Let the members of an n-multiple numerical sequence* $c_\alpha = c_{\alpha_1 \ldots \alpha_n}$ *be represented in the form*

$$c_\alpha = (2\pi i)^{-n} \int_{\Gamma_\rho} \phi(z) \prod_{j=1}^n [f_j^{\alpha_j \beta_j}(z)/z_j^{\beta_j(\alpha_j+1)}] \, dz,$$

where ϕ and the f_j are holomorphic functions in the closed polydisk $\overline{U}_\rho = \{z : |z_j| \leq \rho, j = 1, \ldots, n\}$, $\Gamma_\rho = \{z : |z_j| = \rho, j = 1, \ldots, n\}$. *Then the generating function for the c_α is expressed by the integral*

$$F(t) = \sum_{|\alpha| \geq 0} c_\alpha t^\alpha = (2\pi i)^{-n} \int_{\Gamma_\rho} \frac{\phi(z) \, dz}{\prod_{j=1}^n [z_j^{\beta_j} - t_j f_j^{\beta_j}(z)]}.$$

In particular, if either $f_j(0) \neq 0$ or $\beta_1 = \cdots \beta_n = 1$, then

$$F(t) = \sum_{\mu \in M} \left[\phi(z) \Big/ \frac{\partial w}{\partial z} \right]\Big|_{z = z^{(\mu)}(t)},$$

where

$$w_j = z_j^{\beta_j} - t_j f_j^{\beta_j}(z), \quad j = 1, \ldots, n \tag{21}$$

$M = \{z_{(\mu)}(t), \mu = (\mu_1, \ldots, \mu_n), 1 \leq \mu_j \leq \beta_j\}$ is the set of zeros of the mapping (21) in the polydisk U_ρ.

From this we obtain the following corollary:

Theorem 3.13 (Main Theorem of MacMahon). *Let*

$$X_j = \sum_{k=1}^{n} a_{jk} x_k, \quad j = 1, \ldots, n.$$

Then the coefficient of $x^\alpha = x_1^{\alpha_1} \ldots x_n^{\alpha_n}$ *in the expansion of* $X_1^{\alpha_1} \ldots X_n^{\alpha_n}$ *in powers of* x *is equal to the coefficient of* t^α *in the Taylor series expansion of the function* $F(t) = 1/\det \|\delta_{jk} - a_{jk} t_j\|$, *where*

$$\delta_{jk} = \begin{cases} 1, & j = k, \\ 0, & j \neq k. \end{cases}$$

Example 5. Compute the sum

$$A_{mn} = \sum_{k=0}^{N} \binom{m}{k} \binom{n}{k} \binom{m+n+p-k}{m+n}, \quad N = \max\{m, n, p\}.$$

Since

$$\int_{|z|=\varepsilon} \frac{(1+z)^m \, dz}{z^{k+1}} = \begin{cases} \binom{m}{k}, & k \leq m, \\ 0, & k > m, \end{cases}$$

then the generating function for A_{mn} equals

$F(u, v)$

$$= \sum_{k=0}^{\infty} \frac{u^m v^n}{(2\pi i)^3} \int_\Gamma \frac{(1+z_1)^m (1+z_2)^n (1+z_3)^{m+n+p-k} \, dz_1 \wedge dz_2 \wedge dz_3}{z_1^{k+1} z_2^{k+1} z_3^{m+n+1}}$$

$$= \frac{1}{(2\pi i)^3} \int_\Gamma \frac{(1+z_3)^p \, dz_1 \wedge dz_2 \wedge dz_3}{[z_3 - u(1+z_1)(1+z_3)][z_3 - v(1+z_2)(1+z_3)][z_1 z_2 (1+z_3) - 1]}$$

$$= [(1-u)(1-v)]^{-(p+1)},$$

where $\Gamma = \{|z_1| = |z_2| = 2, |z_3| = 1/2\}$. Hence, after expanding $F(u, v)$ into a Taylor series we obtain $A_{mn} = \binom{m+p}{m} \binom{n+p}{n}$.

Chapter 2
Multidimensional Logarithmic Residues
and Their Applications

L.A. Aizenberg

§ 1. Multidimensional Logarithmic Residues

By the words "*logarithmic residue formula*" one usually understands an integral representation for the sum of the values of a holomorphic function at all the zeros of a holomorphic mapping in a given region (for instance, a formula for the number of these zeroes). Therefore, to begin with we must introduce the concept of the multiplicity of the zero of a holomorphic mapping. We will give the most natural, the so-called "dynamic" definition of multiplicity. We consider a mapping

$$w = f(z) \tag{1}$$

holomorphic in a neighborhood of the point $a \in \mathbb{C}^n$, where $w = (w_1, \ldots, w_n)$, $f = (f_1, \ldots, f_n)$. Let the point a be a zero of this mapping, that is, $f(a) = 0$. If the closure of a neighborhood U_a contains no other zeroes of the mapping (1), then there is an $\varepsilon > 0$ such that for almost all $\zeta = (\zeta_1, \ldots, \zeta_n)$, with $|\zeta| < \varepsilon$, the mapping

$$w = f(z) - \zeta \tag{2}$$

has the property that, at each of its zeroes, the jacobian of the mapping is not zero: $\partial f / \partial z \neq 0$. Such zeroes are called simple zeroes. For a small enough choice of ε, the number of simple zeroes does not depend on the choice of ζ nor on the neighborhood U_a.

This number of zeroes of the mapping (2) is called the *multiplicity of the zero a of the mapping* (1). For the relaton between this multiplicity and the coefficients of the Taylor expansion of the mapping (1) at the point a, see § 4.

Example 1. For the mapping $w_1 = z_1, w_2 = z_2^m + z_1^2$, the point $(0, 0)$ is a zero of multiplicity m, since the mapping $w_1 = z_1 - \zeta_1, w_2 = z_2^m + z_1^2 - \zeta_2$, for small $|\zeta|$ and $\zeta_2 \neq \zeta_1^2$, has m simple zeroes of the form $(\zeta_1, \sqrt[m]{\zeta_2 - \zeta_1^2})$ in a neighborhood of this point.

Let D be a bounded domain in $\mathbb{C}^n$ with piecewise smooth boundary ∂D. Consider a mapping (1), holomorphic on the closed domain $\bar{D}$ and having no zeroes on ∂D. The mapping (1) in this case has only a finite number of zeroes inside D. We consider the function $\phi \in A_c(D)$ (holomorphic in D and continuous in $\bar{D}$) and pose the problem of computing the sum

$$\sum_{a \in Z_f} \phi(a), \tag{3}$$

where Z_f is the set of zeroes of the mapping (1) in D and where the number of times each zero appears in the sum (3) is equal to the multiplicity of the zero. The sum (3) can be written as integrals of various dimensions, similar to the integral representation formula for holomorphic functions of n complex variables (in various circumstances the dimension of the integral can vary from n to $2n - 1$). The formula for the integral representation of the sum (3) (the formula of the multidimensional logarithmic residue) is best-known and has the most applications either when the integration is over the entire $(2n - 1)$-dimensional boundary ∂D or in the case when the integration is over an n-dimensional skeleton. We will present these formulas, assuming throughout that we are given a map $f \in A^n(\bar{D})$ (holomorphic in the closed domain $\bar{D}$) and a function $\phi \in A_c(D)$ and that multiplicities are taken into account in the sum (3).

Consider the following exterior differential form, which will be important in the sequel. The form depends on the mapping f, on a continuous vector function $w^{(0)}$ and on the continuously differentiable vector functions $w^{(1)}, \ldots, w^{(n-1)}$;

$$\Omega(w^{(0)}, w^{(1)}, \ldots, w^{(n-1)}, f)$$

$$= \frac{(-1)^{n(n-1)/2}}{(2\pi i)^n} \frac{\langle w^{(0)}, df \rangle}{\langle w^{(0)}, f \rangle} \wedge d\frac{\langle w^{(1)}, df \rangle}{\langle w^{(1)}, f \rangle} \wedge \cdots \wedge d\frac{\langle w^{(n-1)}, df \rangle}{\langle w^{(n-1)}, f \rangle},$$

where

$$\langle w, f \rangle = \sum_{i=1}^{n} w_i f_i, \quad \langle w, df \rangle = \sum_{i=1}^{n} w_i \, df_i.$$

Theorem 1 (Leray-Koppelman formula (Aizenberg-Yuzhakov (1979)). *If the vector functions* $w^{(j)}, j = 0, 1, \ldots, n - 1$ *satisfy the condition*

$$\langle w^{(j)}(z), f(z) \rangle \neq 0, \quad z \in D, \quad j = 0, 1, \ldots, n - 1,$$

then the following formula holds:

$$\int_{\partial D} \phi \Omega(w^{(0)}, w^{(1)}, \ldots, w^{(n-1/}, f) = \sum_{a \in Z_f} \phi(a). \tag{4}$$

Corollary 1 (Roos (1974)). *If the vector function* $w \in C^{(1)}(\partial D)$ *is such that* $\langle w, f \rangle \neq 0$ *on* ∂D, *then*

$$\int_{\partial D} \phi \omega(w, f) = \sum_{a \in Z_f} \phi(a), \tag{5}$$

where

$$\omega(w, f) = \frac{(n-1)!}{(2\pi i)^n} \frac{\sum_{k=1}^{n} (-1)^{k-1} w_k \, dw[k] \wedge df}{\langle w, f \rangle^n},$$

$$df = df_1 \wedge \cdots \wedge df_n,$$

$$dw[k] = dw_1 \wedge \cdots \wedge dw_{k-1} \wedge dw_{k+1} \wedge \cdots \wedge dw_n.$$

Corollary 2 (Aizenberg-Yuzhakov (1979), Roos (1974)).

$$\int_{\partial D} \phi\omega(\bar{f}, f) = \sum_{a \in Z_f} \phi(a). \qquad (6)$$

Formulas (5) and (6) follow from (4), since

$$\Omega(w, w, \ldots, w, f) = \omega(w, f).$$

On the other hand, the general formula (4) is obtained from formulas (6) and (5), which were already known. This is so because one can show that if the vector functions $w^{(1)}, \ldots, w^{(n-1)}, p^{(1)}, \ldots, p^{(n-1)}$ are in the class $C^{(2)}(\partial D)$, then the difference

$$\Omega(w^{(0)}, w^{(1)}, \ldots, w^{(n-1)}, f) - \Omega(p^{(0)}, p^{(1)}, \ldots, p^{(n-1)}, f)$$

is a $\bar{\partial}$-exact form. Consequently, it is orthogonal to holomorphic functions ϕ with respect to integration over ∂D.

If $f = z - a$, then the logarithmic residue formulas (4)–(6) reduce to, respectively, the general integral representation formula of Leray-Koppelman, the Leray formula, and the Bochner-Martinelli formula.

We observe further that the form $\Omega(w^{(0)}, w^{(1)}, \ldots, w^{(n-1)}, f)$ does not depend on $w^{(0)}$. Thus, the multidimensional logarithmic residue formula for $n > 1$ contains arbitrary choices (the choice of the vector function $w^{(1)}, \ldots, w^{(n-1)}$ in (4) or the vector functions w in (5)). In some cases it is useful to pick the formula for the multidimensional logarithmic residue according to the nature of the problem at hand (see § 3).

To formulate another important multidimensional logarithmic residue formula we will consider special analytic polyhedra

$$D = \{z : z \in G, |f_j| < \rho_j, j = 1, \ldots, n\},$$

where $f \in A^n(G)$, $\bar{D} \subset G \subset \mathbb{C}^n$.

Theorem 2 (Caccioppoli (1949), Martinelli (1955), Bishop (1961), Sorani (1962)). *If D is a special analytic polyhedron, then*

$$\frac{1}{(2\pi i)^n} \int_\Gamma \phi(z) \frac{df(z)}{f(z)} = \sum_{a \in Z_f} \phi(a), \qquad (7)$$

where $\Gamma = \{z : z \in \bar{D}, |f_j(z)| = \rho_j, j = 1, \ldots, n\}$ is the skeleton of this polyhedron.

Using the Stokes Theorem, one can lower the dimension of integration and deduce formula (7) from formula (6), and conversely. This is similar to the way in which the Martinelli-Bochner integral representation is obtained from the Bergmann-Weil integral representation for functions holomorphic in analytic polyhedra (see Volume 7, Chapter II of this series).

For applications the following result is useful; it includes a variant of *Rouché's principle* and a multidimensional logarithmic residue formula.

Proposition 1 (Yuzhakov (Aizenberg-Yuzhakov (1979)). *Let D and $\Gamma = \Gamma_f$ be as in Theorem 2 and let the mappings $f, g \in A^n(\overline{D})$ satisfy the inequalities $|g_j(z)| < |f_j(z)|, j = 1, \ldots, n$, on Γ.*

Then

1) *the cycles Γ_f and*

$$\Gamma_{f+g} = \{z : z \in G, |f_j(z) + g_j(z)| = \varepsilon, j = 1, \ldots, n\}$$

are homologous in the domain $G \setminus \{z : \prod_{j=1}^{n} [f_j(z) + g_j(z)] = 0\}$;

2) *the mappings f and $f + g$ have the same number of zeroes (counting multiplicity) and*

$$\frac{1}{(2\pi i)^n} \int_\Gamma \phi(z) \frac{d(f+g)}{f+g} = \sum_{a \in Z_{f+g}} \phi(a). \tag{8}$$

We observe that in formula (8) the integration is taken over the skeleton of a special analytic polyhedron corresponding to the mapping f (and not to $f + g$).

A number of multidimensional variants of Rouché's Theorem are also known in which the conditions on the mappings are given for the entire boundary ∂D of the domain D. For example, when any one of the following conditions are satisfied, the mappings f and $\phi \in A^n(\overline{D})$ have the same number of zeroes in D (counting multiplicity):

1) on ∂D the inequality $|f - \phi| < |f| + |\phi|$ holds;
2) on ∂D the real part, $\mathrm{Re}(f_1 \bar\phi_1 + \cdots + f_n \bar\phi_n) > -|f||\phi|$;
3) the set

$$\left\{\alpha : \alpha \in \mathbb{C}^1, \alpha = \frac{f_1(z)}{\phi_1(z)} = \cdots = \frac{f_n(z)}{\phi_n(z)}, z \in \partial D\right\}$$

does not separate the points 0 and ∞.

The previous theorem includes the classical case of a discrete set of zeroes for the mapping (1). Let us consider the more general case of a holomorphic mapping of a complex manifold X of complex dimension n to $\mathbb{C}^p$, $1 \le p \le n$. For such a mapping

$$f = (f_1, \ldots, f_p) : X \to \mathbb{C}^p$$

we set

$$Z_f = \{z \in X : f_1(z) = \cdots = f_p(z) = 0\}$$

and introduce the *Martinelli form*

$$\omega_{p,p-1}(f) = \frac{(p-1)!}{(2\pi i)^p} \frac{\sum_{\alpha=1}^{p} (-1)^{\alpha-1} \bar{f}_\alpha \, d\bar{f}[\alpha] \wedge df}{|f|^{2p}},$$

where $|f|^2 = |f_1|^2 + \cdots + |f_n|^2$. Let us consider the case $1 \le p \le n - 1$ under two conditions:

(i) If C_f is the critical set of f, then the analytic set $Z_f \cap C_f$ has at each point complex dimension no greater than $n - p - 1$.

We write $\hat{Z}_f = Z_f \setminus C_f$. If D is a relatively compact open subset of X, then the integral of a continuous $2(n - p)$-form over $Z_f \cap D$ can be defined to be the integral over $\hat{Z}_f \cap D$.

We assume that the boundary ∂D is piecewise smooth and

(ii) the set $\hat{Z}_f \cap \partial D$ has measure 0 in $\hat{Z}_f$.

Now we can formulate a result generalizing formula (6).

Theorem 3 (Lupacciolu (1979)). *Assuming* (i) *and* (ii) *are true, then for any* $\bar{\partial}$-*closed form* $\phi_{n-p,n-p}$ *of type* $(n - p, n - p)$ *which is smooth on* X,

$$\int_{\partial D} \omega_{p,p-1}(f) \wedge \phi_{n-p,n-p} = \int_{Z_f \cap D} \phi_{n-p,n-p}. \qquad (9)$$

For $p = n$ the $\bar{\partial}$-closed form $\phi_{n-p,n-p}$ is actually a holomorphic function and formula (9) reduces to formula (6).

If X is a Kähler manifold with Kähler form Ω, the form $\phi_{n-p,n-p}$ can be taken to be the form $\frac{1}{(n-p)!} \Omega^{n-p}$; we obtain the following corollary.

Corollary 1. *If* (X, Ω) *is a Kähler manifold and if* f *and* D *satisfy conditions* (i) *and* (ii), *then*

$$\frac{1}{(n-p)!} \int_{\partial D} \omega_{p,p-1}(f) \wedge \Omega^{n-p} = V_{2n-2p}(Z_f \cap D), \qquad (10)$$

where V_{2n-2p} *denotes the* $(2n - 2p)$-*dimensional volume.*

Example 2. Let X be an open manifold in $\mathbb{C}^n$ and let Ω be the standard Kähler form

$$\Omega = \frac{i}{2} \sum_{\alpha=1}^{n} dz_\alpha \wedge d\bar{z}_\alpha,$$

then (10) reduces to the following equation

$$\left(\frac{i}{2}\right)^{n-p} \int_{\partial D} \omega_{p,p-1}(f) \wedge \sum_{1 \le \alpha_1 < \cdots < \alpha_{n-p}} dz_{\alpha_1} \wedge d\bar{z}_{\alpha_1} \wedge \cdots \wedge dz_{\alpha_{n-p}} \wedge d\bar{z}_{\alpha_{n-p}}$$

$$= V_{2n-2p}(Z_f \cap D),$$

which for the case $p = 1$ is contained in the work of Wirtinger (1937).

Now we introduce a generalization of formula (7) in the case of a holomorphic mapping $f = (f_1, \ldots, f_p)$ of an open set $D \subset \mathbb{C}^n$ to $\mathbb{C}^p$. We recall (Dolbeault (1985), Sect. 3.4) the inductive definition of an *essential intersection* $C_e(f_1, \ldots, f_p)$:

$$C_e(f_1) = f_1^{-1}(0), \quad C_e(f_1, \ldots, f_s) = \tilde{C}_e(f_1, \ldots, f_{s-1}) \cap f_s^{-1}(0),$$

where $\tilde{C}_e(f_1, \ldots, f_{s-1})$ is the sum of those irreducible components of the cycle $C_e(f_1, \ldots, f_{s-1})$ on which f_s is not identically equal to 0. We consider the following *residue current* $R_f[\phi]$ of Coleff and Herrera (Coleff-Herrera (1978), see also (Dolbeault (1985), Sect. 3.5)); for the form $\phi \in C^\infty$ with compact support in D

$$R_f[\phi] = \lim_{\delta \to 0} \int_{T_\delta(f)} \frac{df}{f_1 \cdots f_p} \wedge \phi,$$

where $T_\delta(f) = \{z : |f_j(z)| = \delta_j, j = 1, \ldots, p\}$ under the condition that $(\delta_j/\delta_{j+1}^k) \to 0$ for all $k > 0$, $1 \leq j \leq p - 1$.

Theorem 4. (Coleff-Herrera (1978), Colomin (1977)) *For any $2(n - p)$-form $\phi \in C^\infty$ with compact support in D,*

$$R_f[\phi] = (2\pi i)^p \int_{C_e(f_1, \ldots, f_p)} \phi.$$

We observe that in this theorem there are no conditions on the intersection of the zero sets of the functions f_s, $s = 1, \ldots, p$ (the intersections are possibly not complete).

Besides the formulas given above for the multidimensional logarithmic residue, there are formulas of another type, but with a similar right-hand side of the equation. They are written in terms of currents, and the corresponding integration is over the whole complex manifold. We will use the following notation.

$$d = \partial + \bar{\partial}, \quad d^c = \frac{i}{4\pi}(\bar{\partial} - \partial)$$

and D_f is the divisor of the meromorphic functon f.

Theorem 5 (Poincaré-Lelong formula (Lelong (1968))). *Let F be a meromorphic functon on a complex manifold X. Then this equation of currents is true:*

$$dd^c \log |f|^2 = D_f. \tag{10}$$

Formula (10) means that for any C^∞-form $\phi = \phi_{n-1,n-1}$ with compact support in X, there is an equality

$$\int_X \log |f|^2 \, dd^c \phi = \int_{D_f} \phi.$$

In the case of a holomorphic function f, the Poincaré-Lelong formula can be generalized to the case when f is a holomorphic mapping from an n-dimensional complex manifold X to $\mathbb{C}^p$. We denote

$$f^* \theta_l = (dd^c \log |f|^2)^l.$$

Theorem 6 (Poincaré-Martinelli formula (King (1971), Griffiths-King (1973))). *The form $f^*\theta_l$ and the form $\log|f|^2 f^*\theta_l$ are locally integrable on X for all l. If $Z_f = f^{-1}(0)$ has dimension $n - p$, then for $l < p$,*

$$dd^c(\log|f|^2 f^*\theta_{l-1}) = f^*\theta_l$$

and

$$dd^c(\log|f|^2 f^*\theta_{p-1}) = Z_f$$

where Z_f is counted with multiplicity. In other words,

$$\int_X \log|f|^2 f^*\theta_{p-1} \wedge dd^c \phi = \int_{Z_f} \phi \qquad (11)$$

for any C^∞-form $\phi = \phi_{n-p,n-p}$ with compact support in X.

From the Poincaré-Lelong and the Poincaré-Martinelli formulas, one can deduce the usual formulas for the multidimensional logarithmic residue. For example, we show how to obtain formula (6) from (11) for $p = n$ and discrete Z_f. In this case the right-hand sides of (6) and (11) are identical. We consider the left side of (11) for $\phi_{0,0} = \phi\chi$, where $\phi \in A(\overline{D})$ and χ is a function of class C^∞ with support in a domain $D_1 \supset \overline{D}$ and $\chi|_{\overline{D}} = 1$ (assuming the boundary ∂D is sufficiently smooth), with $(\overline{D}_1 \setminus D) \cap Z_f = \varnothing$. By the Stokes Formula

$$\int_{D_1} \log|f|^2 f^*\theta_{n-1} \wedge \frac{i}{2\pi} \partial\overline{\partial}\phi_{0,0} = -\frac{i}{2\pi} \int_{D_1 \setminus \overline{D}} \partial \log|f|^2 \wedge f^*\theta_{n-1} \wedge \overline{\partial}\phi_{0,0}$$

$$= -\frac{i}{2\pi} \int_{D_1 \setminus \overline{D}} d(\partial \log|f|^2 \wedge f^*\theta_{n-1}\phi_{0,0})$$

$$+ \int_{D_1 \setminus \overline{D}} f^*\theta_n\phi_{0,0}.$$

Using $f^*\theta_n = 0$ and applying the Stokes Theorem again, we obtain

$$-\frac{i}{2\pi} \int_{\partial(D_1 \setminus \overline{D})} \partial \log|f|^2 \wedge f^*\theta_{n-1}\phi_{0,0} = \int_{\partial D} \phi\omega(\overline{f}, f).$$

To conclude, we will demonstrate versions of Theorems 5 and 6 generalized to the case of line bundles $L \to M$, where M is a complex manifold (for the generalization of the concept of residue to such a case, see Chapter 3, Subsection 3.5). Suppose that the transition functions of L are $\{g_{jk}\}$; a metric on L is given by positive C^∞ functions ρ_j on neighborhoods U_j with relationships

$$\rho_j = |g_{jk}|\rho_k \quad \text{in } U_j \cap U_k.$$

Given this, the $(1, 1)$-form ω given by the equation

$$\omega|_{U_j} = dd^c \log \rho_j,$$

is defined globally and is called the curvature form of the line bundle $L \to M$.

Now take any global holomorphic section $\sigma \in H^0(M, L)$ with divisor D. The function $\log |\sigma|^2$ is locally integrable on M.

Theorem 5' (Griffiths-King (1973)). *On M the following equation of currents is true*:

$$dd^c \log |\sigma|^2 = D - \omega.$$

Intersection in homology is dual to exterior product in de Rham cohomology, so using Theorem 5, we find that Theorem 6 has the following natural analog for holomophic sections $\sigma_1, \ldots, \sigma_r$ of the line bundle $L \to M$.

Theorem 6' (Griffiths-King (1973)). *If the divisors D_{σ_i} intersect in a set of complex codimension r, then this equation of currents is true*,

$$\omega^r - D_{\sigma_1} \cdot D_{\sigma_2} \cdot \cdots \cdot D_{\sigma_r} = dd^c \Lambda,$$

where Λ is the locally integrable form

$$\Lambda = \log \frac{1}{|\sigma|^2} \sum_{k=0}^{r-1} \omega_0^{r-1-k} \wedge \omega^k, \quad \omega_0 = \omega + dd^c \log |\sigma|^2.$$

Moreover, if $\omega \geq 0$, and $|\sigma| \leq 1$, then $\Lambda \geq 0$.

We mention yet another set of directions, related to the generalization or the further study of the multidimensional logarithmic residue (formulas (6) and (7)) (see Aizenberg, Yuzhakov (1979)):

1. In formulas (6) and (7) the inegration is taken over cycles of dimension $2n - 1$ or n. One can also find formulas for the case of cycles of intermediate dimension (Yuzhakov, Kuprikov).

2. The cycle Γ in formula (7) can be replaced by a cycle of a more general nature, called a "separating" cycle. These cycles were studied by Martinelli, Sorani, and Tsikh.

3. One can construct the multidimensional logarithmic residue formula based not on the Bochner-Martinelli integral representation but rather on the more general integral representation of Andreotti-Norguet. In this direction we mention the results of Norguet, Aizenberg and Bolotov.

§2. Series Expansion of Implicit Functions

The easiest corollaries of the Cauchy formula for holomorphic functions of one complex variable are the Taylor expansion and the Laurent expansion for functions holomorphic in a disk or, respectively, in an annulus. Analogously, the easiest corollaries of the logarithmic residue formula for holomorphic functions of one variable are the Lagrange expansion and the Bürmann-Lagrange series. This allows one to represent one holomorphic function in the form of a series in the powers of a second holomorphic function, e.g., the formula for inverting a holomorphic function, etc. In this section, using the multidimensional logarithmic residue (mainly formula (8)), we will introduce some generalizations of these expansions.

Let $\Phi(w, z)$ and $F_j(w, z), j = 1, \ldots, n$, be holomorphic functons of the variables $w = (w_1, \ldots, w_m)$ and $z = (z_1, \ldots, z_n)$ in a neighborhood of a point $(0, 0) \in \mathbb{C}^{m+n}$ such that $F_j(0, 0) = 0$, for $j = 1, \ldots, n$, with $(\partial F/\partial z)|_{(0,0)} \neq 0$. The system of equations

$$F_j(w, z) = 0, \quad j = 1, \ldots, n,$$

defines a system of functions

$$z = \phi(w): z_j = \phi_j(w), \quad j = 1, \ldots, n,$$

holomorphic in a neighborhood of the point $0 \in \mathbb{C}^m$. The problem is to expand the function $\Phi(w, \phi(w))$ in a series. Without loss of generality, we may assume that

$$F_j(0, 0) = 0, \quad \frac{\partial F_j(0, 0)}{\partial z_k} = \delta_{jk}, \quad j, k = 1, \ldots, n.$$

Theorem 7 (Yuzhakov (1975), Aizenberg-Yuzhakov (1979), § 20)). *The function* $\Phi(w, \phi(w))$ *is represented by the following series of functions, which converges absolutely and uniformly in a neighborhood of the origin.*

$$\Phi(w, \phi(w)) = \sum_{\beta \geq 0} \frac{(-1)^{|\beta|}}{\beta!} D_z^\beta \left[\Phi(w, z) g^\beta(w, z) \frac{\partial F}{\partial z} \right]\Bigg|_{z = h(w)}, \qquad (10)$$

where $h = (h_1, \ldots, h_n)$ *is an arbitrary vector-valued function, holomorphic in this neighborhood, with the condition that* $h(0) = 0$;

$$g^\beta = g_1^{\beta_1} \ldots g_n^{\beta_n}, \quad \beta! = \beta_1! \ldots \beta_n!, \quad |\beta| = \beta_1 + \cdots + \beta_n;$$

$$g_j(w, z) = F_j(w, z) - z_j + h_j(w), \quad j = 1, \ldots, n;$$

$$D_z^\beta = \frac{\partial^{|\beta|}}{\partial z_1^{\beta_1} \ldots z_n^{\beta_n}};$$

the notation $\beta \geq 0$ *means that all the* $\beta_j \geq 0, j = 1, \ldots, n$.

Corollary. *Let the mapping* (1) *be holomorphic in a neighborhood of the point* 0 *and satisfy the condition*

$$f_j(0) = 0, \quad \frac{\partial f_j(0)}{\partial z_k} = \delta_{jk}, \quad j, k = 1, \ldots, n,$$

and let the function $\Phi(z)$ *also be holomorphic at* 0. *Then in some neighborhood of the point* 0 *the following expansion is valid*

$$\Phi(\phi(w)) = \sum_{\beta \geq 0} \frac{(-1)^{|\beta|}}{\beta!} D^\beta \left[\Phi(z) \theta^\beta(z) \frac{\partial f}{\partial z} \right]\Bigg|_{z = w}, \qquad (11)$$

where $\theta_j(z) = f_j(z) - z_j, j = 1, \ldots, n$. *For* $\Phi(z) \equiv z_j, j = 1, \ldots, n$, *formula* (11) *represents the inverse of the holomorphic mapping* (1).

If the functions $\Phi(w, z)$ and $F_j(w, z)$, for $j = 1, \ldots, n$, are given in a neighborhood of the origin by their Taylor series, then one can also express the Taylor series of the function $\Phi(w, \phi(w))$. As a corollary of this exapnsion, one obtains the result of Cayley-Sylvester-Sack (see Ajzenberg-Yuzhakov (1983), § 20)). For the series (10) and (11), one can give estimates for the domains of convergence and the remainders (loc. cit.).

Finally, the multidimensional logarithmic residue can also be applied to systems of equations where $\partial F/\partial z|_{(0,0)} = 0$. This permits one in some cases to separate a holomorphic single-valued branch of an implicit vector-valued function.

Example. We separate the holomorphic branches of the complex curve $z^3 - 3wz + w^3 = 0$ in a neighborhood of the point $(0, 0) \in \mathbb{C}^2$. Applying the method described above, one can show that the branch tangent to the complex line $\{(z, w): z = 0\}$ has the form

$$z = \sum_{k=0}^{\infty} c_k w^k,$$

where

$$c_k = \frac{(3k)!}{3^{3k+1} k! (2k + 1)!}.$$

§ 3. Application of the Multidimensional Logarithmic Residue to Systems of Nonlinear Equations

We will investigate the system of algebraic equations

$$Q_j(z) + P_j(z) = 0, \quad j = 1, \ldots, n \tag{12}$$

where the $Q_j(z)$ are homogeneous polynomials with k_j being the highest degree in the variables jointly. We further assume that the only common zero of the polynomials is the origin and that the degree of each P_j is less than $k_j, j = 1, \ldots, n$. It is easy to show that system (12) has a finite number of solutions equal to $N = k_1 k_2 \ldots k_n$. Let $m_1, \ldots, m_n$ be natural numbers such that the equation

$$\sum_{j=1}^{n} w_j Q_j(z) = \sum_{j=1}^{n} |z_j|^{2m_j} \tag{13}$$

has a solution of the form

$$w_j = \sum_{k=1}^{n} a_{jk}(z) \bar{z}_k^{m_k}, \quad j = 1, \ldots, n, \tag{14}$$

where the a_{jk} are polynomials in z. According to the well-known theorem of Macauley (see Chapter 3, Theorem 4.1), this condition is automatically satisfied for $m_j = |k| + 1 - n$, for $j = 1, \ldots, n$ and $|k| = k_1 + \cdots + k_n$; but sometimes m_j

can be chosen smaller. The solution (14) of equation (13), which in general is not unique, can be found by the method of undetermined coefficients. Now one can apply formula (5) to the vector-valued function w in the domain

$$D_\rho = \{z : |z_1|^{2m_1} + \cdots + |z_n|^{2m_n} < \rho\},$$

which results in the following assertion.

Theorem 8 (Aizenberg (Aizenberg-Yuzhakov (1979), § 21)). *Let $R(z)$ be a polynomial of degree μ; then*

$$\sum_e R(z^{(e)}) = \mathcal{M}\left[R\Delta_1\Delta_2 \sum_{j=0}^{\mu} \frac{(-1)^j}{j!} \langle w, p\rangle^j\right], \tag{15}$$

where Δ_1 is the jacobian of system (12) and $\Delta_2 = \det\|a_{jk}\|$. The summation in the left side of (15) is taken over all the zeroes $z^{(e)}$ (counting multiplicity) of system (12), and $\mathcal{M}$ is the linear functional on the polynomials in $z_1, \ldots, z_n, \bar{z}_1^{m_1}, \ldots, \bar{z}_n^{m_n}$ defined by the equation

$$\mathcal{M}\left[z^\beta \bar{z}_1^{m_1\alpha_1} \ldots \bar{z}_n^{m_n\alpha_n}\right] = \begin{cases} \alpha!, & \text{if } \beta_j = m_j\alpha_j + m_j - 1, j = 1, \ldots, n, \\ 0, & \text{otherwise.} \end{cases}$$

Corollary 1.

$$\sum_e R(z^{(e)}) = \frac{t}{s^{\mu+\alpha}},$$

where t is a polynomial in the coefficients of system (12) and the polynomial R, while s is a polynomial only in the coefficients of the polynomials $Q_1, \ldots, Q_n$.

Formula (15) takes an especially simple form in the case of the system

$$z_j^{k_j} + P_j(z) = 0, \quad j = 1, \ldots, n \tag{16}$$

where the degree P_j is less than k_j for $j = 1, \ldots, n$. Here one can consider $m_i = k_i$, $i = 1, \ldots, n$. Then $w_i = \bar{z}_i^{k_i}$, the determinant $\Delta_2 = 1$ and we arrive at the following proposition.

Corollary 2. *For system (16) and a polynomial $R(z)$ of degree μ, the following formula is true:*

$$\sum_e R(z^{(e)}) = \mathcal{N}\left[R\Delta_1 \frac{z_1 \ldots z_n}{z_1^{k_1} \ldots z_n^{k_n}} \sum_{|\alpha|=0}^{\mu} (-1)^{|\alpha|} \left(\frac{P_1}{z_1^{k_1}}\right)^{\alpha_1} \ldots \left(\frac{P_n}{z_n^{k_n}}\right)^{\alpha_n}\right], \tag{17}$$

where $\mathcal{N}$ is the linear functional on the polynomials in $z_1, \ldots, z_n, 1/z_1, \ldots, 1/z_n$ which associates to any such polynomial its free term.

Using formulas (15) and (17) one can compute power series, for example the first coordinates of the roots of

$$\sum_{e=1}^{N} (z_1^{(e)})^j = s_j, \quad j = 1, \ldots, N.$$

The coefficients of the polynomial $\Omega(z_1) = z_1^N + b_1 z_1^{N-1} + \cdots + b_{N-1} z_1 + b_N$, having roots $z_1^{(1)}, \ldots, z_1^{(N)}$ given by Waring's formula or the recurrence formula of Newton

$$jb_j = -s_j - s_{j-1}b_1 - \cdots - s_1 b_{j-1}, \quad j = 1, \ldots, N, \tag{18}$$

are expressed in terms s_j.

Thus formulas (15) and (17) lead to a *new method for eliminating unknowns*, which does not add extra roots and which does not omit any roots. This method seems to us to be simpler than the classical methods of elimination using the resultants of polynomials.

Formula (17) leads to a particularly simple computation in the case when the degree of the polynomial $R(z)$ is small.

Example 1. Consider in $\mathbb{R}^3$ the three surfaces of third order

$$X_1^3 + \sum_{i+j+k \leq 2} a_{ijk} X_1^i X_2^j X_3^k = 0,$$

$$X_2^3 + \sum_{i+j+k \leq 2} b_{ijk} X_1^i X_2^j X_3^k = 0, \tag{19}$$

$$X_3^3 + \sum_{i+j+k \leq 2} c_{ijk} X_1^i X_2^j X_3^k = 0,$$

where a_{ijk}, b_{ijk}, and c_{ijk} are real. Let the surfaces in (19) be in "general position" in the sense that they have 27 points in common in $\mathbb{R}^3$, the maximum possible number. We fix a point $(A, B, C) \in \mathbb{R}^3$ and compute using formula (17) the sum of the squares of the distances from this point to the 27 common points of the surfaces (19). We find that the sum we seek is equal to

$$9(a_{200}^2 + b_{020}^2 + c_{002}^2) - 18(a_{100} + b_{010} + c_{001}) + 6(a_{101}c_{002} + a_{110}b_{020}$$

$$+ a_{200}b_{110} + b_{011}c_{002} + a_{200}c_{101} + b_{020}c_{011}) + 12(a_{002}c_{101} + a_{020}b_{110}$$

$$+ a_{110}b_{200} + b_{002}c_{011} + b_{011}c_{020} + a_{101}c_{200}) + 27(A^2 + B^2 + C^2)$$

$$+ 18(Aa_{200} + Bb_{020} + Cc_{002}).$$

It is curious that the answer does not depend on 12 of the 30 coefficients of the equations of the surfaces (19).

Formula (15) has found an application in the determination of all stationary solutions of certain chemical kinetic equations (see Aizenberg et al. (1983)).

Example 2. In the study of the reaction of the oxidation of hydrogen appear the following equations of stationarity with respect to intermediate substances:

$$2k_1 z^2 - 2k_{-1} x^2 - k_4 xy + k_{-4} zu - k_3 x - k_{-3} z = 0,$$

$$2k_2 z^2 - 2k_{-2} y^2 - k_4 xy + k_{-4} zu - k_5 yu = 0,$$

$$k_4 xy - k_{-4} zu - k_5 yu - 2k_6 u^2 = 0 \tag{20}$$

$$x + y + z + u = 1.$$

After eliminating $u = 1 - x - y - z$ and making the change of variable $z = ty$, we apply formula (15) to the resulting nonlinear system of equations. This allows us to eliminate in the general form all the unknowns except t. One obtains a particularly simple expression when $k_6 = 0$. In this case the desired polynomial is

$$t^2(p_0 t^6 + p_1 t^5 + \cdots p_5 t + p_6),$$

where

$$p_0 = k_2^2 k_{-4}^2 (2k_{-1} + k_3),$$

$$p_1 = k_2 k_{-4}(4k_2 k_5 k_{-1} + 2k_2 k_3 k_5 + k_2 k_3 k_4),$$

$$p_2 = 2k_2^2 k_5^2 k_{-1} + k_2^2 k_3 k_5^2 + k_2^2 k_3 k_4 k_5 + k_2 k_3 k_4 k_5 k_{-4} - k_2 k_4 k_5 k_{-3} k_{-4},$$

and so forth. Moreover, applying the classical methods of Descartes and Budan-Fourier, one can investigate the number of positive roots of system (20) and write down a condition on the parameters (coefficients of system (20)) guaranteeing the uniqueness of the stationary condition or of a certain number of stationary conditions; and this in its turn leads to some information about the chemical reaction in question.

The results cited above have developed in the following directions:

1. Instead of system (16) on can consider a system of more general form

$$z_i^{k_i} + \sum_{j=1}^{i-1} z_j \phi_{ij}(z) + p_i(z) = 0, \quad i = 1, \ldots, n, \tag{21}$$

where the $\phi_{ij}(z)$ are homogeneous polynomials of degree $k_i - 1$ and the $p_i(z)$ are polynomials of degree no greater than $k_i - 1$. Under these conditions, formula (17) remains essentially true. Moreover, instead of system (16) or (21), one can take analogous systems in which the ordinary homogeneous polynomials are replaced by weighted-homogeneous polynomials (Aizenberg-Tsikh, (Aizenberg-Yuzhakov (1983), § 21)). More general systems were investigated by Yuzhakov (in the collection *Some Problems of Multidimensional complex Analysis*, Krasnoyarsk, 1980, 197–214).

2. A multidimensional analog of Waring's formula for systems of the form (21) was found by Bolotov (Aizenberg-Yuzhakov, 1979, § 21).

3. A multidimensional analog of Newton's formula for systems of nonlinear algebraic equations of a different type was proved by Aizenberg and Kytmanov (Siberian Math. Journal 22, No. 2, 19–30 (1981)).

4. On the basis of formulas (15) and (17), Aizenberg proved a relatively simple algorithm which provides an answer to the question of how many real roots the system (12) or (16) has, about the number of roots in a given ellipsoid, etc. (see loc. cit.).

5. The problem of computing the remaining coordinates of the roots if the first coordinates have already been determined, and also the problem of solving nonlinear algebraic systems in radicals has been considered in the work of L.A. Aizenberg, B.A. Bolotov, and A.K. Tsikh (Dokl. Akad. Nauk USSR 252, No. 1, 11–14 (1980)).

§4. Computation of the Zero-Multiplicity of a Holomorphic Mapping

We consider a mapping (1), holomorphic in a neighborhood of a point $a \in \mathbb{C}^n$. Suppose that a is an isolated zero of this mapping with multiplicity $\mu_a(f)$. When $n = 1$ the multiplicity $\mu_a(f)$ only depends on the index of the first non-zero Taylor coefficient of the function f at the point a. For $n > 1$ the situation is significantly more complicated. For example, the systems

$$(z_1 + z_2 + z_1^2, z_1 - z_2 + z_2^2) \text{ and } (z_1 + z_2 + z_1^2, z_1 + z_2 + z_2^2)$$

are made up of the same collection of monomials, but the multiplicity of the zero at $(0, 0)$ differs for the two maps. But for almost all systems (1) the multiplicity $\mu_a(f)$ is nonetheless determined only by the monomials which appear in f_i with non-zero coefficients, in other words, the sets

$$s_i = \operatorname{supp} f_i = \{\alpha \in \mathbb{N}^n, c_{i\alpha} \neq 0\},$$

where

$$f_i = \sum_\alpha c_{i\alpha}(z - a)^\alpha.$$

Next we will study an important concept introduced by Tsikh, *the resultant of a function ψ with respect to a system* (1), denoted by $R(f, \psi)$. If system (1) is holomorphic in a domain $D \subset \mathbb{C}^n$, with a finite set of zeroes $\{z^{(e)}\}$, then $R(f, \psi) = \prod_e \psi(z^{(e)})$, where the number of times each zero appears in the product is equal to its multiplicity. With the help of formula (7), one obtains

Theorem 9. (Tsikh (Aizenberg-Yuzhakov (1979), §22)) *The multiplicity of an isolated zero $z = 0$ of system* (1) *is equal to the multiplicity of the zero $z_n = 0$ of the resultant of the function f_n relative to the system ${}'f = (f_1, \ldots, f_{n-1})$:*

$$\mu_0(f) = \mu_0(R({}'f, f_n)).$$

From this result and Rouché's principle, it is easy to show the theorem of Bezout concerning the number of isolated zeroes of a system of algebraic equations in $\mathbb{CP}^n$.

Moreover, let

$$f_j(z) = \sum_{|\alpha| \geq k_j} c_{i\alpha} z^\alpha, \quad j = 1, \ldots, n.$$

We denote by $d_0(f_j) = k_j$ the order of the zero at $z = 0$ of the function f_j at the point 0 and

$$p_j(z) = \sum_{|\alpha| = k_j} c_{i\alpha} z^\alpha, \quad j = 1, \ldots, n. \tag{22}$$

The most important corollary of Theorem 9 is

Corollary (Tsikh-Yuzhakov). *The multiplicity of an isolated zero $z = 0$ of system* (1) *is equal to the product of the orders of the zero at $z = 0$ of the functions of the system, $\mu_0(f) = d_0(f_1) \ldots d_0(f_f)$, if and only if 0 is an isolated zero of the* *system* (22) *of homogeneous polynomials. It is always true that*

$$\mu_0(f) \geq d_0(f_1)\dots d_0(f_f).$$

This result is easy to generalize to the case when the polynomials are weighted homogeneous polynomials rather than polynomials homogeneous in the ordinary sense (Tsikh, see (Aizenberg-Yuzhakov (1979)).

One can introduce a further generalization of the homogeneous principal part (22) of the system (1) by employing the concept of the *Newton polyhedron* $\Gamma_+(f_1,\dots,f_n)$ *of the system* (1) at the zero. This is the convex hull in $\mathbb{R}^n_+$ of the set

$$\{\alpha : \alpha \in \mathbb{R}^n_+, \alpha \in (\operatorname{supp} f_1 \cup \dots \cup \operatorname{supp} f_n)\backslash\{0\}\}.$$

Let $\Gamma(f_1,\dots,f_n)$ be the *Newton polygon*, the union of the closed faces of the polyhedron $\Gamma_+(f_1,\dots,f_n)$ and let $\Gamma_-(f_1,\dots,f_n)$ be the union of all the segments with beginning point at 0 and end point on $\Gamma(f_1,\dots,f_n)$. We will call the number $n!v(\Gamma_-)$ the *Newton number of the system* (1) at zero; here v is the n-dimensional volume. We include in $f_{j\Gamma}$ only those terms of the form $c_\alpha z^\alpha$ in the Taylor series of f_j for which $\alpha \in \Gamma, j = 1,\dots, n$. The polynomials $f_{j\Gamma}, j = 1,\dots, n$, are called the *principal parts of system* (1) *at zero*. System (1) is called nondegenerate at zero if and only if for any face σ of the Newton polygon the polynomial $f_{j\sigma}$ is not zero on the set $(\mathbb{C}\backslash 0)^n$. Using the methods of §1 we obtain

Theorem 10 (Kushnirenko (1975)). 1) *The multiplicity of the isolated zero $z = 0$ of system* (1) *is no less than the Newton number of the system at zero; if the system is nondegenerate at zero, then equality holds.* 2) *The set of principal parts of systems nondegenerate at zero and having a given Newton polygon, is an open dense subset of the manifold of all principal parts with given Newton polygon.*

§5. Application of the Multidimensional Logarithmic Residue to the Theory of Numbers

Using formula (6) one can obtain an integral formula for the difference between the number of integer points in a domain in the space $\mathbb{R}^n$ and its volume. A whole series of classical problems of number theory are problems of computing the asymptotic behavior of this difference. Thus, these problems reduce to the study of the asymptotics of the integrals below. We will restrict our attention to the cases $n = 2, 3$, which are important in classical problems.

Theorem 11 (Aizenberg (1983)). 1) *Let Q be a bounded domain in $\mathbb{R}^2$ with piece-wise smooth boundary ∂Q, and let $s(Q)$ be the area of Q. If there are no integer points on ∂Q, then*

$$N(Q) - s(Q)$$

$$= \frac{1}{\pi} \int_0^\infty \int_0^\infty dt_1 \wedge dt_2 \int_{\partial Q} \frac{t_2 \sin 2\pi x_1 \, dx_2 - t_1 \sin 2\pi x_2 \, dx_1}{(t_1^2 - 2t_1 \cos 2\pi x_1 + t_2^2 - 2t_2 \cos 2\pi x_2 + 2)^2}.$$

$$(23)$$

2) *Let Q be a bounded domain in $\mathbb{R}^3$ with piecewise smooth boundary. If there are no integer points on ∂Q, then*

$$N(Q) - v(Q)$$

$$= \frac{4}{\pi} \int_0^\infty \int_0^\infty \int_0^\infty dt_1 \wedge dt_2 \wedge dt_3$$

$$\times \int\int_{\partial Q} \frac{t_2 t_3 \sin 2\pi x_1 \, dx_2 \wedge dx_3 - t_1 t_3 \sin 2\pi x_2 \, dx_1 \wedge dx_3 + t_1 t_2 \sin 2\pi x_3 \, dx_1 \wedge dx_2}{(t_1^2 - 2t_1 \cos 2\pi x_1 + t_2^2 - 2t_2 \cos 2\pi x_2 + t_3^2 - 2t_3 \cos 2\pi x_3 + 3)^2},$$

$$\tag{24}$$

where $v(Q)$ is the volume of the domain Q.

The proofs of formulas (23) and (24) reduce to embedding the domain Q in $\mathbb{C}^n$ and considering a suitable extension to a domain $D_a \subset \mathbb{C}^n$ and a suitable mapping which has zeroes only at the integer points of the real subspace, and finally considering the limiting behavior with respect to a in formula (6), applied to this domain and this mapping. If we use the more general formula (5), then we can obtain generalizations of formulas (23) and (24). As the first application of formulas (23) and (24) we mention the following. If in the integrals (23) or (24) the kernel is expanded into a series for the case when Q is a disk (respectively, a ball) of radius t, then after integrating one obtains an expansion of the difference $N(Q) - s(Q)$ (respectively, $N(Q) - v(Q)$) into a series of Bessel functions I_1 (respectively, $I_{3/2}$). This series has several advantages in comparison with previously-known expansions of this kind: its coefficients can clearly be written in terms of Gamma functions and do not contain $r_k(t)$, the number of representations of t as the sum of k squares of integers, $k = 2, 3$, in contrast with the expansion of Voronoy-Hardy ($n = 2$) and Oppenheim ($n = 3$).

Chapter 3
The Grothendieck Residue and its Applications to Algebraic Geometry

A.K. Tsikh

Introduction

From the point of view of the theory of residues in complex analysis, the Grothendieck residue is represented as the integral of a meromorphic form of an extremely general type over a special cycle. It is the most precise multivariable analog of the Cauchy residue at a point for an arbitrary meromorphic function of one complex variable. The Grothendieck residue is defined locally: it is actually associated to what is called a regular sequence of germs $f = (f_1, \ldots, f_n)$ of holomorphic functions at some point $a \in \mathbb{C}^n$. It can be interpreted as a homomorphism

$$\text{res} : \mathcal{O}_a / I_a(f) \to \mathbb{C},$$

where $\mathcal{O}_a$ is the ring of germs of functions holomorphic at a, and $I_a(f)$ is the ideal in this ring generated by the sequence f. In the light of this algebraic interpretation, the Grothendieck residue is a useful and effective instrument in algebraic geometry. In addition, it generalizes the logarithmic residue for a mapping $w = f(z)$. Instead of the terminology "Grothendieck residue," we will also use the phrase "local residue" with respect to the mapping f.

§ 1. Integral Definition and Fundamental Properties of the Local Residue

1.1. Definitions. Let h and $f_1, \ldots, f_n$ be holomorphic functions in a neighborhood U_a of a point $a \in \mathbb{C}^n$ such that the mapping $f = (f_1, \ldots, f_n)$ has an isolated zero at the point $a : f^{-1}(0) \cap U_a = \{a\}$. The "local residue"[7] or "Grothendieck residue" of the meromorphic form $\omega = h\,dz/f \ldots f_n$ at the point a is defined to be the integral (Griffiths-Harris (1978), Tong (1973))

$$\operatorname*{res}_a \omega = \frac{1}{2\pi i} \int_{\Gamma_a} \frac{h(z)\,dz_1 \wedge \cdots \wedge dz_n}{f_1(z)\ldots f_n(z)}, \tag{1}$$

where

$$\Gamma_a = \{z \in U_a : |f_j(z)| = \varepsilon, j = 1, \ldots, n\}.$$

Here ε is taken sufficiently small and the cycle Γ_a is oriented by the condition

$$d(\arg f_1) \wedge \cdots \wedge d(\arg f_n) \ge 0.$$

We will also call the residue (1) the local residue of the function h with respect to the mapping f at the point a and denote it by $\operatorname*{res}_a{}_f(h)$.

We note that in the case $h = \phi \partial f / \partial z$, where $\partial f / \partial z$ is the jacobian of the mapping f, the local residue (1) reduces to the logarithmic residue (see Chapter 2) and is equal to the product $\mu_a(f) \cdot \phi(a)$, where $\mu_a(f)$ is the multiplicity of the zero at a of the mapping f; this coincides (see Palamodov (1967)) with the dimension of the vector space $\mathcal{O}_a / I_a(f)$.

The following proposition then follows from Stokes Formula.

Proposition 1.1 (Griffiths-Harris (1978), Tong (1973)). *The local residue vanishes for functions h which belong to the ideal $I_a(f)$.*

---

[7] In (Dolbeault (1985), Sect. 04) the term "point residue" is used instead of "local residue".

Thus, taking account of the linearity of the local residue with respect to functions $h \in \mathcal{O}_a$, we obtain a well-defined homomorphism

$$\operatorname*{res}_a{}_f : \mathcal{O}_a/I_a(f) \to \mathbb{C},$$

which is the residue mapping of Grothendieck (see Lomadze (1981), Beauville (1971), Carrell (1978), Hartshorne (1966)).

1.2. Representation of the Local Residue by an Integral over the Boundary of a Domain. Analogously to the way that the integral of $h\, df/f$, the logarithmic differential with weight h, can be expressed as an integral over the boundary of a domain of $h\beta(f, \bar{f})$, where β is the Bochner-Martinelli kernel (see Chapter 2), the local residue (1) can be represented in the form of an integral over a $(2n - 1)$-dimensional cycle.

Proposition 1.2 (Griffiths-Harris (1978)). *Let the mapping f be holomorphic in the closure of a neighborhood U_a with piecewise smooth boundary ∂U_a, where $f^{-1}(0) \cap \bar{U}_a = \{a\}$. Then the local residue*

$$\operatorname*{res}_a \omega = \int_{\partial U_a} \eta_\omega,$$

where

$$\eta_\omega = h \cdot \frac{(n-1)!}{(2\pi i)^n} \cdot \frac{\sum_{k=1}^n (-1)^{k-1} \bar{f}_k\, d\bar{f}_1 \wedge \cdots [k] \cdots \wedge d\bar{f}_n \wedge dz_1 \wedge \cdots \wedge dz_n}{(|f_1|^2 + \cdots + (|f_n|^2)^n}.$$

We remark that the form η_ω differs from $h \cdot \beta(f, \bar{f})$ by the absence of the jacobian $J = \partial f/\partial z$ just as the form ω differs from the weighted logarithmic differential $h\, df/f$. Instead of the form η_ω one can take another representative of the Dolbeault cohomology class $[\eta_\omega] \in H_{\bar{\partial}}^{n,n-1}(U_a^*)$, $U_a^* = U_a \backslash \{a\}$, for example the form $h\Omega/J$, where Ω is the form in Theorem 1 of Chapter 2.

1.3. Transformation Formula for the Local Residue. When a system of germs f is changed to a system g, contained in the ideal $I_a(f)$, the local residue is transformed in the following manner. Let

$$g_j = \sum_{k=1}^n \phi_{jk} f_k, \quad j = 1, \ldots, n, \quad \phi_{jk} \in \mathcal{O}_a, \tag{2}$$

where the system $g = (g_1, \ldots, g_n)$, like f, has an isolated zero at the point a. Then the local residue satisfies a transformation formula (Griffiths-Harris (1978), Tong (1973)):

$$\operatorname*{res}_a{}_f (h) = \operatorname*{res}_a{}_g (h \cdot \det \|\phi_{jk}\|). \tag{3}$$

The transformation formula provides a means of computing local residues. For example, inserting the monomial $(z_j - a_j)^{k_j+1}$ into (2) as g_j (by Hilbert's Nullstellensatz any monomial of sufficiently high degree lies in the ideal $I_a(f)$), we arrive at the formula

$$\operatorname*{res}_a f \, (h) = \frac{1}{k_1! \ldots k_n!} \frac{\partial^{k_1 + \cdots + k_n}}{\partial z_1^{k_1} \ldots \partial z_n^{k_n}} [h \cdot \det \|\phi_{jk}\|]_{z=a}.$$

From this formula it is evident that

$$\operatorname*{res}_a f = \sum_{|\alpha| \leq m} C_\alpha D^\alpha \delta(z - a)$$

is an analytic functional which equals a finite linear combination of derivatives of delta functions. (It would be interesting to know if the converse is true: can every such analytic functional be realized as the local residue relative to some mapping?)

We remark that if the only zero of the mapping f in the neighborhood U_a is at the point a, then the mapping g, related to f by formula (2), can have zeroes in U_a in addition to the isolated zero $z = a$. It is not difficult to check that for any such zero $b \neq a$ the determinant $\det \|\phi_{jk}\| \in I_b(f)$. Therefore, according to Proposition 1.1, the local residue $\operatorname*{res}_b g = (h \cdot \det \|\phi_{jk}\|) = 0$. A special case of this fact explains the nature of the *Bergman-Weil integral representation* in special analytic polyhedra (see Ajzenberg Yuzhakov (1979)); it is related to the Cauchy integral representation by the transformation formula (3). To be concrete, let the polyhedron Π be defined by the conditions $|W_j(z)| < \varepsilon$, $j = 1, \ldots, n$, and let $a \in \Pi$. In (3) we set $f = z - a$ and $g = W(z) - W(a)$. Then for the ϕ_{jk} in (3) we can take the coefficients of the *Hefer expansion* for the functions W_j. Consequently, taking account of what was said above and recalling Proposition 1 from Chapter 2, we find that in (3) the residue on the left is equal to the Cauchy integral of the function h while the one on the right is equal to the Bergman-Weil integral of this function (see Volume 7, article 2).

1.4. Local Duality Theorem. According to Proposition 1.1, the symmetric pairing $\operatorname*{res}_a f : \mathcal{O}_a / I_a(f) \otimes \mathcal{O}_a / I_a(f) \to \mathbb{C}$ is well-defined, if we set

$$\operatorname*{res}_a f \, (h, g) = \operatorname*{res}_a f \, (h \cdot g).$$

One of the most remarkable properties of the local residue is reflected in the following:

Theorem 1.3 (Local duality (see Griffiths-Harris (1978), Tong (1973))). *The pairing $\operatorname*{res}_a f$ is nondegenerate, i.e., if for all $g \in \mathcal{O}_a$ the local residue $\operatorname*{res}_a f \, (h \cdot g) = 0$, then $h \in I_a(f)$.*

The nondegeneracy property of the local residue is applied effectively in § 4 to the question of whether or not a polynomial or germ $h \in \mathcal{O}_a$ belongs to a given ideal.

§2. Using the Trace to Express the Local Residue

2.1. Definition of the Trace and its Fundamental Properties. Let $D \subset \mathbb{C}_z^n$ be a bounded domain in the space of the variables z and let $f : \overline{D} \to \mathbb{C}_w^n$ be a holomorphic mapping having no zeroes on the boundary ∂D. Then this mapping has only a finite set Z_f of zeroes in D. We assume that Z_f is not empty. Let G be the connected component of the point 0 in $\mathbb{C}_w^n \setminus f(\partial D)$. Then the systems of equations $f - w = 0$ are homotopic for any $w \in G$, and the number of roots in the domain D is the same for each of these systems. In this case one says that f has finite type.

For every meromorphic function $H(z) = \phi(z)/\psi(z)$ in the domain D we associate its *trace*

$$(\mathrm{Tr}\, H)(w) = \sum_v H(z^{(v)}(w)), \quad w \in G \setminus f(\{\psi = 0\}), \tag{4}$$

where the summation is taken over all roots $z^{(v)}(w)$ in D of the system $f - w = 0$. The trace of the function H is a meromorphic function in G and is holomorphic if H is holomorphic.

Let ε be a noncritical value of the mapping $|f|^2 : D \to \mathbb{R}^n$, $z \mapsto (|f_1(z)|^2, \ldots, |f_n(z)|^2)$. In this case the skeleton $\Gamma_{f,\varepsilon}$, defined by the condition $|f_j|^2 = \varepsilon$, is a smooth manifold. From the definition of the integral comes the following *change of variables formula*: if a polydisk, defined by the condition $|w_j|^2 \leq \varepsilon$, is contained in the domain G and its skeleton $\Gamma_{w,\varepsilon}$ does not touch the image under f of the poles of the meromorphic function H, then

$$(2\pi i)^{-n} \int_{\Gamma_{f,\varepsilon}} H(z)\, dz = (2\pi i)^{-n} \int_{\Gamma_{w,\varepsilon}} (\mathrm{Tr}\, H)(w)\, dw, \tag{5}$$

where $df = df_1 \wedge \cdots \wedge df_n$ and $dw = dw_1 \wedge \cdots \wedge dw_n$.

We apply formula (5) to the case when the left-hand side is the Grothendieck residue, i.e., when $D = U_a$ and $H = h/J \cdot f_1 \ldots f_n$, with $J = \partial f/\partial z$, the jacobian of the mapping f. To do this, we begin by observing that the following assertion is true; it follows directly from Proposition 1 of Chapter 2.

Proposition 2.1 (Khovanskij in Arnold et al. (1982)). *In the polydisk of radius $\sqrt{\varepsilon}$, the trace of the function h/J, meromorphic at the point a, has an integral representation*

$$(\mathrm{Tr}\, h/J)(w) = (2\pi i)^{-n} \int_{\Gamma_{f,\varepsilon}} \frac{h(z)\, dz}{\prod_{j=1}^n [f_j(z) - w_j]}. \tag{6}$$

Consequently it is holomorphic.

Thus $(\mathrm{Tr}\, H)(w) = (\mathrm{Tr}\, h/J) \cdot w_1^{-1} \ldots w_n^{-1}$ and we obtain

Proposition 2.2 (Khovanskij in Arnold et al. (1982)). *The local residue can be expressed using the trace by the formula*

$$\mathop{\mathrm{res}}_a f (h) = (\mathrm{Tr}\, h/J)(0), \tag{7}$$

where $J = \partial f/\partial z$ is the jacobian.

2.2. Algebraic Interpretation. The concept of the trace, which has just been introduced, has an algebraic interpretation in terms of extensions of fields. From this point of view, formulas (5) and (7), which we obtained using the transcendental definition of the residue as an integral over a manifold, are at the foundations of the definition of the residue in algebraic geometry (see Lomadze (1981), Beauville (1971)). Thus if K is a finite extension over the field of formal series E then we set $\mathrm{Res}_K = \mathrm{Res}_E \mathrm{Tr}_{K/E}$, where Res_K is the residue of an object (in our case, of a meromorphic function or form) associated with K while $\mathrm{Res}_E(\psi)$ is the residue of the formal power series $\psi = \sum c(a)w^a$, which is equal to $c(-1, \ldots -1)$.

In our case the algebraic interpretation of the trace $\mathrm{Tr}\, h/J$ allows us to give an improvement of an integral representation of Bishop (Bishop (1961)), which we present here. This concerns an integral representation of the meromorphic function h/J in a polyhedron $\Pi_\varepsilon \subset\subset U_a$, defined by the condition $|f_i| < \varepsilon$. To do this we observe that if we set $w = f(\zeta)$ in (6), then the residue on the right gives the value of the function h/J at the point $z = \zeta$ plus the sum of the values of this function at the preimages $\zeta^{(v)}(f(\zeta))$, $v = 2, \ldots, \mu$ of the point $w = f(\zeta)$ (here μ is the multiplicity of the zero a of the mapping f). In order to avoid summing at these $\mu - 1$ points, we introduce a nontrivial weight $\Omega(\zeta, z)$ into the integrand in (6). This weight has the property that for any fixed $\zeta \in \Pi_\varepsilon$ the function Ω is equal to zero at all of the points $z = \zeta^{(v)}(f(\zeta))$ which are not equal to ζ. Since the analytic set $\{(\zeta, z) \in \Pi_\varepsilon \times \Pi_\varepsilon : f(\zeta) = f(z)\}$ contains the set $\{\zeta = z\}$ as an irreducible component, such a function Ω exists. Its construction goes like this. Let $w^{(0)}$ be a noncritical value of the mapping f and let $\psi(z)$ be a holomorphic function in $\overline{\Pi}_\varepsilon$ for which all the values $\psi(z^{(v)}(w^{(0)}))$ are different. Then

$$\Omega(\zeta, z) = \left\{ \prod_{v=2}^{\mu} [\psi(z^{(v)}(f(z))) - \psi(\zeta)] \right\} \left\{ \prod_{\substack{s>1, k\geq 1 \\ s\neq k}} [\psi(z^{(k)}(f(z))) - \psi(z^{(s)}(f(z)))] \right\}.$$

Theorem 2.3. *At each point $z \in \Pi_\varepsilon$ at which the jacobian J of the system f is different from zero, there is an integral formula for the meromorphic function h/J*

$$\frac{h(z)}{J(z)} \cdot \Omega(z, z) = (2\pi i)^{-n} \int_{\Gamma_{f,\varepsilon}} \frac{h(\zeta)\Omega(z, \zeta)\, d\zeta}{\prod_{j=1}^{n} [f_j(\zeta) - f_j(z)]}, \tag{8}$$

where $\Gamma_{f,\varepsilon}$ is the skeleton of the polyhedron Π_ε.

We remark that $\Omega(z, \zeta)$ is a polynomial of degree $\mu - 1$ in $\psi(z)$, the coefficients of which depend holomorphically on ζ. Besides this, $\Omega(z, z)$ is a function of the form $c(f(z))$, where $c(w)$ is holomorphic in the polydisk of radius ε. Since the integral in (8) is holomorphic with respect to $f(z)$, then from this formula we obtain the result that for any h in the ring of germs $\mathcal{O}_a(z)$ in the variable z there is a representation

$$\frac{h(z)}{J(z)} = c_0(f(z))\psi^0(z) + \cdots c_{\mu-1}(f(z))\psi^{\mu-1}(z), \tag{9}$$

where $c_k(w) \in \mathcal{M}_0(w)$ is meromorphic in a neighborhood of zero. Thus the family of meromorphic germs $\{h/J, h \in \mathcal{O}_a(z)\}$ is contained in the finite extension of the field $\mathcal{M}_0(w)$ generated by the basis $\{\psi^0, \dots, \psi^{\mu-1}\}$. Therefore, the trace of the germ $H = h/J$ is defined as the sum of the diagonal elements of the matrix $\|\alpha_{kj}(w)\|$ which represents H in terms of the basis $\{\psi^k\}$:

$$H\psi^k = \sum_{j=0}^{\mu-1} \alpha_{kj}\psi^j.$$

This trace coincides with the trace defined by formula (4).

It is appropriate to compare Formula (9) with the Weierstrass Preparation Theorem (see (Arnol'd et al. 1982)) according to which, for any $h \in \mathcal{O}_a(z)$, there is a unique representation

$$h(z) = a_1(f(z))e_1(z) + \cdots + a_k(f(z))e_k(z), \quad a_j \in \mathcal{O}_0(w), \tag{10}$$

where $\{e_j(z)\}$ is a basis of the local algebra $\mathcal{O}_a/I_a(f)$ and k is its dimension. From (10) it is easy to conclude that $k \geq \mu$ (μ is the geometric multiplicity, i.e., the degree of the mapping f at the point a). From (9) one derives the opposite inequality. Thus we obtain $\mu = k$ (see Palamodov, 1967), where this equality is obtained from the properties of analytic sheaves; also see Arnol'd et al. (1982).

§3. The Total Sum of Local Residues

3.1. The Total Sum of Residues on a Compact Manifold. The Euler-Jacobi Formula. The local residue is a generalization of the residue of a meromorphic function h/f of one complex variable. We now present the analog of the theorem that the sum of all the residues of a meromorphic function equals zero (the sum is taken over the Riemann sphere or a compact Riemann surface). This analog is a special case of Theorem 6.2 of (Dolbeault (1985)), the case of a complete intersection. Let X be a complex manifold of dimension n and let $D_1, \dots, D_n$ be effective divisors on X, i.e., each D_j can be represented locally as a finite linear combination of irreducible analytic hypersurfaces with nonnegative coefficients. We assume that the intersection $Z = D_1 \cap \cdots \cap D_n$ is discrete. Consider a meromorphic form ω on X of degree n with pole divisor $D = D_1 + \cdots + D_n$. This means that the form can be represented locally as $\omega = h\,dz/f_1 \dots f_n$, where f_j is a function defining the divisor D_j. Thus for every $a \in Z$ a local residue is defined $\operatorname*{res}_a \omega = \operatorname*{res}_a{}_f (h)$. Then the following *theorem on total sum of residues* is true.

Theorem 3.1 (Griffiths (1976)). *If X is compact, then*

$$\sum_{a \in Z} \operatorname*{res}_a \omega = 0.$$

This theorem is proved by representing the local residue as an integral over the boundary ∂U_a (see Proposition 1.2). In fact the result of theorem is a consequence of the homological dependence $\sum_{a \in Z} \Gamma_a \sim 0$ of the cycles Γ_a that appear in the definition of the local residues.

We give some corollaries of Theorem 3.1. Let $X = \mathbb{CP}^n$ be projective space and let $z = (z_1, \ldots, z_n)$ be local coordinates on one of the affine pieces of this space, which we will consider the "finite" part.

Corollary 3.2 (Griffiths (1976)). *Let the system of algebraic equations* $f_j(z) = 0$, $j = 1, \ldots, n$ *have no root at infinity and let* $h(z)$ *be a polynomial of degree* deg $h \le$ deg $f_1 + \cdots +$ deg $f_n - n + 1$. *Then*

$$\sum_{a \in Z_f} \operatorname{res}_f (h) = 0,$$

where Z_f *is the set of roots in* $\mathbb{C}^n$ *of the system* $f = (f_1, \ldots, f_n) = 0$.

The truth of this assertion follows from Theorem 3.1, for the condition on the degree of the polynomial h ensures that the form $\omega = h \, dz/f_1 \ldots f_n$ has no poles along the hyperplane at infinity.

Corollary 3.3 (Euler-Jacobi formula Griffiths-Harris (1978)). *If under the hypotheses of Corollary 3.2 the system of equations* $f = 0$ *has only simple roots (at each of the roots of this system the jacobian* $\partial f/\partial z$ *is different from zero), then*

$$\sum_{a \in Z_f} \frac{h(a)}{\dfrac{\partial f}{\partial z}(a)} = 0.$$

For $n = 1$ the Euler-Jacobi formula reduces to the Lagrange interpolation formula

$$\sum_{a \in Z_f} \frac{h(a)}{f'(a)} = 0, \quad \deg h \le \deg f - 2.$$

The Euler-Jacobi formula admits the following generalization. Let $f = (f_1, \ldots, f_n)$, $f_j = \sum c_{j\alpha} z^\alpha$, be a system of Laurent polynomials with Newton polyhedra $\Delta(f_1), \ldots, \Delta(f_n)$ (see Chapter 2 or Khovanskij (1978)). Each face σ_{jt} with normal vector $t \in \mathbb{R}^n$ of the polyhedron $\Delta(f_j)$ corresponds to the following piece of the polynomial f_j:

$$f_{jt} = \sum_{\alpha \in \sigma_{jt}} c_{j\alpha} z^\alpha.$$

The system f is called nondegenerate if for any $t \in \mathbb{R}^n \backslash \{0\}$ the system $f_t = (f_{1t}, \ldots, f_{nt})$ have only simple zeroes in $(\mathbb{C} \backslash \{0\})^n$.

Corollary 3.4 (Generalized Euler-Jacobi formula (Khovanskij (1978))). *Let* $f = (f_1, \ldots, f_n)$ *be a nondegenerate system of Laurent polynomials and let* Z_f *be the set of zeroes of this system in* $(\mathbb{C} \backslash \{0\})^n$. *If all the zeroes* $z \in Z_f$ *are simple and if the Newton polyhedron of the polynomial* h *lies strictly inside the sum*

$\Delta(f_1) + \cdots + \Delta(f_n)$, then

$$\sum_{a \in Z_f} \left(h/z_1 \ldots z_n \frac{\partial f}{\partial z} \right) \Bigg|_a = 0.$$

Here the inclusion condition $\Delta(h) \subset \Delta(f_1) + \cdots + \Delta(f_n)$ is analogous to the inequality of degrees in Corollary 3.2. This condition permits one to extend a differential form meromorphic on $(\mathbb{C}\setminus\{0\})^n$,

$$\omega = \frac{h}{f_1 \ldots f_n} \frac{dz_1}{z_1} \wedge \cdots \wedge \frac{dz_n}{z_n},$$

to a suitable toroidal compactification without adding any poles.

3.2. Applications to Plane Projective Geometry. The Euler-Jacobi formula show that, in contrast to the one-dimensional case, not every set of points in the projective space of dimension $n > 1$ is the set of solutions of a system of n algebraic equations. As a corollary of the Euler-Jacobi formula, we will introduce some geometric illustrations of this phenomenon in the case of the plane (i.e., $n = 2$).

Theorem 3.5 (Cayley-Bacharach (see Griffiths-Harris (1978))). *Let C and D be curves in $\mathbb{CP}^2$ of degrees l and k; and suppose the curves intersect in $l \cdot k$ points. Then any curve E of degree $l + k - 3$ which passes through $l \cdot k - 1$ of the points of intersection $C \cap D$ must pass through all $l \cdot k$ of these points.*

Theorem 3.6 (Pascal (see Griffiths-Harris (1978))). *The pairs of opposite sides of a hexagon inscribed in a smooth conic Q, a curve of second order, intersect in three collinear points.*

The proof of Pascal's theorem follows from the Cayley-Bacharach theorem: Let $L_1, \ldots, L_6$ be the lines containing the sides of the hexagon. We set $C = L_1 + L_3 + L_5$, $D = L_2 + L_4 + L_5$ and $E = Q + \overline{P_{14}P_{36}}$, where $P_{ij} = L_i \cap L_j$. Then E passes through P_{52}, a point of the intersection $C \cap D$.

The converses of these two theorems are also true (see Griffiths-Harris).

3.3. The Converse of the Theorem on Total Sum of Residues. Assertion 6.4 of (Dolbeault (1985)) shows that if the divisors $D_j = Y_j$ are positive in the sense of Kodaira (see Griffiths-Harris (1978)), then Theorem 3.1 can be reversed. In the case $X = \mathbb{CP}^n$ we give a constructive proof of this fact. For this we take an affine piece of $\mathbb{CP}^n$ with coordinates z such that this affine set contains the discrete set $Z = Y_1 \cap \cdots \cap Y_n$. Then in order to construct the desired form ϕ it is sufficient for any pair of points $a_i, a_j \in Z$ to find a polynomial h such that the following condition is satisfied for the form $\phi = h\, dz/f_1 \ldots f_n$ ($Y_j = \{f_j = 0\}$):

$$\operatorname*{res}_{a_i} \phi = -\operatorname*{res}_{a_j} \phi = 1; \quad \operatorname*{res}_{a} \phi = 0, \quad a \in Z\setminus\{a_i, a_j\};$$

$$\deg h + n < \deg f_1 + \cdots + \deg f_n.$$

Using the Bergman-Weil integral representation, it is not difficult to convince oneself that for h one can take $h(z) = \Omega(z, a_i) - \Omega(z, a_j)$, where $\Omega(z, \zeta)$ is the determinant formed from the coefficients of the Heffer expansion of the polynomials $f_j, j = 1, \ldots, n$.

3.4. Abel's Theorem and its Converse.

It is known that the integral of a rational function $r(x)$ of a variable $x \in \mathbb{C}^1$ is a rational logarithmic function, that is,

$$\int r(x)\, dx = R(x) + \sum a_k \log(x - x_k),$$

where $R(x)$ is a rational function. At the same time, an arbitrary abelian integral

$$u(P) = \int_{P_0}^{P} r(x, y)\, dx,$$

related to the curve $V = \{f(x, y) = 0\}$ and the rational function $r(x, y)$ is not of this type. If one considers the intersections $D_t \cdot V = \{P_\nu(t)\}$ of the curve V with a family of curves $D_t = \{\theta(x, y, t) = 0\}$, which depend rationally on t, then according to the classical theorem of Abel, the sum $u(t) = \sum_\alpha u(P_\nu(t))$ is rational logarithmic. Thus the differential $du(t)$ is equal to the sum

$$\sum_\nu \psi(P_\nu(t)), \quad \psi = r(x, y)\, dx|_V, \tag{11}$$

is rational.

We will consider sum (11) in the multidimensional case, when $V = \{f(x_1, \ldots, x_n, y) = 0\}$ is a hypersurface in $\mathbb{CP}^{n+1}$, defined in local coordinates $x_1, \ldots, x_n, y$ as the zeroes of a polynomial f without multiple factors. In this case ψ is a meromorphic n-form on V (for a precise definition of a meromorphic form on an analytic set, see Griffiths (1976); here, for simplicity, we may assume that V is smooth). The family D_t is the family $\{L\}$ of all complex lines in $\mathbb{CP}^{n+1}$, in other words the Grassmannian $G(1, n + 1)$.

Consider in $V \times G(1, n + 1)$ the submanifold of flags (the pairs "point and a line passing through the point"):

$$I = \{(P, L) : P \subset L\}.$$

If the line L is described by an equation, $L = \{l_1 = x_1 - a_1 y - b_1 = 0, \ldots, l_n = x_n - a_n y - b_n = 0\}$, then local coordinates are defined on the product $\mathbb{CP}^{n+1} \times G(1, n + 1)$ by $x = (x_1, \ldots, x_n), y, a = (a_1, \ldots, a_n), b = (b_1, \ldots, b_n)$. Then we have

$$I = \{x = ay + b, f(x, y) = 0\}.$$

Let $\pi_1 : I \to V$ and $\pi_2 : I \to G(1, n + 1)$ be the projections (such a pair of projections in integral geometry is called a "dual fibration"). Then the analog of the sum (11) is the following concept of the trace of a form

$$\text{Tr } \psi = (\pi_2)_*(\pi^* \psi),$$

which for a line L in "general position" (that is, for which $L \cdot V$ consists of d different points P_ν, where d is the degree of f) is written in the form,

$$\text{Tr } \psi = \psi(P_1(L)) + \cdots + \psi(P_d(L)).$$

We will show a formula for the trace of a holomorphic form ψ on V. According to Chapter 1, Subsection 1.2, every such form is the Poincaré residue of some rational form, i.e.,

$$\psi = \frac{p(x, y)}{\dfrac{\partial f}{\partial y}(x, y)} \, dx_1 \wedge \cdots \wedge dx_n \bigg|_V. \tag{12}$$

For a general line $L = L(a, b)$ we set

$$L \cdot V = \{P_1(a, b), \ldots, P_d(a, b)\},$$

where the $P_\nu(a, b) = (x_\nu(a, b), y_\nu(a, b))$ have distinct coordinates $y_\nu = y_\nu(a, b)$. Let

$$P(y) = p(ay + b, y), \quad F(y) = f(ay + b, y);$$

then by a direct computation we obtain

$$\text{Tr } \psi = \sum_A \pm \left\{ \sum_\nu \frac{y_\nu^{|A|} P(y_\nu)}{F'(y_\nu)} \right\} da_A \wedge db_{A^c},$$

where $A = (i_1, \ldots, i_k) \subset \{1, \ldots, n\}$, $da_A = da_{i_1} \wedge \cdots \wedge da_{i_k}$; A^c is the complement to A in the set of indices, and $|A| = k$ is the cardinality of A.

Since on L the form $dl_1 \wedge \cdots \wedge dl_n \wedge df = F'(y) \, dx_1 \wedge \cdots \wedge dx_n \wedge dy$, then from the Euler-Jacobi formula applied to the system $(l_1, \ldots, l_n, f)$ we have

$$\sum_\nu \frac{y_\nu^{|A|} P(y_\nu)}{F'(y_\nu)} = 0, \quad |A| = 0, 1, \ldots, n,$$

if $\deg P \leq d - n - 2$. Consequently, for the given bounds on the degree, the trace $\text{Tr } \psi \equiv 0$. One can show (Griffiths (1976)) that every meromorphic form ψ on V whose trace is holomorphic on $G(1, n + 1)$ has the form (12), where $\deg p \leq \deg f - n - 2$ (the trace equals zero, since on $G(1, n + 1)$ there are no nontrivial forms). Forms of this type are called forms of the first kind. Thus, Abel's Theorem for forms of the first kind becomes the statement that

$$\psi(P_1(L)) + \cdots + \psi(P_d(L)) \equiv 0.$$

We remark that the left side of this identity has a local character; that is, it can be defined for arbitrarily small pieces V_ν of a hypersurface V, with each piece intersecting some fixed line at a single point. Moreover, this identity characterizes those collections of pieces of analytic sets $_\nu$ which extend to a global algebraic hypersurface V. Namely, let some given pieces $V_1, \ldots, V_d$ be irreducible n-dimensional complex analytic sets in $\mathbb{CP}^{n+1}$, along with meromorphic forms $\psi_\nu \not\equiv 0$ on V_ν. We assume that there exists a line L_0 intersecting each V_ν in a single simple point which is not a pole for ψ_ν. In such a case, for lines L in a neighborhood $U(L_0) \subset G(1, n + 1)$, one can define the trace

$$\text{Tr}\{\psi\} = \psi(P_1(L)) + \cdots + \psi(P_d(L)), \quad P_v(L) = L \cdot V_v.$$

Then the following is true:

Theorem 3.7 (Griffiths (1976)). *If the trace* $\text{Tr}\{\psi\} \equiv 0$, *then there exists an* n-*dimensional algebraic set* $V \subset \mathbb{CP}^{n+1}$ *and a rational* n-*form* ψ *of the first kind such that* $V_v \subset V$ *and* $\psi|_{V_v} = \psi_v$.

Here is an outline of the proof of this theorem. Let f_v be the defining functions for V_v; then $\psi_v = \text{Res}_{V_v} \Psi_v$ is the Poincaré residue of the form $\Psi_v = g_v \, dx \wedge dy/f_v$. We denote by $I_W = \{(P, L): P \subset L \subset W\}$ the part of the flag manifold lying over the neighborhood $W \subset \mathbb{CP}^{n+1}$ of the line L_0. On I_W we consider the form

$$\Phi = \sum_A \pm \left\{ \sum_v \left[\underset{P_v}{\text{res}} \, \omega_v \right] \frac{dy}{y - y_v} \right\} y^{|A|} \, da_A \wedge db_{A^c},$$

where $P_v = (x_v(a, b), y_v(a, b))$ are the points of intersection $L(a, b) \cdot V_v$, $l_k = x_k - a_k y - b_k$, and $\underset{P_v}{\text{res}} \, \omega_v$ is the local residue of the form $\omega_v = \Psi_v / l_1 \dots l_n$.

The condition $\text{Tr}\{\psi_v\} \equiv 0$ implies that Φ descends to W under the projection $\pi_1 : I_W \to W$; this means that $\Phi = \pi_1^* \Psi$, where Ψ is a meromorphic $(n + 1)$-form on W. Moreover, using the fact that every function which is meromorphic in a neighborhood of a line in $\mathbb{CP}^{n+1}$ is a rational function, we conclude that Ψ is rational. Its pole set is the desired hypersurface V, and the form ψ is the Poincaré residue $\text{Res}_V \Psi$.

3.5. Residue Theorem for Vector Bundles.

It is known that to every divisor D on X with local data $f_\alpha \in \mathcal{M}^*(U_\alpha)$ there corresponds a line bundle $L = [D]$ with transition functions $\{g_{\alpha\beta} = f_\alpha/f_\beta\}$. The local residue $\text{res} \, \omega$ of the form ω with pole divisor $D = D_1 + \cdots + D_n$ at the point $a \in Z = \overset{a}{D} = D_1 \cap \cdots \cap D_n$ can be viewed as the residue with respect to the holomorphic section $f = f_1(z)e_1 + \cdots + f_n(z)e_n$ of the vector bundle $E = [D_1] \oplus \cdots \oplus [D_n]$, where $\{e_k\}$ is a frame. Proceeding with this analogy, we define the local residue relative to a holomorphic section with an isolated zero for any vector bundle of rank n. Let X be a complex analytic manifold of dimension n and let $E \to X$ be a vector bundle of rank n. Consider the vector space $H^0(X, \mathcal{O}(K \otimes \det E))$, where $K = \bigwedge^n T_X^{*\prime}$ is the highest exterior power of the holomorphic cotangent bundle. An element of this space can be written in terms of a local holomorphic frame e_1, $\dots$, e_n for E and local holomorphic coordinates $z = (z_1, \dots, z_n)$ in the form $\psi = h(z)(dz_1 \wedge \cdots \wedge dz_n) \otimes (e_1 \wedge \cdots \wedge e_n)$. To the element ψ and the section $f = f_1(z)e_1 + \cdots + f_n(z)e_n$ we associate the differential form

$$\frac{\psi}{f} = \frac{h(z) \, dz_1 \wedge \cdots \wedge dz_n}{f_1(z) \dots f_n(z)}. \tag{13}$$

Under the assumption that the section f has an isolated zero at the point $a \in X$, we define the residue

$$\operatorname*{res}_{a} \left(\frac{\psi}{f} \right) = \operatorname*{res}_{a} {}_{f} (h). \tag{14}$$

Although the right side of (13) depends on the choice of frame and coordinates, according to the transformation formula (3), the residue (14) does not depend on these choices. A theorem about the total sum of residues is also true.

Theorem 3.8 (Griffiths-Harris (1978)). *Let $E \to X$ be a holomorphic vector bundle of rank n over a compact complex analytic manifold of dimension n, and let $f \in H^0(X, \mathcal{O}(E))$ be a holomorphic section having a discrete set of zeroes Z. If for $\psi \in H^0(X, \mathcal{O}(K \otimes \det E))$ and $a \in Z$ we define the residue $\operatorname*{res}_{a} (\psi/f)$ by formula (14), then*

$$\sum_{a \in Z} \operatorname*{res}_{a} \left(\frac{\psi}{f} \right) = 0.$$

Griffiths and Harris proved that for residues relative to holomorphic sections the converse of the theorem on total sums of residues is also true (Ann. Math. *108*, No. 3, 461–508 (1978)).

3.6. The Total Sum of Residues Relative to a Polynomial Mapping in $\mathbb{C}^n$.

Let $h, f_1, \ldots, f_n$ be polynomials in the ring $\mathbb{C}[z] = \mathbb{C}[z_1, \ldots, z_n]$ such that the set $Z_f = \{z \in \mathbb{C}^n : f_1(z) = \cdots = f_n(z) = 0\}$ is discrete. We consider the total sum of residues (*the global residue*) of the polynomial h relative to the polynomial mapping $f = (f_1, \ldots, f_n)$:

$$\operatorname{Res}_f(h) = \sum_{a \in Z_f} \operatorname*{res}_{a} {}_{f} (h).$$

For the global residue Res_f the transformation formula and duality remain true.

Proposition 3.9 (Transformation formula for the global residue (Tsikh, Yuzhakov (1984))). *If $g = (g_1, \ldots, g_n)$, with $g_j = \sum \phi_{jk} f_k$, $\phi_{jk} \in \mathbb{C}[z]$, and if $g^{-1}(0)$ is discrete, then*

$$\operatorname{Res}_f(h) = \operatorname{Res}_g(h \cdot \det \|\phi_{jk}\|). \tag{15}$$

Proposition 3.10 (Global duality (Tsikh, Yuzhakov (1984))). *Let $I(f)$ be the ideal in the ring $\mathbb{C}[z]$ generated by a system of polynomials $f = (f_1, \ldots, f_n)$ with a discrete set of zeroes in $\mathbb{C}^n$. Then the polynomial F belong to the ideal $I(f)$ if and only if $\operatorname{Res}_f(F\phi) = 0$ for every $\phi \in \mathbb{C}[z]$.*

We remark that from the transformation formula for the global residue one can deduce formula (15) of Chapter 2, by applying it to a system f with no zeroes at infinity. In Tsikh Yuzhakov (1984) Yuzhakov used the transformation formula (15) and elimination theory to produce an algorithm for the computation of an arbitrary global residue. Global duality is applied in §4 to the problem of the inclusion of $F \in I(f)$.

§4. Application of the Grothendieck Residue to the Algebra of Polynomials and to the Local Ring $\mathcal{O}_a$

In this section we will apply the Grothendieck residue (principally local and global duality) to the problem of determining whether or not a given polynomial $F(z)$ belongs to the ideal $I(f)$ generated in the ring of polynomials $\mathbb{C}[z] = \mathbb{C}[z_1, \ldots, z_n]$ by the system of polynomials $f = (f_1(z), \ldots, f_n(z))$. In other words, when can F be represented in the form

$$F = q_1 f_1 + \cdots + q_n f_n, \tag{16}$$

where the q_j are polynomials? We will also consider the analogous question in the local ring $\mathcal{O}_a$. In the polynomial case we will examine two approaches: in the first approach it will be assumed that the zeroes of the system f are known (Subsection 4.1, Subsection 4.2) and in the second we will not assume they are known (Subsection 4.4).

4.1. Macauley's Theorem. The application of local duality to the question of equation (16) is most transparent in the case when the f_j are forms (homogeneous polynomials).

Theorem 4.1 (Macauley). *If the system of forms $f = (f_1(z), \ldots, f_n(z))$ has an isolated zero at the coordinate origin, then every form of degree greater than $\deg f_1 + \cdots + \deg f_n - n$ belongs to the ideal $I(f)$.*

The proof follows immediately from local duality if we observe that every differential n-form $P\,dz/Q$, where P and Q are homogeneous polynomials, is exact if $\deg P + n \neq \deg Q$.

4.2. Noether-Lasker Theorem in $\mathbb{CP}^n$. The well-known Noether-Lasker theorem in $\mathbb{C}^n$ asserts that for a polynomial F equation (16) is satisfied with polynomial coefficients q_j (i.e., $F \in I(f)$) if this equation is satisfied at every point $a \in Z_f = f^{-1}(0)$ for some holomorphic germs $q_j = q_{ja}$ (that is, if the *local Noether condition* is fulfilled: $F \in I_a(f)$, $a \in Z_f$, where $I_a(f)$ is the ideal generated by the system f in the ring of germs $\mathcal{O}_a$). It turns out that in the case when the system f has no zeroes at infinity in the projective space, then the local Noether condition ensures that equation (16) is valid for q_j with $\deg q_j + \deg f_j \leq \deg F, j = 1, \ldots, n$. This assertion can be formulated in the language of forms in homogeneous coordinates on projective space. Namely, let $\zeta = (\zeta_0, \zeta_1, \ldots, \zeta_n)$ be homogeneous coordinates on the projective space $\mathbb{CP}^n$ and let $\tilde{f} = (\tilde{f}_1(\zeta), \ldots, \tilde{f}_n(\zeta))$ be a system of homogeneous forms which has in this space a discrete set of zeroes $Z_{\tilde{f}}$. For each $a \in Z_{\tilde{f}}$ (let $a = (a_0, a_1, \ldots, a_n)$ and for definiteness, assume that $a_0 \neq 0$) denote by $I_a(f)$ the ideal in $\mathcal{O}_a$, the ring of germs in the variables $z_j = \zeta_j/\zeta_0$ generated by the system $f(z) = \tilde{f}(1, z_1, \ldots, z_n)$. In this case, the following theorem is true.

Theorem 4.2 (Noether-Lasker in $\mathbb{CP}^n$ (see Griffiths, Harris (1978) and Tsikh (1984)). *If a homogeneous form $\tilde{F}(\zeta)$ satisfies the local Noether condition, $F(z) \in I_a(f)$, at each point $a \in Z_{\tilde{f}}$ in the local coordinates z, then $\tilde{F} \in I(f)$.*

We remark that the system $\tilde{f}$ does not have isolated zeroes in the space $\mathbb{C}^{n+1}$ of the variables ζ. Therefore, local duality does not apply to it directly. However, the local Noether conditions ensure that the global residue is equal to zero (we assume that z is a local coordinate system in an affine piece that contains $Z_{\tilde{f}}$)

$$\operatorname{Res}_f(F \cdot g) = \sum_{a \in Z_{\tilde{f}}} \operatorname{res}_a (F \cdot g), \quad g \in \mathbb{C}[z].$$

This residue admits "localization" (see Tsikh (1984), which explains the link between arbitrary residues in $\mathbb{CP}^n$ and local residues in $\mathbb{C}^{n+1}$) in the form of a local residue in $\mathbb{C}^{n+1}$ with respect to the extended system $\Phi = (\zeta_0^m, \tilde{f}_1(\zeta), \ldots, \tilde{f}_n(\zeta))$:

$$\operatorname{Res}_f(F \cdot g) = \operatorname{res}_{0} (\tilde{F} \cdot \tilde{g}),$$

where m is some number depending on the degree of the polynomials F and g. Thus the proof of Theorem 4.2 reduces to local duality. In Griffiths, Harris (1978), in the proof of this theorem in the case $n = 2$ this "localization" is replaced by the Kodaira Vanishing Theorem and Kodaira-Serre duality.

4.3. Verification of the Local Noether Condition. In order to apply the Noether-Lasker Theorem, it is useful to have criteria for the fulfillment of the local Noether condition. We will now state several criteria in terms of the multiplicity $\mu_a(f)$ at the zero a of the mapping f and $d_a(F)$, the order of the zero at the point a of the function F under consideration. This number is defined to be the order of the lowest-order derivative of the function F which does not vanish at a.

The crudest of these criteria is the following:

Proposition 4.3 (see Arnold et al. (1982)). *If $d_a(F) \geq \mu_a(f)$, then $F \in I_a(f)$.*

A more precise criterion is the following generalization of a result of Bertini (1889).

Theorem 4.4 (Tsikh (1984)). *Let $f_j = P_j + \cdots, j = 1, \ldots, n$, where P_j is the homogeneous polynomial of lowest degree in the Taylor expansion of the function f_j at the point a, with $d_j = \deg P_j$ the order of the zero of the function f_j at a. If for some $j \in \{1, \ldots, n\}$ the set of solutions of the system of equations*

$$P_1(z) = \cdots = P_{j-1}(z) = P_{j+1}(z) = \cdots = P_n(z) = 0$$

consists of a finite number of rays in $\mathbb{C}^n$ and the order

$$d_a(F) > \mu_a(f) - d_1 \ldots d_n + d_1 + \cdots + d_n - n,$$

then $F \in I_z(f)$.

As a corollary of this theorem we state the following assertion:

Proposition 4.5. *Just as in Theorem 4.4, let*

$$f_j = P_j + \cdots, \quad j = 1, \ldots, n,$$

where the system $P = (P_1, \ldots, P_n)$ has a unique zero $z = a$. If $d_a(F) > d_1 + \cdots + d_n - n$, then $F \in I_a(f)$.

Proposition 4.5 follows from Theorem 4.4, since in this case it is a consequence of Theorem 9 of Chapter 2 that the multiplicity $\mu_a(f)$ is equal to the product of the orders $d_1 \ldots d_n$. In turn, this proposition generalizes Macauley's Theorem.

The proof of Theorem 4.4 is obtained by using local duality and a sigma process at the point a, as a result of which the system f is transformed into a system f^* with a finite number of zeroes $\{a_\nu\}$ on the "blown-up" hyperplane. Then one applies Proposition 4.1 to the system f^* at each zero a_ν for which the multiplicity $\mu_{a_\nu}(f^*) \leq \mu_a(f) - d_1 \ldots d_n$.

4.4. A Consequence of Global Duality. Now let us consider the problem of representing F in the form (16) in the case when the zeroes of the system f are unknown. In this case it is appropriate to use global duality (see Section 3.6). It turns out that to verify whether the polynomial F belongs to the ideal $I(f)$ it is sufficient to compute the global residue for only a finite set of polynomials $\phi \in \mathbb{C}[z]$. The following assertion explains how to choose these polynomials. Let

$$f_j(\zeta) - f_j(z) = \sum_{k=1}^{n} P_{jk}(\zeta, z) \cdot (\zeta_k - z_k), \quad j = 1, \ldots, n,$$

be the Hefer expansion of the polynomials f_j and let $\Omega(\zeta, z) = \det \| P_{jk} \|$. The determinant Ω can be represented in the form

$$\Omega(\zeta, z) = \sum_{k=1}^{L} g_k(\zeta) \cdot h_k(z). \tag{17}$$

Theorem 4.6. *If the system of polynomials $f = (f_1, \ldots, f_n)$ has a discrete set of zeroes, then the polynomial F belongs to the ideal $I(f)$ if and only if $\operatorname{Res}_f(F \cdot h_k) = 0$, $k = 1, \ldots, L$, where the h_k are the polynomials in the expansion (17). If the mapping $w = f(z)$ is proper, then the family of polynomials $\{h_k(z)\}_{k=\overline{1,L}}$ generates the ring $\mathbb{C}[z]$ viewed as a module over $\mathbb{C}[w]$, i.e., for every $h \in \mathbb{C}[z]$ there is a representation*

$$h(z) = c_1(f(z))h_1(z) + \cdots + c_L(f(z))h_L(z), \quad c_k(w) \in \mathbb{C}[w];$$

moreover, this family forms a basis of the $\mathbb{C}[w]$-module $\mathbb{C}[z]$ if and only if

$$\det \| \operatorname{Res}_f(g_j \cdot h_k) \| \neq 0.$$

Bibliography

The results of § 5 of the second chapter are contained in Aizenberg (1983). The monograph by Aizenberg-Yuzhakov (1979) contains an exposition, from one point of view, of integral representations of holomorphic functions of several complex variables and the theory of multidimensional residues, including the logarithmic residue and its applications, in particular the results of §§ 1, 2 and subsections 3 and 4 of Chapter 1 and §§ 2–4 of Chapter 2 of this survey. In the first volume of the book by Arnol'd et al. (1982, 1984) there is a proof of the equality of the algebraic multiplicity of a holomorphic mapping and its topological degree. This fact was proved earlier in Palamodov (1967) using the theory of analytic sheaves. Morever, in Arnold et al. (op. cit.) local duality for the Grothendieck residue is proved; propositions 2.1 and 2.2 of Chapter 3 and also proposition 4.5 of Chapter 3 are stated.

The second volume of the book Arnol'd et al. (op. cit.), examines integrals depending on a parameter and works out a technique for computing monodromy and the asymptotics of integrals. In Varchenko (1983) the asymptotics of integrals are investigated by the saddle-point method. This is equivalent to the study of the asymptotics of integrals over vanishing cycles of holomorphic forms which depend continuously upon parameters; it gives a definition of the mixed Hodge structure on the cohomology vanishing at a critical point of a holomorphic function. The survey by Golubeva (1976) is devoted to the study of the analytic properties of Feynmann integrals, their singularities (Landau manifolds) and their branching properties. The monograph by Egorychev (1977) develops a method of determining generating functions and the computation of combinatorial sums using integral representations and residues. The work by Krasnov (1972) studies the cohomology of complexes of meromorphic forms on a Kähler manifold, forms whose poles lie on a fixed submanifold; here one investigates the link between the order of the pole of these forms and the type of the class-residue depending of the infinitesimal neighborhood of the submanifold. Theorem 1 of Chapter 2 was taken from Kushnirenko (1975), which also includes other results about the multiplicity of a zero of a holomorphic mapping and the Newton polyhedron. In Lomadze (1981) and Beauville (1971) the algebraic concept of residue is introduced and examined. The generalized Euler-Jacobi formula of Chapter 3 is stated in Khovanskij (1978). Theorems 4.2 and 4.4 in Chapter 3 are proved in Tsikh (1984) where, also, more general local residues are studied than in Chapter 3. The algorithm for computing the total sum of local residues relative to a polynomial mapping is stated in Tsikh-Yuzhakov (1984).

In Yuzhakov (1975), using the multiple logarithmic residue, formulas are given for the expansion of implicit functions into power series; this generalizes the expansion of Bürmann-Lagrange. The article by Aizenberg et al. (1983) is devoted to applying the results of §3 of Chapter 2 to some mathematical problems of chemical kinetics. In Andreotti-Norguet (1964) there is a formula generalizing the Bochner-Martinelli integral representation, the so-called formula of Andreotti-Norguet. The work by Bertini (1889) contains Theorem 4.4 of Chapter 3 in the case $n = 2$. The article by Bishop (1961) is about holomorphic mappings of analytic spaces; it uses several variants of formula (8) of Chapter 3 and also formula (7) of Chapter 2. The latter formula was also in (Caccioppoly (1949); Martinelli (1955); Sorani (1962).) The goal of the work by Carrell (1978) is to obtain a direct proof of the representability of several integrals on a compact complex manifold X by means of Grothendieck residues with respect to a meromorphic vector field with isolated zeroes on X and then to apply this to prove the theorem of Bott and Baum on residues. The monograph by Coleff-Herrera (1978) is devoted to residue-currents generalizing the Grothendieck residue considered in Chapter 3. Moreover, in particular, there is an assertion close to Theorem 4 of Chapter 2, though the stated formulation of the theorem was taken from Solomin (1977). In the work by Fotiadi et al. (1965) the theorem of Froissard on expansions is presented (see Chapter 1) as well as an application of the Thom isotopy theorem to the study of integrals depending on parameters. The Thom theorem itself is proved in the work of Mather (see the footnote on page 17). In Gordon (1974) the concept of geometrical residue is introduced as an element of the kernel

$$\text{Ker}[H_p(X \setminus S) \to H_p(X)],$$

where S is an analytic subset of arbitrary codimension of a manifold X; this generalizes the exact sequence of Leray. The article by Griffiths (1969) is the fundamental work on the residues of rational functions; it describes a procedure for bringing these integrals to canonical form, and it studies problems of algebraic cycles on projective manifolds and inversion of generalized abelian integrals. In Griffiths (1976) the generalized theorem on the total sum of residues is stated (see Chapter 3) and this is applied to the multidimensional Abel's theorem. The book by Griffiths-Harris (1978) contains a systematic exposition of local residues and their application to algebraic geometry. In particular, it contains the algebraization of local and global duality in the language of the Ext functor. As a result of this, the analytic definition of the local residue appears in invariant form (not depending on local coordinates and the choice of generators of the ideal). Moreover, in this book are described many of the conceptual and technical methods useful in working with multidimensional residues.

Theorem 6 of Chapter 2 is contained in the paper by King (1971), and Theorem 5 is contained in Lelong (1968). All these results and their generalizations (Theorems 5, 6, 5', 6' of Chapter 2) are found in the book by Griffiths-King (1973), which is about Nevanlinna theory for holomorphic mappings of algebraic manifolds.

The work by Hartshorne (1966) is devoted to the algebraic computation of residues as morphisms of the ring $\mathcal{O}_X$ associated to an algebraic set X. It contains a proof of (Grothendieck) duality for proper schemes over fields, with arbitrary singularities; this generalizes the local and global duality of Chapter 3 and also Serre duality.

Leray (1959) developed the theory of residues on a complex analytic manifold; he introduced the concept of form-residue and the exact (co)homological sequence associated with a submanifold of codimension 1 and proved the existence of the class-residue and the residue formula (simple and multiple).

Theorem 4 of Chapter 2 is in Lupacciolu (1979); formulas (5) and (6) of this chapter are in Roos (1974); example 3 of § 1, Chapter 2, is taken from Wirtinger (1937).

Norguet (1959 and 1971) generalized the theory of residues of Leray to the case of submanifolds of codimension $k > 1$ and also to closed subsets of a locally compact space. He introduced the concept of partial derivatives for differential forms.

In the book by Pham (1967), integrals depending on a parameter are investigated. The majority of the results of § 3 of Chapter 2 were taken from this book: the isotopy theorem, the concept of a Landau set, the Picard-Lefschetz theorem, etc.

Although individual fragments relating to multiple residues are encountered in works by Jacobi and Picard, one should consider the true beginning of the theory of multidimensional residues to be the memoir of Poincaré (1887), where he generalized the Cauchy integral theorem to functions of two complex variables and investigated several cases of the residues of rational functions of two variables and proved that they can be expressed as the periods of abelian integrals. Moreover, using double residues he established a generalization of the Lagrange expansion for functions of two variables that had been stated by Stieltjes.

In Tong (1973), apparently for the first time, the Grothendieck residue was considered as the integral (1) of Chapter 3 and an integral proof was given for the transformation formula and local duality.

References*

Aizenberg, L.A. (1983): Application of multidimensional logarithmic residue to represent the difference between the number of integer points in a domain and its volume in the form of an integral. Dokl. Akad. Nauk SSSR *270*, No. 3, 521–523. Engl. transl.: Sov. Math., Dokl. *27*, 615–617 (1983), Zbl.535.10047

---

* For the convenience of the reader, references to reviews in Zentralblatt für Mathematik (Zbl.), compiled using the MATH database, and Jahrbuch über die Fortschritte der Mathematik (Jbuch) have, as far as possible, been included in this bibliography.

Aizenberg, L.A., Yuzhakov, A.P. (1979): Integral Representations and Residues in Multidimensional Complex Analysis. Novosibirsk: Nauka, 335 pp. Engl. transl.: Transl. Math. Monogr. 58. Providence: Am. Math. Soc., 283 pp. 1983, Zbl.445.32002

Aizenberg, L.A., Bykov, V.I., Kytmanov, A.M., Yablonskij, G.S., (1983): Search for all steady-states chemical kinetic equations with the modified method of elimination. Chem. Eng. Sci. 38, No. 9, 1555–1567

Andreotti, A., Norguet, F. (1964): Problème de Levi pour les classes de cohomologie. C. R. Acad. Sci., Paris, Sér. A 258, No. 3, 778–781, Zbl.124,388

Arnol'd, V.I., Varchenko, A.N. Gusein-Zade S.M. (1982, 1984): Singularities of Differentiable Mappings. I, Moscow: Nauka. Engl. transl.: Boston: Birkhäuser. 382 pp (1985). Zbl.513.58001 II, Moscow: Nauka. 336 pp. Engl. transl. Boston: Birkhäuser 1988, Zbl.545.58001

Beauville, A. (1971): Une Notion de Résidue en Geométrie Analytique. Lect. Notes Math. 205, 183–203, Zbl.236.32004

Bertini, E. (1889): Zum Fundamentalsatz aus der Theorie der algebraischen Funktionen. Math. Ann. 34, 447–449, Jbuch21,426

Bishop, E. (1961): Mappings of partially analytic spaces. Am. J. Math. 83, No. 1, 209–242, Zbl.118,77

Caccioppoli, R. (1949): Residui di integrali doppi e intersezioni di curve analitiche. Ann. Mat. Pura Appl., IV. Ser. 29, No. 4, 1–14, Zbl.40,192

Carrell, J.B. (1978): A remark on the Grothendieck residue map. Proc. Am. Math. Soc. 70, No. 1, 43–48, Zbl.409.32005

Coleff, N.R., Herrera, M.E. (1978): Les Courants Résiduels Associés à une Forme Méromorphe. Lect. Notes Math. 633, 211 pp., Zbl.371.32007

Dolbeault, P. (1985): General Theory of Multidimensional Residues. Itogi Nauki Tekh., Ser. Sovrem. Probl. Mat., Fundam. Napravlenija 7, 227–251. Engl. transl. in: Encyclopaedia of Math. Sc. 7, Berlin, Heidelberg, New York: Springer-Verlag 1990, 215–242

Egorychev, G.P. (1977): Integral Representation and Computation of Combinatorial Sums. Novosibirsk: Nauka. 285 pp. English transl.: Transl. Math. Monogr. 59, Providence 1984, Zbl.453.05001

Fotiadi, D., Froissart, M., Lascoux, J., Pham, F. (1965): Applications of an isotopy theorem. Topology 4, No. 2, 159–191, Zbl.173,93

Golubeva, V.A. (1976): Some questions from the analytic theory of Feynman integrals. Usp. Mat. Nauk 31, No. 2, 135–202. Engl. transl.: Russ. Math. Surv. 31, No. 2, 139–207 (1976), Zbl.334.28008

Gordon, G. (1974): The Residue Calculus in Several Complex Variables. Lect. Notes Math. 409, 430–438, Zbl.297.32009

Griffiths, P.A. (1969): On the periods of certain rational integrals. Ann. Math. II. Ser. 90, No. 3, 460–541, Zbl.215,81

Griffiths, P.A. (1976): Variations on a theorem of Abel. Invent. Math. 35, No. 3, 321–390, Zbl.339.14003

Griffiths, P.A., Harris, J. (1978): Principles of Algebraic Geometry. New York: John Wiley & Sons. 813 pp., Zbl.408.14001

Griffiths, P.A., King, J. (1973): Nevanlinna theory and holomorphic mappings between algebraic varieties. Acta Math. 130, No. 3–4, 145–220, Zbl.258.32009

Hartshorne, R. (1966): Residues and Duality. Lect. Notes Math. 20, 423 pp., Zbl.212,261

King, J. (1971): The currents defined by analytic varieties. Acta Math. 127, No. 1–2, 185–220, Zbl.224.32008

King, J. (1972): A residue formula for complex subvarieties (preprint). See also: Proc. Conf. Chapel Hill 1970, 43–56 (1970), Zbl.224.32009

Krasnov, V.A. (1972): Cohomology of complexes of meromorphic forms and residues. Izv. Akad. Nauk SSSR, Ser. Mat. 36, 1237–1268. Engl. transl.: Math. USSR, Izv. 6, 1217–1250 (1974), Zbl.248.32007

Khovanskij, A.G. (1978): Newton polyhedra and the Euler-Jacoby formula. Usp. Mat. Nauk 33, No. 6, 237–238. Engl. transl.: Russ. Math. Surv. 33, No. 6, 237–238 (1978), Zbl.442.14020

Kushnirenko, A.G. (1975): Newton polyhedra and the number of solutions for systems of k equations with k unknowns. Usp. Mat. Nauk 30, No. 2, 266–267 (Russian)

Lelong, P. (1968): Fonctions Plurisousharmoniques et Formes Differentielles Positives. New York: Gordon and Breach. 79 pp., Zbl.195,116

Leray, J. (1959): Le calcul différentiel et intégral sur une variété analytique complexe (Problème de Cauchy, III). Bull. Soc. Math. Fr. *87*, 81–180, Zbl.199,412

Lomadze, V.G. (1981): On residues in algebraic geometry. Izv. Akad. Nauk SSSR, Ser. Mat. *45*, No. 6, 1258–1287. Engl. transl.: Math. USSR, Izv. *19*, 495–520 (1982), Zbl.528.14003

Lupacciolu, G. (1979): On the argument principle in multidimensional complex manifolds. Atti Accad. Naz. Lincei, VIII. Ser. Rend., Cl. Sci. Fis. Mat. Nat. *66*, No. 5, 323–330, Zbl.507.32003

Martinelli, E. (1955): Contributi alla teoria dei residue per le funzioni di due variabili complesse. Ann. Mat. Pura Appl., IV. Ser. *39*, No. 4, 335–343, Zbl.66,62

Norguet, F. (1959): Dérivées partielles et résidus de formes différentielles sur une variété analytique complexe. Sémin. Anal. P. Lelong 2, Paris 1958–1959, No. *10*, Zbl.197,69

Norguet, F. (1971): Introduction à la Théorie Cohomologique des Résidus. Lect. Notes Math. *205*, 34–55, Zbl.218.32004

Palamodov, V.P. (1967): On the multiplicity of a holomorphic mapping. Funkts. Anal. Prilozh. *1*, No. 3, 54–65. Engl. transl.: Funct. Anal. Appl. *1*, 218–226 (1967), Zbl.164,92

Pham, F. (1967): Introduction à l'Étude Topologique des Singularités de Landau. Mém. Sci. Math. *164*, 143 pp., Paris: Gauthier-Villars, Zbl.157,275

Poincaré, H. (1887): Sur les résidus des intégrales doubles. Acta Math. *9*, 321–380, Jbuch 19, 275

Roos, G. (1974): L'intégrale de Cauchy dans $\mathbb{C}^n$. Lect. Notes Math. *409*, 171–195, Zbl.304.32005

Solomin, J.E. (1977): Le résidu logarithmique dans les intersections non complètes. C. R. Acad. Sci., Paris, Sér. A *284*, No. 17, 1061–1064, Zbl.354.32005

Sorani, G. (1963): Sui residue delle forme differenziali di una varieta analitica complessa. Rend. Mat. Appl., V. Ser. *22*, No. 1–2, 1–23, Zbl.124,389

Tong, Y.L. (1973): Integral representation formulae and Grothendieck residue symbol. Am. J. Math. *95*, No. 4, 904–917, Zbl.291.32008

Tsikh, A.K. (1984): Local residues in $\mathbb{C}^n$. Algebraic applications. Mat. Sb., Nov. Ser. 123, No. 2, 230–242. Engl. transl.: Math. USSR, Sb. *51*, 225–237 (1985)

Tsikh, A.K., Yuzhakov, A.P. (1984): Properties of the complete sum of residues with respect to a polynomial mapping, and their applications. Sib. Mat. Zh. *25*, No. 4, 207–213. Engl. transl.: Sib. Math. J. *25*, 677–682 (1984), Zbl.561.32003

Varchenko, A.N. (1983): Asymptotics of integrals and Hodge structures. Itogi Nauki Tekh., Ser. Sovrem. Probl. Mat. *22*, 130–166. Engl. transl.: J. Sov. Math. *27*, 2760–2784 (1984), Zbl.543.58008

Wirtinger, W. (1937): Ein Integralsatz über analytische Gebilde im Gebiete von mehreren komplexen Veränderlichen. Monatsh. Math. Phys. *45*, 418–431, Zbl.16,408

Yuzhakov, A.P. (1975): On an application of the multiple logarithmic residue to the expansion of implicit functions in power series. Mat. Sb., Nov. Ser. *97*, No. 2, 177–192. Engl. transl.: Math. USSR, Sb. *26*, 165–179 (1976), Zbl.326.32002

II. Plurisubharmonic Functions

A. Sadullaev

Translated from the Russian
by P.M. Gauthier

Contents

Introduction

Plurisubharmonic functions play a major role in the theory of functions of several complex variables. The extensiveness of plurisubharmonic functions, the simplicity of their definition together with the richness of their properties and, most importantly, their close connection with holomorphic functions have assured plurisubharmonic functions a lasting place in multidimensional complex analysis. (Pluri)subharmonic functions first made their appearance in the works of Hartogs at the beginning of the century. They figure in an essential way, for example, in the proof of the famous theorem of Hartogs (1906) on joint holomorphicity.

Defined at first on the complex plane $\mathbb{C}$, the class of subharmonic functions became thereafter one of the most fundamental tools in the investigation of analytic functions of one or several variables. The theory of subharmonic functions was developed and generalized in various directions: subharmonic functions in Euclidean space $\mathbb{R}^n$, plurisubharmonic functions in complex space $\mathbb{C}^n$ and others.

Subharmonic functions and the foundations of the associated classical potential theory are sufficiently well exposed in the literature, and so we introduce here only a few fundamental results which we require. More detailed expositions can be found in the monographs of Privalov (1937), Brelot (1961), and Landkof (1966). See also Brelot (1972), where a history of the development of the theory of subharmonic functions is given.

The theory of plurisubharmonic functions is connected with analytic functions of several variables. The foundations of the theory were laid in the 40's and 50's while studying properties of domains of holomorphy and entire and meromorphic functions. Of primary importance were the works of Oka (1942, 1953), Lelong (1941, 1957, 1966), and Bremermann (1956, 1959). These results are presented fairly completely in the monograph of Vladimirov (1964).

The subsequent development of the theory of plurisubharmonic functions was connected with the elaboration of a complex potential theory encompassing capacitary properties of plurisubharmonic functions. Interest in this direction has increased particularly in recent years thanks to the connection, established by Bedford and Taylor, between extremal plurisubharmonic functions and solutions of the complex Monge-Ampère equation $(dd^c u)^n = 0$. At the present time there are already a large number of papers devoted to both complex potential theory itself as well as to its applications.

The aim of the present paper is to, firstly set forth the foundations of the theory of plurisubharmonic functions and, in particular, the complex potential theory (Chap. 2), and secondly, to exhibit some of its applications to multidimensional complex analysis (Chap. 3). Our exposition of complex potential theory is, as much as possible, self-contained. To this end, we, first of all in Chap. 1, introduce some necessary results from the theory of subharmonic functions. The further development proceeds without invoking more profound theorems of classical potential theory.

Fundamental applications of complex potential theory are related to rational and polynomial approximation and problems of analytic continuation. As in the classical case $n = 1$, at the core of such applications lie estimates connected with extremal plurisubharmonic functions. The multidimensional problems which are considered in this context have deep connections with the classical works of Bernstein, Walsh, Keldysh-Lavrentiev (1937), Gonchar (1972, 1974) and others.

Chapter 1
Elementary Theory of Plurisubharmonic Functions

§ 1. Subharmonic Functions

1.1. Definition and Basic Properties. A function u, $-\infty \leq u(x) < +\infty$, defined in a domain D of Euclidean space $\mathbb{R}^n$, is said to be subharmonic in D if

a) u is upper semicontinuous in D, i.e. $\overline{\lim\limits_{x \to x^0}} u(x) \leq u(x^0)$ for an arbitrary point $x^0 \in D$;

b) for each point $x^0 \in D$, for sufficiently small $r > 0$ ($r \leq r(x^0)$), the value $u(x^0)$ does not exceed the average value of u on the sphere $S(x^0, r)$:

$$u(x^0) \leq \frac{1}{\sigma_n r^{n-1}} \int_{S(x^0, r)} u(x) \, d\sigma, \tag{1}$$

where σ_n is the area of the unit sphere $S(0, 1)$ in $\mathbb{R}^n$.

From condition a) it follows that the set $\{x \in D : u(x) < t\}$ is open for each $t \in \mathbb{R}$. Moreover, on each compact subset $K \subset D$, the function u attains its maximum. From this together with condition b), we easily conclude the following:

Theorem (maximum principle). *Suppose u is subharmonic in a domain $D \subset \mathbb{R}^n$ and attains a maximum at some point $x^0 \in D : u(x^0) = \sup\limits_{x \in D} u(x)$. Then $u \equiv const$ in D.*

The following properties follow immediately from the definition of subharmonic functions:

1) A linear combination $a_1 u_1 + \cdots + a_m u_m$ of subharmonic functions u_j with nonnegative constant coefficients ($a_j \geq 0$) is a subharmonic function.

2) The maximum of finitely many subharmonic functions, $u(x) = \sup\{u_1(x), \ldots, u_m(x)\}$, is also subharmonic.

3) The limit of a uniformly convergent or monotonically decreasing sequence of subharmonic functions is subharmonic.

We remark that for a harmonic function v, the functions v and $-v$ are both subharmonic. The converse is also true: if v and $-v$ are both subharmonic functions, then v is necessarily harmonic. From this and from the maximum principle, it follows that if u is subharmonic and v is harmonic, then if on the boundary of the domain D, we have $u|_{\partial D} \leq v|_{\partial D}$, then we have the same inequality, $u(x) \leq v(x)$, everywhere in D. This relationship to harmonic functions actually characterizes subharmonic functions and in some textbooks is given as the definition of a subharmonic function.

1.2. The Poisson Integral. If $u(x)$ is harmonic in the ball

$$B(x^0, r): |x - x^0| < r$$

and continuous in the closure $\bar{B}$, then the values of the function u in the interior B can be expressed in terms of the values of u on the boundary $\partial B = S(x^0, r)$ via the *Poisson integral*

$$u(x) = \int_{S(x^0, r)} u(y)\mathscr{P}(x, y)\, d\sigma(y) \tag{2}$$

where

$$\mathscr{P}(x, y) = \frac{r^2 - |x - x^0|^2}{\sigma_n r |x - y|^n}$$

is the Poisson kernel. Formula (2) can also be used for the harmonic extension into the ball, of a continuous function φ given on the boundary ∂B. The integral

$$u(x) = \int_{S(x^0, r)} \varphi(x, y)\mathscr{P}(x, y)\, d\sigma(y)$$

is the solution to the classical Dirichlet problem in the ball B, since u is harmonic in B and continuous on $\bar{B}$ and $u|_{\partial B} = \varphi$.

In the sequel we shall repeatedly make use of the following technique for studying subharmonic (and also plurisubharmonic) functions. Consider, in a domain D, a subharmonic function $u(x)$ and take its trace $u|_S$ on the boundary S of some ball $B \subset D$. Since $u|_S$ is upper semicontinuous (on S), there exists a decreasing sequence of continuous functions $\varphi_j \downarrow u|_S$. For each j, let v_j be the Poisson integral of the function φ_j. Then the v_j, being a decreasing sequence (as $j \to \infty$), converge to the Poisson integral of u,

$$v(x) = \int_S u(y)\mathscr{P}(x, y)\, d\sigma(y).$$

The function $v(x)$ is harmonic in B and its boundary values agree with $u|_S$ almost everywhere on S.

From this follows the interesting consequence that for an arbitrary subharmonic function u, the average

$$\frac{1}{\sigma_n r^{n-1}} \int_{S(x^0, r)} u(x)\, d\sigma$$

is an increasing function of r (for $r < \text{dist}(x^0, \partial D)$.

Indeed, if v is the Poisson integral for $u|_{S(x^0, r)}$ and $r_1 < r$, then

$$\frac{1}{\sigma_n r_1^{n-1}} \int_{S(x^0, r_1)} u(x) \, d\sigma \le \frac{1}{\sigma_n r_1^{n-1}} \int_{S(x^0, r_1)} v(x) \, d\sigma = v(x^0)$$

$$= \frac{1}{\sigma_n r^{n-1}} \int_{S(x^0, r)} v(x) \, d\sigma = \frac{1}{\sigma_n r^{n-1}} \int_{S(x^0, r)} u(x) \, d\sigma.$$

From this monotonicity it is not difficult to obtain also the monotonicity of the averages over balls

$$\frac{1}{V_n r^n} \int_{B(x^0, r)} u(x) \, dV,$$

where V_n is the volume of the unit ball $B(0, 1)$.

Thus, from condition b) on subharmonic functions, we have the inequality

$$u(x^0) \le \frac{1}{V_n r^n} \int_{B(x^0, r)} u(x) \, dV \le \frac{1}{\sigma_n r^{n-1}} \int_{S(x^0, r)} u(x) \, d\sigma, \tag{3}$$

and moreover, in view of the semicontinuity, both integrals in (3) converge to $u(x^0)$ as $r \to 0$.

1.3. Polar Sets. A set $E \subset D$, on which some function u, subharmonic with $u \not\equiv -\infty$ in D, takes the value $-\infty$, is called a *polar set*. In classical potential theory, it is shown that a set E is polar if and only if it has outer *Newtonian capacity* zero. In particular, for $n > 2$, metric dimension $(n - 2)$ separates polar and non-polar sets (for smaller dimension, polarity; for larger, non-polarity). The proofs of these facts are non-elementary and we shall neither present them nor refer to them. From (3) easily follows the weaker (but sufficient for us) assertion:

Theorem. *Polar sets have zero Lebesgue measure.*

Indeed, from (3) it follows that if $u \not\equiv -\infty$ is subharmonic in a domain D, then u is locally integrable. Consequently, the set $E : u(x) = -\infty$ has zero Lebesgue measure.

1.4. Approximation of Subharmonic Functions. For approximation, it is usual to make use of some kernel $K(x) \ge 0$ of class C^∞[1], whose support lies in the unit ball $B(0, 1)$, normalized such that $\int K(x) \, dV = 1$. As $K(x)$ it is convenient for us to take, for example, the function

$$K(x) = \begin{cases} \alpha_n e^{1/(|x|^2 - 1)}, & \text{for } |x| \le 1, \\ 0, & \text{for } |x| > 1 \end{cases}$$

---

[1] Here and subsequently, C^k, $0 \le k \le \infty$, denotes the class of k times continuously differentiable functions ($C^0 = C$ is the class of continuous functions).

with an appropriate constant α_n. For a function u subharmonic in D, we define
$$u_\delta(x) = \int u(x + \delta y)K(y)\,dV = \frac{1}{\delta^n}\int u(y)K\left(\frac{y-x}{\delta}\right)dV, \ \delta > 0.$$ The first integral
clearly represents a subharmonic, and the second, an infinitely smooth function.
Consequently, $u_\delta(x)$ is a subharmonic function of class C^∞.

In contrast to the situation for an arbitrary summable function, for a subharmonic function u, the sequence u_δ monotonically decreases as $\delta \downarrow 0$, and converges to u at each point $x \in D$. This follows easily from what was said at the end of 1.2.

1.5. The Riesz Representation. For functions of class C^2, subharmonicity is equivalent to positivity of the *Laplacian* $\Delta u = \dfrac{\partial^2 u}{\partial x_1^2} + \cdots + \dfrac{\partial^2 u}{\partial x_n^2}$; $u \in C^2(D)$ is subharmonic in a domain D if and only if $\Delta u \geq 0$ in D. Indeed, substituting the Taylor formula for $u(x)$ (about x^0) in the integral $\dfrac{1}{\sigma_n r^{n-1}}\displaystyle\int_{S(x^0,r)} u(x)\,d\sigma$, we obtain

$$\frac{1}{\sigma_n r^{n-1}}\int_{S(x^0,r)} u(x)\,d\sigma = u(x^0) + C\cdot r^2[\Delta u|_{x=x^0} + o(1)],$$

where C is a constant independent of r. The validity of the criterion $\Delta u \geq 0$ clearly follows from this formula and property b).

If $u \not\equiv -\infty$ is an arbitrary (not necessarily twice smooth) subharmonic function in a domain D, then u is locally summable in D (see 1.3) and, consequently, is a distribution. Let Δu be its Laplacian in the distributional sense. By approximating u by infinitely smooth subharmonic functions, we see that $\Delta u \geq 0$. Since every positive distribution is a measure, the distributional Laplacian Δu is a (positive) measure on D. This measure is called the *measure associated* with u.

The associated measure Δu allows us to associate, with a subharmonic function u, a corresponding potential with respect to the Newtonian (Logarithmic for $n = 2$) kernel

$$K(x) = \begin{cases} \dfrac{1}{2\pi}\ln|x|, & \text{for } n = 2, \\[3mm] -\dfrac{\Gamma(n/2 - 1)}{4\pi^{n/2}}\cdot\dfrac{1}{|x|^{n-2}}, & \text{for } n > 2. \end{cases}$$

We remark that if μ is any finite Borel measure in $\mathbb{R}^n$, then its *potential*

$$U^\mu(x) = \int K(x - y)\,d\mu(y)$$

is a subharmonic function with Laplacian $\Delta u = \mu$. Consequently, if we consider the restriction μ_G of the associated measure $\mu = \Delta u$ to some subdomain $G \subset\subset$

D, we obtain that the difference

$$u(x) - \int K(x - y)\, d\mu_G(y)$$

is a harmonic function in *G*. Thus, we arrive at the following Riesz Representation Theorem.

Theorem. *Let u be a function subharmonic in a domain $D \subset \mathbb{R}^n$. Then on any relatively compact domain $G \subset\subset D$, u has a representation of the form*

$$u(x) = U^\mu(x) + h(x),$$

where U^μ is the potential of some measure μ and h is a function harmonic in G.

Corollary. *If a distribution u is such that $\Delta u \geq 0$, then it is given by a subharmonic function.*

1.6. Hartog's Lemma. *Suppose in a domain D that $g(x)$ is a continuous function and $u_j(x)$ is a sequence of locally uniformly upper-bounded subharmonic functions such that*

$$\varlimsup_{j \to \infty} u_j(x) \leq g(x) \tag{4}$$

at each point $x \in D$. Then, on any compact set $K \subset D$, inequality (4) holds uniformly; that is, for each $\varepsilon > 0$ there exists an integer j_0 such that $u_j(x) \leq g(x) + \varepsilon$, for each $x \in K$ and each $j \geq j_0$.

The proof of this very important lemma is not difficult (cf. e.g. Shabat (1976)). It follows easily from the following inequalities which result from (3) and (4):

$$\varlimsup_{j \to \infty} u_j(x^0) \leqslant \lim_{j \to \infty} \frac{1}{V_n r^n} \int_{B(x^0, r)} u_j(x)\, dV \leq \frac{1}{V_n r^n} \int_{B(x^0, r)} g(x)\, dV, \quad x^0 \in K, \quad r > 0.$$

§2. Plurisubharmonic Functions and Their Elementary Properties

The concept of a plurisubharmonic (psh) function is connected with the complex structure of $\mathbb{C}^n$: a function $u(z)$ is said to be *plurisubharmonic* in a domain $D \subset \mathbb{C}^n$ if

a) it is upper semicontinuous in *D* and
b) the restriction $u|_l$ is subharmonic on $l \cap D$ for every complex line *l*.

It follows immediately from the definition that a psh function is simultaneously a subharmonic function. Hence, properties 1), 2), and 3) from 1.1 for subharmonic functions are valid also for psh functions. Rather than repeat these properties, we pass directly to properties which are inherent only to psh functions.

2.1. Approximation. If u is a plurisubharmonic function in a domain $D \subset \mathbb{C}^n$, then the functions u_δ, $\delta > 0$, constructed in 1.4, are also plurisubharmonic. Hence a function $u \in \mathrm{Psh}(D)$ can always be approximated by psh functions of class C^∞ on compact subsets of D. The functions u_δ will be defined on all of D only in the case $D = \mathbb{C}^n$.

As the following example of J.E. Fornaess and J. Wiegerinck (1989) shows, for an arbitrary domain D, it is not in general possible to approximate on all of D.

Example. On the plane $\mathbb{C}_w$ consider the subharmonic function

$$v(w) = \sum_{k=2}^{\infty} \frac{2^{-k}}{\ln k} \ln \left| w - \frac{1}{k} \right|.$$

Note, $v(1/k) = \infty$, but $v(0) = -1/2$.

We surround the point $w = 1/k$ by a disc $U_k : |w - 1/k| < r_k$ of radius $r_k > 0$ so small that $v(w) < -1$ in U_k, $k = 2, 3, \ldots$. For such a choice, the function

$$u(z, w) = \begin{cases} \max\{v(w), -1\}, & \text{for } |z| < 1, \\ -1, & \text{for } |z| > 1, \end{cases}$$

is plurisubharmonic in the disconnected set $\mathbb{C}^2(z, w) \setminus \{|z| = 1, w \in \mathbb{C}\}$ and extends to be psh through the holes

$$Q = \bigcup_{k=2}^{\infty} \{\{|z| = 1\} \times U_k\}$$

by setting $u(z, w) = -1$ on Q. In other words, the extended function $u(z, w)$ is psh in the domain $\Omega = \{(z, w) : |z| \neq 1\} \cup Q$. This function $u(z, w)$ cannot be globally approximated by a sequence of continuous psh functions $u_j \downarrow u$. Indeed, if it could, then the sequence u_j would converge to -1 on the circles $\{|z| = 2, w = 0\}$ and $\{|z| = 2, w = 1/k\}$, $k = 2, 3, \ldots$. Since the discs $\{|z| < 2, w = 1/k\}$ belong entirely to the domain Ω, by the maximum principle, the convergence $u_j \downarrow -1$ also holds (uniformly) on the union of these discs. But this is impossible since $u(z, 0) = -1/2$, $|z| < 1$, and the u_j are continuous.

The domain Ω in the preceding example is not a domain of holomorphy in $\mathbb{C}^2$. It turns out that on domains of holomorphy, global approximation by psh functions of class C^∞ is nevertheless possible. This assertion easily follows from the following theorem on continuation of psh functions from submanifolds.

Theorem (Sadullaev (1982)). *If $M \subset \mathbb{C}^n$ is a closed complex submanifold, then any function u psh on M^2 psh-extends to all of $\mathbb{C}^n$; i.e. there exists a psh function w in $\mathbb{C}^n$ such that $w|_M = u$.*

To construct global approximations $u_j \downarrow u$ on a Stein manifold M embedded in $\mathbb{C}^n$, we approximate the extension w in $\mathbb{C}^n$ and consider the restrictions of these approximating functions on M.

---

[2] Psh on a manifold is defined by local coordinates.

2.2. The Operator dd^c. We suppose that the reader is acquainted with the theory of *currents* and, in particular, with the notion of a positive current. The necessary facts are well exposed in Harvey (1977) (see the supplement in the Russian translation by Chirka).

We recall that $d = \partial + \bar{\partial}$, $d^c = i(\bar{\partial} - \partial)$, and hence $dd^c = 2i\partial\bar{\partial}$. Let u be a psh function of class C^2 in a neighborhood of 0. Then if $l : z_j = a_j\omega, j = 1, 2, \ldots, n$ is any complex line, with parameter $\omega \in \mathbb{C}$, the restriction $u|_l = u(a\omega)$ is subharmonic in ω.

Therefore $\Delta_\omega u|_l \geq 0$ which means that the quadratic form

$$\sum_{j,k} \frac{\partial^2 u}{\partial z_j \partial \bar{z}_k} a_j \bar{a}_k,$$

is positive for all $a \in \mathbb{C}^n$, i.e. this form is positive definite. From this it follows that $dd^c u$ is a positive differential form of bidegree $(1, 1)$. For an arbitrary psh function u we use smooth approximations $u_j \downarrow u$ to see that $dd^c u$ is also a positive current, i.e.

$$\int u \, dd^c \alpha \geq 0 \tag{5}$$

for any positive differential test form α of bidegree $(n - 1, n - 1)$.

Formula (5) may be taken as the definition of a psh function: an upper semicontinuous function u is psh if and only if $dd^c u \geq 0$ as a current of bidegree $(1, 1)$.

We remark that every positive current is a current of measure type, i.e. is a generalized differential form whose coefficients are Borel measures (cf. Harvey (1977)).

2.3. The Upper Envelope of psh Functions. It was mentioned above that the maximum of finitely many psh functions is again plurisubharmonic. In practice, we often encounter the upper envelope $u = \sup u_\alpha$ of an infinite number of psh functions $u_\alpha, \alpha \in \Lambda$. In this situation, we require, first of all, that u be locally upper-bounded (i.e. the family $\{u_\alpha\}$ be locally uniformly upper-bounded). However, this is insufficient for plurisubharmonicity of u; u could turn out not to be a semicontinuous function. For this reason, we employ the regularization u^* of a function u (the least upper semicontinuous majorant):

$$u^*(z) = \lim_{\varepsilon \to 0} \sup_{|w-z| < \varepsilon} u(w) = \varlimsup_{w \to z} u(w).$$

The function u^* is plurisubharmonic. This follows easily from the definition of a psh function.

The situation is analogous for the upper limit

$$u(z) = \varlimsup_{j \to \infty} u_j(z)$$

of a locally uniformly upper-bounded sequence $\{u_j\}$: the regularization u^* is psh in the domain of plurisubharmonicity of the functions u_j. Later (see Chap. 2) it will be shown that the set $\{z : u(z) < u^*(z)\}$, where u differs from its regularization u^*, is negligible: it is a pluripolar set.

2.4. Connection with Holomorphic Functions. If a function f is holomorphic in a domain $D \subset \mathbb{C}^n$, then $\ln|f|$ is a psh function. Conversely, every psh function can be represented by holomorphic functions in the following way.

Theorem (Bremermann (1956)). *Any psh function u on a Stein manifold D can be represented in the form*:

$$u(z) = \left[\overline{\lim_{j\to\infty}} \; \alpha_j \ln|f_j| \right]^*, \quad \alpha_j \geq 0, \quad f_j \in \mathcal{O}(D). \tag{6}$$

For a plane domain $D \subset \mathbb{C}$, this theorem was proved by Lelong (1941). His proof is based on methods of classical potential theory and is rather long. However, if we invoke a famous theorem of Oka on holomorphic nonextendability for pseudoconvex domains, then the proof becomes simpler and carries over to the general case.

Indeed, consider the domain

$$D^* = \{(z, w) \in D \times \mathbb{C} : \ln|w| + u(z) < 0\}$$

in the product $D \times \mathbb{C}_w$. Since $\ln|w| + u$ is a plurisubharmonic function in $D \times \mathbb{C}$ and D is pseudoconvex, the domain D^* is also pseudoconvex. By a theorem of Oka (1942, 1953), D^* is a domain of holomorphy. Consequently, there exists a function $F \in \mathcal{O}(D^*)$ which is nonextendable holomorphically outside of D^*. Expanding F in a Hartog's series: $F(z, w) = \sum\limits_{j=1}^{\infty} f_j(z)w^j$, we find that the radius of convergence of this series

$$R(z) = \left[\overline{\lim_{j\to\infty}} \; |f_j(z)|^{1/j} \right]^{-1}$$

almost everywhere agrees with e^{-u}. Thus, $u(z) = \left[\overline{\lim_{j\to\infty}} \; \dfrac{\ln|f_j|}{j} \right]^*$.

Remark. Bremermann showed the uniform proximity of functions of the type $\alpha \ln|f|$ to continuous psh functions. More precisely, if u is a continuous psh function in D, then for each compact $K \subset D$ and each $\varepsilon > 0$, there exist functions $f_1, f_2, \ldots, f_N$ and positive constants $\alpha_1, \alpha_2, \ldots, \alpha_N$ such that, on K,

$$\max\{\alpha_1 \ln|f_1|, \ldots, \alpha_N \ln|f_N|\} \leq u \leq \max\{\alpha_1 \ln|f_1|, \ldots, \alpha_N \ln|f_N|\} + \varepsilon. \tag{7}$$

Thus, using an approximation $u_j \downarrow u$, we get that for any psh function u, we may omit the regularization symbol $(*)$ in (6).

Now, we introduce an example showing that the requirement that D be holomorphically convex (a Stein manifold) is essential in the theorem of Bremermann.

Example. Let $D = \mathbb{C}^2 \backslash S^+$, where S^+ is the half-sphere

$$S^+ = \{z = (z_1, z_2) : |z| = 1, \operatorname{Im} z_2 \geq 0\}.$$

Then the function

$$u(z) = \begin{cases} 0, & \text{for } |z| > 1 \text{ or } \operatorname{Im} z_2 < 0, \\ \operatorname{Im} z_2, & \text{for } |z| < 1 \text{ and } \operatorname{Im} z_2 \geq 0 \end{cases} \tag{8}$$

is plurisubharmonic in D. However u does not have a representation in the form (6), because every function f_i would be holomorphic in $\mathbb{C}^2$, but the function u is not psh in $\mathbb{C}^2$.

Remark. The theory of subharmonic functions and potential theory are so intertwined in their contemporary state of development that they are often identified. However, previous to the connection between subharmonic functions and potentials of Borel measures (F. Riesz (1926), cf. 1.5), these two theories developed in parallel: subharmonic functions, as a part of function theory, and potential theory in connection with the Dirichlet problem for equations of mathematical physics. In the theory of psh functions of classical potential theory, certain capacitary properties (continuity almost everywhere in capacity, structure of the set $\{u < u^*\}$ for the upper envelope etc.) correspond to these functions. We now turn to such properties.

Chapter 2
Complex Potential Theory

The following investigation of psh functions is connected with extremal problems and capacitary properties. In these questions, an important role is played by the complex Monge-Ampère operator $(dd^c)^n$, where n is the dimension of the given complex manifold. The operator $(dd^c)^n$ will play a role here similar to the role played by the Laplacian in classical potential theory.

In contrast to the classical situation, for $n > 1$, this equation is not linear. For this reason, the associated "complex potential theory" is also called non-linear potential theory. Research in this area was initiated by Chern, Levin, and Nirenberg, who showed that the operator $(dd^c)^n$ is bounded in the mean for the class of uniformly bounded psh functions of class C^2, and by the fundamental paper of Bedford and Taylor (1976), in which the operator $(dd^c)^n$ is defined for bounded (not necessarily smooth) psh functions as a certain positive measure.

In the years 1976–1983, interesting studies in multi-dimensional complex potential theory were carried out. These studies laid the foundations for the central concepts of this theory (Bedford-Taylor (1976, 1982), Sadullaev (1981)). During this same period, methods were developed for applying this theory to problems of several complex variables, especially to questions of approximation

(Zakharyuta (1974, 1976), Siciak (1962, 1969), Sadullaev (1982b, 1984), Nguyen Thanh Van et Zeriahi Ahmed (1983, 1985)). At the present time the elaboration of the foundations of complex potential theory is essentially completed. A remarkable result of this theory has been the solution of two problems of Lelong concerning pluripolar sets posed by him in the early 60's. The first of these problems, global pluripolarity of a locally pluripolar set, was solved by Josefson (1978). The solution of the second problem concluded a series of studies by Bedford-Taylor (1982) and the author (1982b) (see 3.3 below); the first solution to the second problem is due to Bedford-Taylor (see also Cegrell (1978)).

In this chapter we give a brief presentation of the foundations of complex potential theory.

Our approach is to construct this theory based on certain integral estimates and somewhat differs from the path taken by Bedford-Taylor.

§ 1. The Complex Monge-Ampère Operator

1.1. The Operator $(dd^c)^n$. For a function u of class C^2, we define $(dd^c u)^k = dd^c u \wedge \cdots \wedge dd^c u$ (k times). This represents a differential form of bidegree (k, k). It is easy to see that

$$(dd^c u)^n = \pi^n n! \det\left(\frac{\partial^2 u}{\partial z_j \partial \bar{z}_k}\right) \beta_n,$$

where $\beta_n = \left(\frac{i}{2}\right)^n \prod_{j=1}^{n} dz_j \wedge d\bar{z}_j$ is the volume form on $\mathbb{C}^n$.

The operator $(dd^c u)^k$ for an arbitrary bounded (not necessarily twice smooth) psh function can be defined in the distributional sense. Let u be a bounded psh function in a domain $G \subset \mathbb{C}^n$ and denote by $D^{(n-k,n-k)}$ the space of differential forms of bidegree $(n - k, n - k)$, of class C^∞ and of compact support in G. Then, the recurrence relation

$$\int (dd^c u)^k \wedge \varphi = \int u(dd^c u)^{k-1} \wedge dd^c \varphi, \quad \varphi \in D^{(n-k,n-k)} \tag{1}$$

defines $(dd^c u)^k$ as a positive current of bidegree (k, k).

Indeed, for $k = 1$, this was shown in 2.2 of Chap. 1. Suppose, by induction, that the assertion is true for $k - 1$ and let us prove it for k. Since by assumption $(dd^c u)^{k-1}$ is positive, it is a current of measure type. Consequently, $u(dd^c u)^{k-1}$ is also a current of measure type and hence the right side of (1) is well-defined (recall that u is bounded). Thus, $(dd^c u)^k$ is a current. In order to show the positivity of $(dd^c u)^k$, we use an approximation $u_j \downarrow u$. We have

$$\int u(dd^c u)^{k-1} \wedge dd^c \varphi = \lim_{j \to \infty} \int u_j(dd^c u)^{k-1} \wedge dd^c \varphi.$$

By definition of the current $(dd^c u)^{k-1}$,

$$\int (dd^c u)^{k-1} \wedge u_j dd^c \varphi = \int (dd^c u)^{k-2} \wedge dd^c u_j \wedge dd^c \varphi = \int (dd^c u)^{k-1} \wedge dd^c u_j \wedge \varphi.$$

Hence, $\int u_j (dd^c u)^{k-1} \wedge dd^c \varphi \geq 0$ for all positive forms φ, since the current $(dd^c u)^{k-1}$ is positive by the inductive hypothesis and the form $dd^c u_j \wedge \varphi$ is positive, being the exterior product of two positive forms. It follows that the current $(dd^c u)^k$ is positive.

1.2. Integral Estimates. The basic difficulty in making practical use of the operator $(dd^c u)^k$ consists in establishing the convergence of the currents $(dd^c u_j)^k$ for a monotone sequence of psh functions u_j. In this section we show such a convergence only for a uniformly convergent sequence u_j. If u is a continuous psh function, then its approximating sequence $u_j \downarrow u$ of C^∞-functions converges uniformly on compacta. In this case, it follows that $(dd^c u_j)^k \to (dd^c u)^k$ in the sense of currents, for $k = 1, 2, \ldots, n$. The convergence of $(dd^c u_j)^k$ for an arbitrary monotone sequence u_j will be shown in 3.2. Till then, it will be necessary for us to work only with continuous functions.

Lemma. *If $G = \{\rho(z) < 0\}$ is a strictly pseudoconvex domain, $\rho \in C^2(G)$, $\sigma = \min\limits_{G} \rho(z)$, and u is a continuous psh function in G, then for each r and k, $\sigma < r < 0, 1 \leq k \leq n$,*

$$\int_\sigma^r dt \int_{\rho(z) \leq t} (dd^c \rho)^{n-k} \wedge (dd^c u)^k \leq (M - m) \int_{\rho(z) \leq r} (dd^c \rho)^{n-k+1} \wedge (dd^c u)^{k-1}, \quad (2)$$

where $M = \max\limits_{G} u(z), m = \min\limits_{G} u(z)$.

For a C^2-function, the lemma follows from Stokes' Theorem and Fubini's Theorem

$$\int_\sigma^r dt \int_{\rho(z) \leq t} (dd^c \rho)^{n-k} \wedge (dd^c u)^k = \int_{\rho \leq r} d\rho \wedge d^c u \wedge (dd^c \rho)^{n-k} \wedge (dd^c u)^{k-1}$$

$$= \int_{\rho = r} u d^c \rho \wedge (dd^c \rho)^{n-k} \wedge (dd^c u)^{k-1}$$

$$- \int_{\rho \leq r} u (dd^c \rho)^{n-k+1} \wedge (dd^c u)^{k-1}. \quad (3)$$

Since $d^c \rho \wedge (dd^c \rho)^{n-k} \wedge (dd^c u)^{k-1} = \lambda s$, where $\lambda \geq 0$ and s is the Euclidean volume form on $\{\rho = r\}$ and since the form $(dd^c \rho)^{n-k+1} \wedge (dd^c u)^{k-1}$ is positive, we obtain (2) by estimating u and invoking once again Stokes' Theorem. In case u is an arbitrary continuous psh function, we use an approximation $u_j \downarrow u$. In order to pass to the limit, it is sufficient to establish the compactness of the family $(dd^c u_j)^k$ (cf. Theorems 1 and 2 below).

Let u be a psh function of class C^2 in G. Applying the inequality (2) successively for $k = p, p - 1, \ldots, 1$, we obtain:

$$\int_\sigma^0 dt_1 \int_\sigma^{t_1} dt_2 \cdots \int_\sigma^{t_{p-1}} dt_p \int_{\rho \le t_p} (dd^c\rho)^{n-p} \wedge (dd^c u)^p \le (M - m)^p \int_{\rho \le 0} (dd^c\rho)^n. \tag{4}$$

For fixed r, $\sigma < r < 0$, the left side of (4) can be estimated from below by the quantity

$$\frac{|r|^p}{p!} \int_{\rho \le r} (dd^c\rho)^{n-p} \wedge (dd^c u)^p.$$

Consequently, for each $r < 0$, the following mean estimate holds

$$\int_{\rho \le r} (dd^c\rho)^{n-p} \wedge (dd^c u)^p \le C_r (M - m)^p(u), \tag{5}$$

where C_r is a constant independent of u. Since $(dd^c u)^p$ is a strictly positive current and ρ is strictly psh, the left member of (5) is greater than $C \cdot \|(dd^c u)^p\|$, where the constant $C > 0$ depends only on ρ and $\|\cdot\|$ denotes the mass of a current of measure type (see Harvey (1977)). Consequently, for any test form $\alpha \in D^{(n-p, n-p)}$, we have the estimate

$$\int_{\rho \le r} \alpha \wedge (dd^c u)^p \le C_r \cdot \|\alpha\|_G (M - m)^p(u), \tag{6}$$

expressing the weak compactness of the family $(dd^c u)^p$ for the class of bounded (above and below) psh functions u. From this estimate and the ones shown earlier, the assertion of the lemma follows easily for the general case.

Let us now state separately the compactness result and the passage to the limit which we have obtained (see Bedford-Taylor (1976)).

Theorem 1. *Let $\{u\}$ be a locally uniformly bounded family of psh functions of class C^2 in a domain $G \subset \mathbb{C}^n$. Then, the family of currents $\{(dd^c u)^p\}$ is weakly compact in G for any p, $1 \le p \le n$.*

Theorem 2. *If a sequence of psh functions $u_j \in C^2(G)$ converges uniformly to u in a domain G, then $(dd^c u_j)^p \to (dd^c u)^p$, $1 \le p \le n$, weakly.*

Indeed, for $p = 1$, the theorem is obvious. We suppose the theorem true for $p - 1$ and show it for p. By definition, for any test form $\alpha \in D^{(n-p, n-p)}$,

$$\int_G (dd^c u_j)^p \wedge \alpha = \int_G u_j (dd^c u_j)^{p-1} \wedge (dd^c\alpha) = \int_G u(dd^c u_j)^{p-1} \wedge dd^c\alpha$$

$$+ \int_G (u_j - u)(dd^c u_j)^{p-1} \wedge (dd^c\alpha).$$

By the inductive hypothesis, the first integral on the right side tends to

$$\int_G u(dd^c u)^{p-1} \wedge dd^c\alpha := \int_G (dd^c u)^p \wedge \alpha,$$

and the second integral to zero, in view of the boundedness of $(dd^c u_j)^{p-1}$ and the uniform convergence of $u_j - u$ to zero.

Remark. The estimate (6), which was shown for psh functions of class C^2, is valid also for continuous functions. Thus Theorem 1, and consequently Theorem 2, are also true for the class of continuous psh functions (see also below, 3.2).

1.3. The Dirichlet Problem. Plurisubharmonic functions satisfying the Monge-Ampère equation $(dd^c u)^n = 0$ (for $n = 1$ we have the classical harmonic functions) play an important role for $n > 1$. These functions, in general, are not C^∞-smooth; they can even be discontinuous.

Consider the following Dirichlet problem. Given a continuous function φ on the boundary ∂G of a domain $G \subset \mathbb{C}^n$, we seek a solution ω the homogeneous Monge-Ampère equation with boundary data φ.

$$(dd^c \omega)^n = 0, \quad \omega|_{\partial G} = \varphi. \tag{7}$$

For an arbitrary domain G, problem (7) in general has no solution. However, if G is strictly pseudoconvex, then a solution exists and can be found, for instance, by the Perron method (see Brelot (1961)). Let $\mathscr{U}(\varphi)$ be the family of all psh functions u in G such that the boundary values along the normal are dominated by the boundary data: $u^*|_{\partial G} \leq \varphi$. Set

$$\omega(z) = \sup_{u \in \mathscr{U}} u(z).$$

Then the regularization ω^* satisfies the condition $\omega^*|_{\partial G} = \varphi$ (Bremermann) and is continuous in $\overline{G}$ (Walsh). Bedford and Taylor (1976) showed that ω^* is a solution of (7), i.e.

$$(dd^c \omega)^n = 0 \quad \text{and} \quad \omega^*|_{\partial G} = \varphi.$$

Just as for harmonic functions, the solutions u of the homogeneous Monge-Ampère equation satisfy a maximum principle: for any psh function v in G and for any domain $D \subset\subset G$, $v|_{\partial D} \leq u|_{\partial D}$ implies $v \leq u$ in all of D. The proof of this statement is connected with the following proposition.

Theorem. *If $u, v \in L^\infty \cap \mathrm{Psh}(G)$ and the set*

$$U = \{z \in G : u(z) < v(z)\}$$

is precompact in G, then

$$\int_G (dd^c u)^n \geq \int_G (dd^c v)^n.$$

For infinitely smooth functions, the proof is not difficult. Using C^∞-approximations and Theorem 2, 1.2, it is easy to prove the theorem for continuous functions u, v. The general case can be proved as a consequence of the convergence property of $(dd^c u_j)^n$ for a monotone sequence u_j, which will be given in 3.2 of Chap. 2.

§2. Extremal Functions and Capacity

In this section we introduce the notions of $\mathscr{P}$-measure and capacity of a set E with respect to a domain $G \supset E$. These are plurisubharmonic analogs of the classical notions of harmonic measure and capacity of condensers. For simplicity, we suppose that $G = \{\rho(z) < 0\}$ is a strictly pseudoconvex domain in $\mathbb{C}^n$, though all results hold in the more general situation.

2.1. $\mathscr{P}$-measure. Let E be a subset of G and $\mathscr{U}(E, G)$ the class of psh functions u in G such that $u|_G \leq 0$, $u|_E \leq -1$. Set

$$\omega(z, E, G) = \sup\{u(z) : u \in \mathscr{U}(E, G)\}.$$

The regularization $\omega^*(z, E, G)$ is called the $\mathscr{P}$-measure of the set E with respect to the domain G.

Note that the topological lemma of Choquet (cf. Brelot (1961)) implies that there is a countable family $\mathscr{U}' \subset \mathscr{U}(E, G)$ such that

$$[\sup\{u(z) : u \in \mathscr{U}'\}]^* = \omega^*(z, E, G).$$

As a corollary, it follows that the $\mathscr{P}$-measure ω^* can be represented as a limit

$$\omega^* = \left[\lim_{j \to \infty} u_j(z) \right]^*, \quad \text{where } u_j \in \mathscr{U}(E, G)$$

and $\{u_j\}$ is an increasing sequence.

We state the following simple properties of $\mathscr{P}$-measure (cf. Sadullaev (1981)).

a) *Monotonicity.* If $E_1 \subset E_2$, then

$$\omega(z, E_1, G) \geq \omega(z, E_2, G).$$

b) *Approximation.* If $U \subset G$ is an open set and $U = \bigcup_{j=1}^{\infty} K_j$, where $K_j \subset K_{j+1}$ are compact, then

$$\omega^*(z, K_j, G) \downarrow \omega^*(z, U, G).$$

If E is an arbitrary set, then there exists a decreasing sequence of open sets $U_j \supset E$, $U_j \supset U_{j+1}$ ($j = 1, 2, \ldots$) such that

$$\left(\lim_{j \to \infty} \omega(z, U_j, G) \right)^* = \omega^*(z, E, G).$$

In analogy to polar sets (Chap. 1, 1.3), a subset $E \subset G$ is said to be *pluripolar* in G if there is a psh function $u \not\equiv -\infty$ in G such that $u|_E \equiv -\infty$. Since psh functions are also subharmonic, every pluripolar set is polar as well. In particular, pluripolar sets have Lebesgue measure zero. It is easy to prove that a countable union of pluripolar sets is pluripolar.

c) *$\mathscr{P}$-measure and pluripolarity.* $\omega^*(z, E, G)$ is either nowhere 0 or identically 0. The latter holds if and only if E is pluripolar in G.

In fact, if $\omega^* = 0$ at some point $z^0 \in G$, then by the maximum principle, $\omega^* = 0$ in G. In this case, there is a sequence $z^j \in G(z^j \to z^0 \in G)$ such that $\omega(z^j, E, G) \to 0$ as $j \to \infty$. It follows that $\omega(z, E, G) = 0$ almost everywhere in a neighborhood of z^0, because ω satisfies the inequality (see 1.2):

$$\omega(z^j, E, G) \le \frac{1}{V_{2n} r^{2n}} \int_{B(z^j, r)} \omega(z, E, G) \, dV.$$

Fix a point z' such that $\omega(z', E, G) = 0$ and take a sequence $u_j \in \mathcal{U}(E, G)$ with the property $u_j(z') \ge -2^{-j}$. The sum $u(z) = \sum_{j=1}^{\infty} u_j(z)$ is psh in G; moreover, $u(z') \ge -1$ and $u|_E \equiv -\infty$. On the other hand, if u is a psh function in G such that $u \le 0$, $u \not\equiv -\infty$ in G, but $u|_E \equiv -\infty$, then $\varepsilon u \in \mathcal{U}(E, G)$ for every $\varepsilon > 0$. This means that $\omega(z, E, G) = 0$ for every $z \in G$ such that $u(z) \ne -\infty$, i.e., for almost every $z \in G$. Hence, $\omega^*(z, E, G) \equiv 0$ in G.

Although the function ω is equal to -1 on E, the regularization ω^* is not necessarily -1 on E. In the process of regularization, it can jump at some points at which E is sufficiently thin. A compact set $K \subset G$ is said to be *pluriregular* if $\omega^*(z, K, G)|_K \equiv -1$. For more information about pluriregular sets, see Sadullaev (1981).

d) *Continuity of $\mathscr{P}$-measure* (Zakharyuta (1974)). If a compact set K is pluriregular, then the function $\omega^*(z, K, G)$ is continuous in G. In fact, using the function ρ defining the domain G, we can continue ω^* to some neighborhood $D \supset \bar{G}$ as a psh function. Then, in the neighborhood D, we construct approximations $u_j \downarrow \omega^*$ by functions u_j of class C^∞. Applying twice the Hartogs Lemma (see 1.6, Chap. 1) to the sequence u_j (first for K in some neighborhood, and then for $\bar{G}$ and D), we find, for every $\varepsilon > 0$, a j_0 such that

$$u_j|_K \le -1 + \varepsilon \quad \text{and} \quad u_j|_{\bar{G}} \le \varepsilon, \quad j \ge j_0.$$

The functions $u_j - \varepsilon$ belong to $\mathcal{U}(K, G)$ and thus $u_j - \varepsilon \le \omega^*(z, K, G), j \ge j_0$. On the other hand, $u_j \ge \omega^*(z, K, G)$, which means that u_j converges uniformly to $\omega^*(z, K, G)$. Thus, ω^* is continuous.

e) *For a pluriregular compact set $K \subset G$, the function $\omega^*(z, K, G)$ satisfies the homogeneous Monge-Ampère equation in $G \backslash K$.* In fact, consider a ball $B \subset G \backslash K$ and let v be the solution of the Dirichlet problem in B:

$$(dd^c v)^n = 0, \quad v|_{\partial B} = \omega^*|_{\partial B}.$$

By the maximum principle (1.3), $v(z) \ge \omega^*(z, K, G)$, $z \in B$. On the other hand, the function

$$w(z) = \begin{cases} v(z), & z \in B, \\ \omega^*(z, K, G), & z \in G \backslash B, \end{cases}$$

is in $\mathcal{U}(K, G)$ and consequently, $w(z) \le \omega^*(z, K, G)$. In particular, $v \le \omega^*$ in B and thus $\omega^* = v$, q.e.d.

In conclusion, we state a two-constants theorem which easily follows from the definition of $\mathscr{P}$-measure.

Theorem. *If u is a plurisubharmonic function in G, $u \leq M$ and $u|_E \leq m$ on some subset $E \subset G$, then we have the estimate*

$$u(z) \leq M(1 + \omega^*(z, E, G)) - m\omega^*(z, E, G), \quad z \in G.$$

2.2. Condenser Capacity. The integral

$$C(K, G) = \int_K (dd^c\omega^*(z, K, G))^n$$

is called the *capacity* of the compact set $K \subset G$ with respect to the domain G. For $n = 1$ the capacity $C(K, G)$ is the well known capacity of the condenser with plates K and ∂G. For an open set $U \subset G$, we set

$$C(U, G) = \sup\{C(K, G) : K \subset U, K \text{ pluriregular and compact}\}.$$

The constraint of pluriregularity of the compact sets in the definition of $C(U, G)$ is needed for technical reasons. At the end of this section, this constraint will be dropped. We recall that the $\mathscr{P}$-measure of a pluriregular compact set is continuous and Theorems 1 and 2 of 1.2 as well as Theorem 3 of 1.3, so important in the theory of capacity, were proved for continuous functions.

For an arbitrary set $E \subset G$, we can define its outer capacity as

$$C^*(E) = C^*(E, G) = \inf\{C(U, G) : U \text{ open}, E \subset U \subset G\}.$$

We state several important properties of capacity (see Sadullaev (1980, 1982b) and Bedford-Taylor (1982)).

a) For any increasing sequence of open sets $U_j \subset U_{j+1}, j = 1, 2, \ldots$, we have[3]

$$C\left(\bigcup_{j=1}^{\infty} U_j\right) = \lim_{j \to \infty} C(U_j) \text{ (obvious)}.$$

b) For any open set $U \subset G$,

$$C(U) = \sup\left\{\int_U (dd^c u)^n : u \in C(G) \cap \text{Psh}(G), -1 \leq u \leq 0\right\}. \tag{8}$$

Indeed, taking as psh function u the $\mathscr{P}$-measure of a pluriregular compact set $K \subset U$, we have that the left side of (8) is no greater than the right side. On the other hand, fix a number $\varepsilon > 0$ and a function $u \in C(G) \cap \text{Psh}(G)$, $-1 \leq u \leq 0$. Then there exists a pluriregular compact set $K \subset U$ with

$$\int_U (dd^c u)^n - \int_K (dd^c u)^n < \varepsilon. \tag{9}$$

It is clear that the set

$$D = \{(1 + 2\varepsilon)\omega^*(z, K, G) + \varepsilon < u(z)\}$$

---

[3] When the containing domain G does not play an important role, we shall drop it from the notation.

is precompact in G. Thus, using Theorem 1.3, and noting that $\omega^* = \omega^*(z, K, G)$ is continuous in G, we have

$$\int_D \{dd^c[(1 + 2\varepsilon)\omega^* + \varepsilon]\}^n \geq \int_D (dd^c u)^n \geq \int_K (dd^c u)^n.$$

By property d) of $\mathscr{P}$-measure,

$$\int_D (dd^c \omega^*)^n = \int_K (dd^c \omega^*)^n \leq C(U).$$

From this and from (9), we have that the right side of (8) does not exceed $C(U)$.

□

If the open set U in b) is precompact in G, $U \subset\subset G$, then we can replace the functions u in (8) by $\max\{u(z), Mp(z)\}$, where $M = M(U)$ is a constant, such that $M \cdot \rho|_U < -1$. Such a function can be continued plurisubharmonically to some neighborhood $D \supset \bar{G}$, such that $|u| < 1$ in D. Approximating such a u by infinitely smooth functions, we see that in (8) we may replace the class $C(G) \cap \text{Psh}(G)$ by the class

$$L = \{u \in C^\infty(D) \cap \text{Psh}(D): -1 \leq u \leq 0 \text{ in } G\} \tag{10}$$

where $|u| \leq 1$ in D, and $D \supset G$ is a fixed neighborhood.

c) The set function $C(E)$ is countably subadditive, i.e., for any collection of sets $E_j \subset G, j = 1, 2, \ldots$, we have

$$C^*(E) \leq \sum_{j=1}^\infty C^*(E_j), \quad E = \sum_{j=1}^\infty E_j.$$

For open sets E, this property follows from (8). In the case of arbitrary E, for any $\varepsilon > 0$ we may construct open sets $U_j \supset E_j$ such that $C(U_j) - C^*(E_j) < \frac{\varepsilon}{2^j}$, $j = 1, 2, \ldots$. Thus

$$C^*(E) \leq C\left(\bigcup_{j=1}^\infty U_j\right) \leq \sum_{j=1}^\infty C^*(E_j) + \varepsilon.$$

d) $C^*(E) = 0$ if and only if E is pluripolar in G. Moreover, we have the uniform estimate:

$$\alpha(r) \cdot C^*(E) \leq \int_G |\omega^*(z, E, G)|(dd^c \rho)^n \leq \beta(r)[C^*(E)]^{1/n}, \tag{11}$$

where E is an arbitrary subset of $G_r = \{\rho(z) \leq r\}$, $r < 0$, and $\alpha(r)$, $\beta(r)$ – are constants depending only on r.

It is sufficient to show (11) for an arbitrary pluriregular compact set E. We invoke the integral estimate (2). Applying (2) $(n - 1)$ times to the function $\omega^*(z, E, G)$, we obtain

$$\int_\sigma^0 dt_1 \int_\sigma^{t_1} dt_2 \cdots \int_\sigma^{t_{n-1}} dt_n \int_{\rho \leq t_n} (dd^c \omega^*)^n \leq \int_\sigma^0 dt_1 \int_{\rho \leq t_1} dd^c \omega^* \wedge (dd^c \rho)^{n-1}. \tag{12}$$

According to the middle estimate in (3), the right side of (12) is no greater than $\int_{\rho \leq 0} |\omega^*| (dd^c \rho)^n$. As in 1.2 of Chap. 2, the left side of (12) is estimated below by the quantity

$$\frac{|r|^n}{n!} \int_{\rho \leq r} (dd^c \omega^*)^n.$$

Thus,

$$C^*(E) \leq \frac{n!}{|r|^n} \int_G |\omega^*| (dd^c \rho)^n,$$

which is the first inequality in (11). To prove the second inequality, we set

$$M = \int_{G_r} |\omega^*| (dd^c \rho)^n.$$

Then there exists a constant $\gamma > 0$ such that $\inf_{G_r} |\omega^*| \geq \gamma M$. Thus, the open set $D = \left\{ \omega^* < \dfrac{\gamma M}{|\sigma|} (\rho + \sigma) \right\}$, where, as above, $\sigma = \inf_G \rho(z)$, is compact in G and contains the compact set G_r. Consequently (cf. 1.3),

$$\int_D (dd^c \omega^*)^n \geq \left(\frac{\gamma M}{|\sigma|} \right)^n \cdot \int_D (dd^c \rho)^n \geq \left(\frac{\gamma}{|\sigma|} \right)^n \cdot \int_{G_r} (dd^c \rho)^n \cdot M^n$$

$$= C(r) \left[\int_{G_r} |\omega^*| (dd^c \rho)^n \right]^n.$$

Since $\int_D (dd^c \omega^*)^n = \int_E (dd^c \omega^*)^n = C^*(E)$, the second inequality in (11) follows.

2.3. Solution of the First Lelong Problem. An important role in the study of pluripolar sets is played by the following problem of Lelong (1957). Let E be a pluripolar set in a domain $G \subset \mathbb{C}^n$. Is E pluripolar in all of $\mathbb{C}^n$?

The properties of the capacity $C^*(E)$ given in 2.2 yield a simple solution to this problem (first solved by Josefson (1978)). It is based on the following reasoning: without loss of generality, in place of G we may consider the unit ball $B_1 = B(0, 1)$. If E is pluripolar in $B(0, 1)$, then it has zero capacity, $C^*(E, B_1) = 0$ (property d)). It is not difficult to show that if $R > 1$, then $C^*(E, B_1) \geq C^*(E, B_R)^4$. Thus, E is pluripolar with respect to any ball B_R of radius $R > 1$. But then it is easy to construct a psh function $u \not\equiv -\infty$ in all of $\mathbb{C}^n$ such that $u|_E \equiv -\infty$ (for further details, see Sadullaev (1981)).

---

[4] We remark that if a compact set K is pluriregular with respect to the ball B_1, then it is also pluriregular with respect to the ball B_R, $R > 1$. Thus in the definitions of the outer capacities $\mathbb{C}^*(E, B_1)$ and $\mathbb{C}^*(E, B_R)$, we may use the same compacta.

§ 3. Capacitary Properties of psh Functions

3.1. C-Property of psh Functions. Every measurable function is "almost continuous" (C-property of Luzin). The capacitary analog of this property for subharmonic functions in $\mathbb{R}^n$ was proved by H. Cartan. For psh functions we have the following variant.

Theorem. *If u is plurisubharmonic in a domain G, then for each $\varepsilon > 0$ there exists an open set $U \subset G$ such that $C(U) < \varepsilon$ and the restriction of u to $G\backslash U$ is continuous.*

It is sufficient to prove the theorem in the simple case when $G = B$, the unit ball. We establish first an integral inequality for the difference of psh functions in the ball. Let $\mathscr{L}$ – be the class of functions (10) for $G = B$. Consider functions $u, v, \varphi_1, \varphi_2, \ldots, \varphi_n \in \mathscr{L}$ such that $\varphi_0 = u - v \geq 0$ in B and $\varphi_0 \equiv$ const on the sphere $S = \partial B : |z| = 1$. Then

$$\int_B \varphi_0 dd^c \varphi_1 \wedge \cdots \wedge dd^c \varphi_n = \int_S \varphi_0 d^c \varphi_1 \wedge dd^c \varphi_2 \wedge \cdots \wedge dd^c \varphi_n$$

$$- \int_B d\varphi_0 \wedge d^c \varphi_1 \wedge dd^c \varphi_2 \wedge \cdots \wedge dd^c \varphi_n. \quad (13)$$

The surface integral here is no greater than

$$\|\varphi_0\|_s \cdot \int_B dd^c \varphi_1 \wedge dd^c \varphi_2 \wedge \cdots \wedge dd^c \varphi_n \leq C \cdot \|\varphi_0\|_s,$$

where the constant on the right side is independent of $\varphi_1, \varphi_2, \ldots, \varphi_n$ (cf. the argument in 1.2). The last integral in (13) can be estimated with the help of the following *Cauchy-Bunyakovskij inequality*. Let α – be a positive $(n - 1, n - 1)$-form and let φ, ψ be smooth real-valued functions in the neighborhood of $\bar{B}$. Then the expression

$$(\varphi, \psi) = \int_B d\varphi \wedge d^c \psi \wedge \alpha$$

defines a scalar product for which the following inequality holds:

$$|(\varphi, \psi)|^2 \leq (\varphi, \varphi) \cdot (\psi, \psi).$$

Applying this inequality, we obtain:

$$\left| \int_B d\varphi_0 \wedge d^c\varphi_1 \wedge dd^c\varphi_2 \wedge \cdots \wedge dd^c\varphi_n \right|$$

$$\leq \left[\int_B d\varphi_1 \wedge d^c\varphi_1 \wedge dd^c\varphi_2 \wedge \cdots \wedge dd^c\varphi_n \right]^{1/2}$$

$$\times \left[\int_B d\varphi_0 \wedge d^c\varphi_0 \wedge dd^c\varphi_2 \wedge \cdots \wedge dd^c\varphi_n \right]^{1/2}$$

$$\leq C \left[\int_S \varphi_0 d^c\varphi_0 \wedge dd^c\varphi_2 \wedge \cdots \wedge dd^c\varphi_n \right.$$

$$\left. - \int_B \varphi_0 dd^c\varphi_0 \wedge dd^c\varphi_2 \wedge \cdots \wedge dd^c\varphi_n \right]^{1/2}$$

$$\leq C_1 \left(\|\varphi_0\|_S + \int_B \varphi_0 dd^c\varphi_0^+ \wedge dd^c\varphi_2 \wedge \cdots \wedge dd^c\varphi_n \right)^{1/2},$$

where $\varphi_0^+ = \dfrac{u + v}{2} \in \mathcal{L}$ (since $\varphi_0 \geq 0$, the form $2\varphi_0 dd^c\varphi_0^+ + \varphi_0 dd^c\varphi_0 = 2\varphi_0 dd^c u$ is positive). Repeating this procedure n times, we obtain the inequality

$$\int_B \varphi_0 dd^c\varphi_1 \wedge \cdots \wedge dd^c\varphi_n \leq \alpha \left(\|\varphi_0\|_S + \int_B \varphi_0 (dd^c\varphi_0^+)^n \right)^\gamma \tag{14}$$

where $\alpha > 0$, $\gamma > 0$ are some constants independent of $\varphi_0, \varphi_1, \ldots, \varphi_n$.

Using this estimate, we sketch the proof of the theorem. Involving the countable subadditivity of $C^*(E)$ and the local character of the theorem, it is sufficient to show that for each $\varepsilon > 0$, there exists an open set $U \subset B$ such that $C(U \cap B') < \varepsilon$ and the restriction of u is continuous on $B' \backslash U$, where B' is the ball $|z| < 1/2$.

Since $C(F_M)$ tends to zero, for the set $F_M = \{z : u(z) < -M\}$, as $M \to \infty$ (cf. 2.2), then, replacing u by an appropriate combination $a \cdot \max\{u, -M\} + b$, where $a > 0$ and $M > 0$, we may assuame that $-1 \leq u \leq 0$. Then, replacing u by $\max\{u(z), v(z)\}$, where $v(z) = 2(|z|^2 - 3/4)$, we render $u \equiv v$ in some neighborhood of the sphere S.

Consequently, the average (see 1.4)

$$u_p(z) = \int u \left(z + \frac{\xi}{p} \right) K(\xi) \, dV(\xi),$$

decreases to u, and, moreover, $u_p \equiv v_p$ in a neighborhood of S, where v_p is the average of v. Passing to a subsequence, if necessary, we may suppose that the sequence of numbers $\int_B u_p (dd^c u_p)^n$ (bounded, by Theorem 1, 1.2) has a limit.

Set

$$U_{p,m}(\delta) = \{z \in B' : u_p(z) > u_{p+m}(z) + \delta\}, \quad \delta > 0.$$

In view of the monotonicity, we have

$$U_{p,m}(\delta) \subset U_{p,m+1}(\delta), \quad m = 1, 2, \ldots,$$

and

$$\bigcup_{m=1}^{\infty} U_{p,m}(\delta) = U_p(\delta) = \{z \in B' : u_p(z) > u(z) + \delta\}.$$

From the properties of capacity, it follows that

$$C(U_p(\delta)) = \lim_{m \to \infty} C(U_{p,m}(\delta)) \tag{15}$$

Since the set $U_{p,m}(\delta)$ is open,

$$C(U_{p,m}(\delta)) = \sup\left\{\int_{U_{p,m}(\delta)} (dd^c w)^n : w \in \mathscr{L}\right\}$$

$$\leq \sup\left\{\frac{1}{\delta}\int_{U_{p,m}(\delta)} (u_p - u_{p+m})(dd^c w)^n : w \in \mathscr{L}\right\}$$

$$\leq \sup\left\{\frac{1}{\delta}\int_{B} (u_p - u_{p+m})(dd^c w)^n : w \in \mathscr{L}\right\}.$$

By (14), it follows, then, that

$$C(U_{p,m}(\delta)) \leq \frac{\alpha}{\delta}\left[\|v_p - v\|_S + \int_B \varphi_{p,m}(dd^c \varphi_{p,m}^+)^n\right]^\gamma, \tag{16}$$

where $\varphi_{p,m} = u_p - u_{p+m}$.

From the convergence of $\int_B u_p(dd^c u_p)^n$ as $p \to \infty$, it follows that the integral on the right side of (16) tends to zero. Thus, $C(U_{p,m}) \to 0$ as $p \to \infty$, and the convergence is uniform in m. From this and from (15), $\lim_{p \to \infty} C(U_p(\delta)) = 0$, from which the assertion of the theorem follows easily. $\square$

3.2. Convergence of $(dd^c)^k$

Theorem. *If $\{u_j\}$ is a montonically decreasing sequence of functions psh in G, such that the limit $u(z) = \lim_{j \to \infty} u_j(z)$ is locally bounded (from below), then, weakly,*

$$u_j(dd^c u_j)^k \to u(dd^c u)^k, \quad as \ j \to \infty, \ 0 \leq k \leq n.$$

We prove this by induction on k. For $k = 0$, it is obvious. If the assertion is valid for $k - 1$, then, weakly,

$$(dd^c u_j)^k \to (dd^c u)^k,$$

since, by definition,

$$\int (dd^c u_j)^k \wedge \alpha = \int u_j(dd^c u_j)^{k-1} dd^c \alpha,$$

for an arbitrary fundamental form $\alpha \in D^{n-k,n-k}(G)$. Thus, it is sufficient to prove the theorem for k under the hypotheses that it is true for $k-1$ and that $(dd^c u_j)^k \to (dd^c u)^k$ weakly.

Let us fix an $\varepsilon > 0$ and a fundamental form α. By Theorem 3.1 there exists an open set $U \subset G$, $C(U) < \varepsilon$, outside of which the restrictions of all u_j and u are continuous. If $\tilde{u}$ is a continuous function in G coinciding with u outside of U, and E_α is the support of α, then

$$\left| \int_G u_j (dd^c u_j)^k \wedge \alpha - \int_G u(dd^c u)^k \wedge \alpha \right|$$

$$\leq \left| \int_{E_\alpha \setminus U} (u_j - u)(dd^c u_j)^k \wedge \alpha \right| + \left| \int_G \tilde{u}[(dd^c u_j)^k - (dd^c u)^k] \wedge \alpha \right|$$

$$+ \left| \int_{E_\alpha \cap U} (u_j - u)(dd^c u_j)^k \wedge \alpha \right| + \left| \int_{E_\alpha \cap U} (u - \tilde{u})[(dd^c u_j)^k - (dd^c u)^k] \wedge \alpha \right|.$$

$$\tag{17}$$

The integrals on the sets $E_\alpha \setminus U$ and G tend to zero as $j \to \infty$: the first, because $u_j \to u$ uniformly on $E_\alpha \setminus U$, and the second, because $(dd^c u_j)^k \to (dd^c u)^k$ and $\tilde{u}$ is continuous on G. The last two integrals in (17) are small because of the smallness of $\varepsilon > 0$ (since $C(E_\alpha \cap U) < \varepsilon$) and the uniform boundedness of u_j on E_α. $\square$

Remark. If u_j is a locally uniformly bounded monotonically increasing sequence of psh functions and $u = \lim\limits_{j \to \infty} u_j$, then in order to prove the weak convergence

$$u_j (dd^c u_j)^k \to u^* (dd^c u^*)^k,$$

we need the pluripolarity of the set $\sigma : \{u < u^*\}$. If σ is pluripolar, then for any $\varepsilon > 0$, there exists an open set $U \subset G$ such that $C(U) < \varepsilon$ and u_k converges uniformly in $G \setminus U$. Precisely this property, for a decreasing sequence, was used in proving the theorem of this section. The pluripolarity of the set σ will be shown in the next section.

3.3. The Structure of the Irregular Points. Let $K \subset G$ be compact. A point $z^0 \in K$ is said to be *irregular* for K, if $\omega^*(z^0, K, G) > -1$ (cf. 2.1). The collection of such points is denoted by I_K. The following theorem of Bedford and Taylor (1982) is an analog of the famous theorem of Kellogg.

Theorem 1. *The collection I_K of irregular points of a compact set K has zero capacity:* $C(I_K) = 0$.

It is clear that the proof of the theorem reduces to the assertion:

a) If a compact set $K \subset G$ is such that $\omega^*(z, K, G) \geq \delta$, $\delta > -1$ on K then $\omega^*(z, K, G) = 0$.

In turn, a) follows from the following assertion:

b) For each compact set K, its $\mathscr{P}$-measure $\omega^*(z, K, G)$ satisfies, the Monge-Ampère equation $(dd^c \omega^*)^n = 0$ in $G \setminus K$.

The following theorem, which is related to Theorem 1 gives a positive solution to the *second problem of Lelong* for psh functions.

Theorem 2. *Let $\{u_j\}$ be an increasing sequence of locally uniformly bounded psh functions in G and $u(z) = \lim_{j \to \infty} u_j(z)$. Then the set $\sigma = \{z: u(z) < u^*(z)\}$ is pluripolar in G.*

The proofs of Theorems 1 and 2 as well as assertions a) and b) proceed here simultaneously, by induction on n, according to the following scheme, introduced by Bedford and Taylor.

Step I. From b) follows a), and hence, Theorem 1.

Step II. From Theorem 1 follows Theorem 2.

Step III. From Theorem 2 in $\mathbb{C}^n$ follows b) in $\mathbb{C}^{n+1}$.

The correctness of b) for $n = 1$, i.e., the harmonicity of the function $\omega^*(z, K, G)$ in $G \backslash K$, follows easily from property e) of 2.1 of Chap. 2 and from the Harnack Theorem for harmonic functions: if $K_j, j = 1, 2, \ldots$ are approximations of K by regular compacta (cf. 2.1), then $\omega^*(z, K_j, G) \uparrow \omega^*(z, K, G)$ on $G \backslash K$.

Step I. Suppose, for the sake of contradiction, that $\omega^*(z, K, G) \geq \delta > -1$ on K but $\omega^* \not\equiv 0$. If

$$\omega^*(z, K, G) \geq \delta \tag{18}$$

everywhere in the domain G, then in the class

$$\{\sqrt{|\delta|}\, u(z) : u \in \mathcal{U}(K, G)\},$$

there is a function $v(z)$ such that $v > \omega^*$ at some point $z^0 \in G \backslash K$. But this contradicts the fact that $v \leq \omega^*$ on the boundary $\partial(G \backslash K)$ and $(dd^c \omega^*)^n = 0$ in $G \backslash K$ (cf. 1.3). A priori, the inequality (18) may not hold on all of G, although it holds on K. In this case we use the same argument, only for the chosen function v satisfying the condition $v \leq \omega^*$ on the boundary $\partial(G \backslash K)$; we consider the class

$$\{\sqrt{|\delta|}\, u(z) + \omega^*(z, U_\varepsilon, G) : u \in \mathcal{U}(K, G), \varepsilon > 0\},$$

where U_ε is the open set from the theorem in 3.1 : $C(U_\varepsilon) < \varepsilon$ and the restriction of $\omega^*(z, K, G)$ is continuous outside of U_ε.

Step II. Theorem 1 $\Rightarrow$ Theorem 2. Suppose, first of all, that u_j are continuous in G. Then u is lower semicontinuous and the set $u \leq \alpha$ is closed in G for any number α. From this it follows that for any $\alpha < \beta$ the set $K_{\alpha\beta} : u(z) \leq \alpha < \beta \leq u^*(z)$ is also closed in G. From Theorem 1 it follows that $K_{\alpha\beta}$ has zero capacity in any ball $B \Subset G$, since

$$\omega(z, K_{\alpha\beta} \cap B, B) \geq \frac{u(z) - \sup_B u(z)}{\sup_B u(z) - \alpha},$$

and, consequently, $K_{\alpha\beta} \subset I_{K_{\alpha\beta}}$. The union $\bigcup K_{\alpha\beta}$ over all rationals α, β, and hence the set $\sigma \subset \bigcup K_{\alpha\beta}$ also have zero capacity.

In the case where the u_j are arbitrary, there exists, for any $\varepsilon > 0$, according to the C-property of psh functions, an open set $U \subset G$ such that $C(U) < \varepsilon$ and the restrictions of all u_j are continuous on $G \backslash U$. From the previous reasoning, it follows that $C^*(\sigma \setminus U) = 0$ and consequently, $C^*(\sigma) \le C^*(\sigma \setminus U) + C^*(\sigma \cap U) < \varepsilon$. Since $\varepsilon > 0$ is arbitrary, it follows that $C^*(\sigma) = 0$.

Step III. Theorem 2 in $\mathbb{C}^n \Rightarrow$ b) in $\mathbb{C}^{n+1}$. Let $K \subset G \subset \mathbb{C}^{n+1}$ and K_j be pluriregular compacta such that $K_j \supset K_{j+1}$, $\bigcap_{j=1}^{\infty} K_j = K$. Then $\omega_j^* = \omega^*(z, K_j, G)$ satisfies the Monge-Ampère equation in $G \setminus K_j$ (cf. 2.1), is monotonically increasing, and $\omega^* = \omega^*(z, K, G) = \left(\lim_{j \to \infty} \omega_j^* \right)^*$. Hence for the proof of b) it is sufficient to show that $(dd^c \omega_j^*)^{n+1} \to (dd^c \omega^*)^{n+1}$ weakly as $j \to \infty$. For this we must establish the convergence of

$$\lim_{j \to \infty} \int_G \omega_j^* (dd^c \omega_j^*)^n \wedge dd^c \alpha = \int_G \omega^* (dd^c \omega^*)^n \wedge dd^c \alpha, \qquad (19)$$

for any test function α. Using the obvious identity

$$dz_p \wedge d\bar{z}_q = \frac{1}{2} [d(z_p + z_q) \wedge d\overline{(z_p + z_q)} + \mathrm{i} d(z_p + \mathrm{i} z_q) \wedge d\overline{(z_p + \mathrm{i} z_q)}]$$

$$- (1 + \mathrm{i}) dz_p \wedge \overline{dz_p} - (1 + \mathrm{i}) dz_q \wedge \overline{dz_q},$$

we obtain that the differential form $dd^c \alpha$ in (19) can be represented as the sum of forms $\beta \, d\varphi \wedge d\bar{\varphi}$, where φ is a linear function of $z_1, \ldots, z_{n+1}$. Thus it is sufficient to show (19) for $dd^c \alpha = \beta \, dz_{n+1} \wedge d\bar{z}_{n+1}$. In this situation, by Fubini's Theorem, the integrals in (19) are transformed into iterated integrals of the form

$$\int dz_{n+1} \wedge d\bar{z}_{n+1} \int_{G[z_{n+1}]}$$

on the section $G[z_{n+1}^0] = G \cap \{z_{n+1} = z_{n+1}^0\}$. For (almost) every section $G[z_{n+1}]$

$$\lim_{j \to \infty} \int_{G[z_{n+1}]} \beta \omega_j^* (dd^c \omega_j^*)^n = \int_{G[z_{n+1}]} \beta \omega^* (dd^c \omega^*)^n,$$

by Theorem 2 and Remark 3.2. From this, the convergence in (19) follows. $\square$

3.4. Capacitability of Borel Sets. From Theorem 2 of 3.3 it follows that for any compact set $K \subset G$ its capacity $C(K, G)$ coincides with its outer capacity $C^*(K, G)$. Moreover, the capacity introduced above sastifies the capacitability axiom of Choquet: for any increasing sequence $E_j \subset E_{j+1}$, we have

$$\lim_{j \to \infty} C^*(E_j, G) = C^*(E, G),$$

where $E = \bigcup_{j=1}^{\infty} E_j$. This property follows from the analogous identity for $\mathscr{P}$-measure:

$$\lim_{j \to \infty} \omega^*(z, E_j, G) = \omega^*(z, E, G). \qquad (20)$$

Let us prove (20). By Theorem 2 of 3.3, the sets

$$P_j: \omega(z, E_j, G) < \omega^*(z, E_j, G), \quad j = 1, 2, \ldots,$$

are pluripolar in G. Consequently, their union $P = \bigcup_{j=1}^{\infty} P_j$ is also pluripolar, and hence, there exists in G a psh function $u \not\equiv -\infty$, $u \leq 0$, such that $u|_P \equiv -\infty$. Then, the function $w(z) + \varepsilon u(z)$, where w is the limit on the left side of (20), belongs to the class $\mathcal{U}(E, G)$ for each $\varepsilon > 0$. Thus, $w(z) + \varepsilon u(z) \leq \omega^*(z, E, G)$ and consequently, $w(z) \leq \omega^*(z, E, G)$. From this and the obvious inequality $w(z) \geq \omega^*(z, E, G)$, we obtain (20).

From the preceding result, Choquet's Theorem (cf. Landkoff (1966)) yields

Theorem. *For any Borel set $E \subset G$, its outer capacity $C^*(E, G)$ agrees with its inner capacity $C_*(E, G) = \sup\{C(K, G): K \subset E, K \text{ compact}\}$.*

Corollary. *Let $\{U_\alpha\}$ – be a locally uniformly bounded family of psh functions in a domain G and set $u(z) = \sup U_\alpha(z)$. Then the set $\sigma: u(z) < u^*(z)$ is pluripolar in G.*

Remarks. The method of integral estimates (1.3) for the elaboration of a complex potential theory is due to the author. Bedford and Taylor (1982) have used, for this purpose, the convergence of the sequence of currents $(dd^c u_j^{(1)}) \wedge \cdots \wedge dd^c u_j^{(p)}$ for a monotone bounded sequence $\{u_j^{(k)}\}$, $k = 1, 2, \ldots, p$.

For an unbounded psh function the operator $(dd^c u)^n$ may not be a bounded Borel measure. To see this, consider, for example, the function

$$u(z_1, z_2) = (|z_1|^2 - 1) \ln^{1/2}(1/|z_2|^2),$$

which is plurisubharmonic in a neighborhood of the origin. For an arbitrary function $u \in \text{Psh}(G)$ Kiselman has proposed the following definition of $(dd^c u)^k$: in the domain $G \times \{w \in \mathbb{C}: |\text{Im } w| < 1/2\}$ consider the auxilary function $F(z, w) = (u(z) - \text{Re } w)^+$. Then the measure $(dd^c F)^{n+1}$ is concentrated on the graph $\text{Re } w = u(z)$, and we call its projection on G the measure $(dd^c u)^n$. This measure is, in general, not finite, but for a bounded function, it coincides with the well-known measure $(dd^c u)^n$ of Bedford-Taylor (1.1).

$\mathcal{P}$-measures (2.1) were first employed in the works of Zakharyuta (1974, 1976) and the author (1976, 1980). With the help of $\mathcal{P}$-measures, a capacitary quantity has also been constructed which is close to condenser capacity.

The capacity of a condenser in $\mathbb{C}^n$ in terms of $(dd^c \omega^*)^n$ (2.2) was introduced practically simultaneously by Bedford and the author.

The C-property of psh functions and the structure of irregular points (3.1, 3.2) were established by Bedford and Taylor (1982).

If u is a twice continuously differentiable function in G and $(dd^c u)^n = 0$, then through each point $z^0 \in G$ there is an analytic curve on which u is harmonic. Lempert (1982*)[5] (see also Slodkowski (1990)) has established an interesting

---

[5] Translator's note. A date with an asterisk refers to a reference added (by the author) in translation

connection between such analytic curves and extremal surfaces for the Kobayashi and Carathéodory metrics. It turns out that the extremal surfaces for these metrics coincide with the analytic curves on which the solutions of the Monge-Ampère equation, with logarithmic singularity in the given fixed point, are harmonic (see Poletskij and Shabat, in Vol. 9 of this series). For more information on the Monge-Ampère equation see also Cegrell (1986, 1990), Demailly (1987), Levenberg (1985).

Chapter 3
Applications of Complex Potential Theory

In this chapter we give applications of extremal plurisubharmonic functions and capacity to various problems of multidimensional complex analysis. In Section 1, we present the results of the author on the structure of the singularity set of functions admitting rapid approximation by rational functions. In Section 2, we present the results of Chirka and Kazaryan (1983) on holomorphic extension in a fixed direction. In the third and last section, we present a multidimensional analog of the Bernstein-Walsh Theorem (Siciak (1962)) and the holomorphic extension of separately analytic functions (Zakharyuta (1976), see also Siciak (1969)). Besides, the third section contains results on holomorphic extension from circled sets (Forelli, cf. Rudin (1980), Alexander (1981)).

§ 1. Rational Approximation and Pluripolar Sets

Gonchar (1974) showed that if the complement of a domain $D \subset \mathbb{C}^n$ is pluripolar, then, in D, any holomorphic function f can be rapidly approximated with respect to Lebesgue measure in D by rational functions, i.e., $|f - r_k|^{1/k} \to 0$ in measure on D, for some sequence of rational functions r_k, deg $r_k \le k$ (denoted $r_k \xrightarrow{*} f$ in D). Such rapid approximations have the following transmission property (Gonchar (1972, 1974)): if $r_k \xrightarrow{*} f$ in a domain U, then the function f is single valued in its entire Weierstrass domain of existence W_f[6], moreover $r_k \xrightarrow{*} f$ in W_f.

Let us denote by R^0 the class of all functions which admit such a rapid approximation in the vicinity of 0 in $\mathbb{C}^n$. Gonchar's theorem may be formulated as follows:

$$0 \in D, \quad \mathbb{C}^n \setminus D \text{ pluripolar} \Rightarrow \mathcal{O}(D) \subset R^0.$$

---

[6] Let f be holomorphic in a domain $U \subset \mathbb{C}^n$. The set of holomorphic elements (g, V), which can be obtained from the element (f, U) by holomorphic continuation, when appropriately glued together, is called the Weierstrass domain of existence of the element (f, U).

From the preceding, it follows that the envelope of holomorphy $\hat{D}$ of the domain D is such that $\mathcal{O}(D) \subset R^0$ without ramification, that is, lies in $\mathbb{C}^n$. Below, it will be shown that

$$\mathcal{O}(D) \subset R^0 \Rightarrow \mathbb{C}^n \backslash \hat{D} \text{ is pluripolar.}$$

(The set $\mathbb{C}^n \backslash D$ for $n > 1$ can be very thick, but all function in $\mathcal{O}(D)$ can be holomorphically continued to $\hat{D}$ and $\mathbb{C}^n \backslash \hat{D}$ is the "natural" set of non-removeable singularities for $\mathcal{O}(D)$.) Thus, we obtain the following criterion:

$$\mathcal{O}(D) \subset R^0 \Leftrightarrow 0 \in D, \quad \mathbb{C}^n \backslash \hat{D} \text{ is pluripolar,}$$

which is well known in the classical case (for $n = 1$, $D = \hat{D}$)

The complement of a pseudoconvex domain is said to be *pseudoconcave*; we shall begin the proof of the pluripolarity of $\mathbb{C}^n \backslash \hat{D}$ by a study of the general properties of pseudoconcave sets, which are themselves of independent interest.

1.1. The Maximum Principle for Pseudoconcave Sets. A set $S \subset \mathbb{C}^n$ is said to be *pseudoconcave* if for each point $z^0 \in S$ there is a neighbourhood $U \ni z^0$ such that the set $U \backslash S$ is open and pseudoconvex in $\mathbb{C}^n$.

Lemma. *Let D be a bounded convex[7] domain in $\mathbb{C}^n$ and $S \subset D$ a closed pseudoconcave subset of D. Then the Shilov boundary δS of the compact set S (with respect to polynomials) is contained in $\bar{S} \cap \partial D$.*

Suppose, to obtain a contradiction, that we can find a polynomial P such that $\|P\|_{\bar{S}} = P(z^0) = 1$ at some point $z^0 \in S$, but $\|P\|_{\partial S} < 1$ (here $\partial S = \bar{S} \backslash S = \bar{S} \cap \partial D$). Let $D' \Subset D'' \Subset D$ be convex domains such that $|P| < 1$ also on $\bar{S} \backslash D'$. The domain $D'' \backslash S$ is pseudoconvex (being locally pseudoconvex) and the algebraic sets

$$A_k = \{z \in D' : P(z) = 1 + 1/k\}, \quad k = 1, 2, \ldots,$$

which belong to $D'' \backslash S$ along with their boundaries, are separated from S uniformly in k. Since the limit surface and S have the common point z^0, we obtain a contradiction to the continuity principle (cf. Vladimirov (1964), Shabat (1976)). $\square$

1.2. Pluripolarity of Pseudoconcave Sets

Theorem. *Let S be a pseudoconcave set in $\mathbb{C}^n$ such that $0 \notin S$ and for almost each complex line l passing through 0, the section $l \cap S$ is polar (in l). Then S is a pluripolar subset of $\mathbb{C}^n$.*

We outline the method of proof in the case $n = 2$ for simplicity. The theorem is local and so, with the aid of a fractional linear transformation mapping the family of lines passing through 0 (in the neighborhood of a fixed direction) to

---

[7] The convexity plays no role.

a family of parallel lines, we may view the problem in the following form: S is a pseudoconcave closed subset of the unit polydisc $U = U_1 \times U_2$ with $\bar{S} \cap \{\bar{U}_1 \times \partial U_2\} = \emptyset$ and such that the intersection of S with $\{z_1 = z_1^0\}$ is polar for almost every z_1^0. Shrinking U_1, if necessary, we may, without loss of generality, assume that $S \cap \{z_1 = z_1^0\}$, is polar for almost all z_1^0, $|z_1^0| = 1$, with respect to linear measure on the circle.

According to the lemma in 1.1 and Bremermann's approximation theorem (2.4, Chap. 1), the $\mathscr{P}$-measure $\omega^* = \omega^*(z, S, U) = \omega^*(z, \partial S, U)$ and hence $(dd^c \omega^*)^2 = 0$ in U. Moreover, from the conditions of the theorem, it follow that the boundary values of ω^* are almost everywhere equal to zero on ∂U, and on that part of the boundary $\Gamma_2 = \bar{U}_1 \times \partial U_2$, where there are no points of $\bar{S}$, ω^* is identically zero. From the maximality of ω^* in U,

$$\int_0^r dt \int_{\|z\| \leq t} (dd^c \omega^*)^2 = 0,$$

where $\|z\| = \max\{|z_1|, |z_2|\}$, $r < 1$. On the other hand, such an integral can be transformed into

$$\int_{\|z\| = r} \omega^* d^c \|z\| \wedge dd^c \omega^* - \int_{\|z\| \leq r} \omega^* dd^c \|z\| \wedge dd^c \omega^*$$

(cf. 1.2, Chap. 2). As $r \to 1$ the surface integral, here, tends to zero. From this we obtain that

$$\omega^* dd^c \|z\| \wedge dd^c \omega^* = 0 \quad \text{in } U$$

from which it easily follows that $\omega^* \equiv 0$. $\square$ For further details, cf. Sadullaev (1982b).

1.3. Some Properties of the Class R^0.

We formulate a criterion for a function f to belong to the class R^0 at a point $0 \in \mathbb{C}$, in terms of the Taylor coefficients of this function. The first such criterion was obtained by Gonchar (in terms of the Padé table of the function).

Lemma. Let $f(z) = \sum_{k=0}^{\infty} a_k z^k$ be holomorphic in a neighborhood of the closed unit disc $\bar{U} : |z| \leq 1$ and let $A_{j_1, \ldots, j_k}$ be the absolute value of the determinant $(a_{j_\nu}, a_{j_\nu + 1}, \ldots, a_{j_\nu + k - 1})$, $\nu = 1, 2, \ldots, k$. We set

$$V_k = \sup_{j_1, \ldots, j_k} A_{j_1, \ldots, j_k}.$$

Then

$$f \in R^0 \Leftrightarrow \lim_{k \to \infty} V_k^{1/k^2} = 0.$$

From this lemma, the following assertion follows in $\mathbb{C}^n$.

Theorem 1. If a function f, holomorphic in a neighborhood of $0 \in \mathbb{C}^n$, belongs to the class R^0, then for each complex line $l \ni 0$, the restriction $f|_l$ also belongs to R^0.

Indeed, with the help of an appropriate linear transformation of $\mathbb{C}^n$, we may assume that f is holomorphic in a neighborhood of the unit polydisc $U = 'U \times U_n$ and $l = \{'z = 0\}$, where $'z = (z_1, \ldots, z_{n-1})$. Let us expand f in a Hartogs series $f(z) = \sum_{j=0}^{\infty} a_j('z)z_n^j$. Corresponding to the coefficients $a_j('z), j = 0, 1, \ldots$, we obtain a sequence $V_k('z)$. The functions $\frac{1}{k^2} V_k('z)$ are then plurisubharmonic in $'U$.

From the holomorphy of f in the vicinity of the closure of $'U$, it follows that the coefficients $a_j('z)$ are bounded by some constant C, i.e., $|a_j('z)| \leq C$ for all $j \geq 0$. From this it follows easily that the sequence $\frac{1}{k^2} \ln V_k('z)$ is upper bounded in $'U$.

Moreover, from the condition $f \in R^0$ it follows that

$$\lim_{k \to \infty} \int_U \frac{1}{k^2} \ln V_k('z) \, dV = -\infty.$$

From the property of plurisubharmonicity, it follows, then, that

$$\frac{1}{k^2} \ln V_k(0) \leq \frac{1}{\text{mes}'U} \int_U \frac{1}{k^2} \ln V_k('z) \, dV \to -\infty,$$

as $k \to \infty$. Thus $\lim_{k \to \infty} V_k^{1/k^2}(0) = 0$, and therefore, by the lemma, $f(0, z_n)$ belongs to the class R^0. $\square$

With the help of Theorem 1, we establish the fundamental assertion mentioned at the beginning of this section.

Theorem 2. *Let $D \subset C^n$ be a domain such that $\mathcal{O}(D) \subset R^0$. Then the set $\mathbb{C}^n\backslash\hat{D}$ is pluripolar in $\mathbb{C}^n$.*

Indeed, let us fix a complex line $l \ni 0$ and consider the function $f \in \mathcal{O}(l \cap \hat{D})$. Since $\hat{D}$ is a domain of holomorphy, the function f extends holomorphically to $\hat{D}$, i.e., there exists a function $\tilde{f} \in \mathcal{O}(\hat{D})$ such that $\tilde{f}|_l \equiv f$. From the hypotheses of the theorem, $f \in R^0$. But then by Theorem 1, $\tilde{f}|_l \equiv f$ also belongs to R^0. Thus, each function holomorphic in the plane domain $l \cap \hat{D}$ belongs to the class R^0 and, consequently, the set $l\backslash\hat{D}$ is polar in $\mathbb{C}^n$. By the theorem in 1.2, $\mathbb{C}^n\backslash\hat{D}$ is a pluripolar set in $\mathbb{C}^n$. $\square$

1.4. Further Properties of Pseudoconcave Sets. Let S be a closed pseudoconcave set in the unit polydisc $U = 'U \times U_n$ such that the closure (in $\mathbb{C}^n$)$\bar{S}$ does not intersect the face $'U \times \partial U_n$. If the intersection $S \cap \{'z = 'z^0\}$ is polar for almost every $'z^0 \in 'U$, then S is pluripolar in U (Theorem in 1.2). However, an even stronger assertion is valid, which characterizes pseudoconcavity of a set via its sections.

Theorem 1. *Let $S \subset U$ be a pseudoconcave set such that $\bar{S} \cap \{'U \times \partial U_n\} = \emptyset$, and let E be a set of positive capacity in $'U$. If the intersections $l_{,z^0} \cap S$ are polar for all lines $l_{,z^0} = \{'z = 'z^0\}$, $'z^0 \in E$, then they are polar for all lines $l_{,z}$, $'z \in 'U$, and consequently, S is pluripolar in U.*

Indeed, since $\bar{S} \cap \{'U \times \partial U_n\} = \varnothing$, it follows that S is also pseudoconcave in the domain $'U \times \mathbb{C}$. Supposing, without loss of generality, that S does not meet the plane $\{z_n = 0\}$, we prove the theorem by contradiction. Suppose for some point $'\alpha \in 'U$ the intersection $l_{,\alpha} \cap S$ is not polar. Then there exists, in $l_{,\alpha} \backslash S$, a function f which does not belong to the class R^0. Now, as in Theorem 1 of the previous section, we extend f from $l_{,\alpha} \backslash S$ to the domain $('U \times \mathbb{C}) \backslash S$, expand it in a Hartogs series and construct the sequence $V_k('z)$ corresponding to the coefficients. Then

$$\lim_{k \to \infty} \frac{1}{k^2} \ln V_k('z) = -\infty \tag{1}$$

for each fixed $'z \in E$. Since E is of positive capacity, (1) holds throughout $'U$. In particular, $\frac{1}{k^2} \ln V_k('\alpha) = -\infty$, and this contradicts the fact that $f('\alpha, z_n), \notin R^0$.

$\square$

Remark. We may also give a different proof of Theorem 1, without using the class R^0, by invoking a theorem of Pólya. If, for some point $'\alpha \in 'U$ the intersection $l_{,\alpha} \cap S$ is not polar, then, there exists, in $l_{,\alpha} \backslash S$, a holomorphic function f, such that $\lim_{k \to \infty} |\mathscr{A}_k|^{1/k^2} \neq 0$, where $\mathscr{A}_k$ are the Hankel determinants (cf., e.g., Goluzin (1966)). Extending f to the domain $('U \times \mathbb{C}) \backslash S$, we obtain, by the theorem of Pólya, that

$$\lim_{k \to \infty} \frac{1}{k^2} \ln |\mathscr{A}_k('z)| = -\infty, \quad z \in E,$$

where $\mathscr{A}_k('z)$ are the Hankel determinants for the function $\tilde{f}$, $\tilde{f}|_{l'\alpha} \equiv f$. Since E is not pluripolar, $\lim_{k \to \infty} \frac{1}{k^2} \ln |\mathscr{A}_k('\alpha)| = -\infty$. This contradicts the fact that $|\mathscr{A}_k|^{1/k^2} = |\mathscr{A}_k('\alpha)|^{1/k^2}$ does not tend to zero as $k \to \infty$.

Properties of pseudoconcave sets were studied also in the works of Oka, Nishino, and Slodkowski. The following theorem is due to Oka (cf. Nishino (1962)) and Levi-Hartogs (cf., e.g., Shabat (1976)).

Theorem 2. *Under the hypotheses of Theorem 1, if the intersections $l_{,z_0} \cap S$ are finite (or discrete) for all $'z^0 \in E$, then they are also finite (respectively, discrete) for all lines $l_{,z}$, $'z \in 'U$, and S itself is an analytic subset of U.*

We present also, without proof, the following connection between polar and pluripolar sets. A pluripolar set in $\mathbb{C}^n$ is polar in $\mathbb{R}^{2n} \simeq \mathbb{C}^n$ (Sect. 2, Chap. 2), but the converse is, in general, not true. The metric dimension of non-pluripolar sets can range over the interval $[0, 2n]$, while the metric dimension of non-polar sets is necessarily $\geq 2n - 2$. However, in the class of pseudoconcave sets, these two notions coincide (cf. Sadullaev (1982b)).

Theorem 3. *If S is a pseudoconcave polar set in $\mathbb{C}^n \simeq \mathbb{R}^{2n}$, then it is pluripolar in $\mathbb{C}^n$.*

For other properties and applications of pseudoconcave sets cf. Alexander and Wermer (1985, 1989), Berndtsson and Ransford (1986), Aupetit (1984) and Slodkowski (1986, 1990).

§ 2. Holomorphic Extension in a Fixed Direction

In this section we shall prove the following

Theorem 1. *Let f be a function holomorphic in the polydisc*

$$U = 'U \times U_n = 'U \times \{|z_n| < r\}$$

which, for each fixed z^0 in some non-pluripolar set $E \subset 'U$, extends to a function holomorphic in $\mathbb{C}_{z_n}$ except for a finite number of singularities. Then f extends holomorphically to $('U \times \mathbb{C}) \backslash A$ where A is some analytic set.

The difficulty in this theorem consists in showing that the union of singular sets of $f('z, z_n)$ is closed (under the additional assumption that this set is closed, we have the theorem of Oka, stated in 1.4, Chap. 3). This difficulty is overcome with the help of Jakobi series.

2.1. Analyticity of the Singularity Set. Let us consider, in the plane $\mathbb{C}$, a rational *lemniscate G_r*, more precisely, a connected component of the set $|g(z)| < r$, determined by some rational function g. Let f be holomorphic in some neighborhood of $\overline{G}_r$. Then the integral

$$\frac{1}{2\pi i} \int_{\partial G_r} \frac{f(\xi)}{g(\xi) - w} \cdot \frac{g(\xi) - g(z)}{\xi - z} \, d\xi$$

defines a holomorphic function $F(z, w)$ in the domain $G_r \times \{|w| < r\}$, and $F(z, g(z)) \equiv f(z)$ by Cauchy's integral formula. Expanding $F(z, w)$ in its Taylor series in w and setting $w = f(z)$, we obtain an expansion of f in a Jacobi series (cf. Chirka (1976)):

$$f(z) = \sum_{k=0}^{\infty} C_k(z) g^k(z), \tag{2}$$

where the coefficients C_k are defined by the formula

$$C_k(z) = \frac{1}{2\pi i} \int_{\partial G_r} f(\xi) \frac{g(\xi) - g(z)}{g^{k+1}(\xi)(\xi - z)} \tag{3}$$

and, consequently, are rational functions having poles at the poles of g, and $\deg C_k \leq \deg g$.

We remark that the series (2) converges uniformly in the interior of the lemniscate $|g(z)| < R$, where

$$\frac{1}{R} = \varlimsup_{k \to \infty} \|C_k\|_{|g| \leq r_0}^{1/k}, \tag{4}$$

and $r_0 > 0$ is arbitrary, and diverges almost everywhere outside of this lemniscate (analog of the Cauchy-Hadamard formula).

Proof of Theorem 1. Let f be holomorphic in the polydisc $'U \times \{|z_n| < r\}$ and extendable to $\{'z\} \times (\mathbb{C} \setminus E('z))$, as in the hypotheses, for arbitrary $'z \in 'U$. For fixed $'z \in 'U$, we develop this function in a Jacobi series with respect to the rational function

$$g(z_n) = \frac{z_n^{m+1}}{(z_n - a_1)\ldots(z_n - a_m)},$$

where $a_1, a_2, \ldots, a_m$ are "rational" numbers such that $|a_j| \geq r, j = 1, 2, \ldots, m$. In the domain $'U \times \{|g(z_n)| < 2\varepsilon\}$, which is contained in $'U \times \{|z_n| < r\}$, for small $\varepsilon < 0$, we have

$$f('z, z_n) = \sum_{k=0}^{\infty} c_k('z, z_n) g^k(z_n). \tag{5}$$

If $R('z)$ is the radius of convergence of (4), then the function $(-\ln R('z))^*$ is psh in $'U$, and the series (4) converges uniformly in the domain

$$\{'z \in 'U, |g(z_n)| < R_*('z)\},$$

where $R_*('z) = \exp\{-(-\ln R('z))^*\}$ is the regularization. Consequently, the function f is holomorphic in the union G of domains $\{'z \in 'U, |g(z_n)| < R_*('z)\}$ over all functions g corresponding to "rationals" $a_1, a_2, \ldots, a_m, |a_j| \geq 1, j = 1, 2, \ldots, m, m = 1, 2, \ldots$.

For a fixed point $'z \in E$, the union of lemniscates $|g(z_n)| < R_*('z)$, over such g, is the plane minus a discrete set. Thus, for points $'z^0 \in E$ such that $R('z^0) = R_*('z^0)$ for all g, the set $\{'z = 'z^0\} \setminus G$ is discrete. In view of the plurisubharmonicity of the functions

$$\ln \sup_{|g(z_n)| \leq \varepsilon} |c_k('z, z_n)|^{1/k}$$

the set of points $'z$, for which $R('z) \neq R_*('z)$ is pluripolar (cf. 3.3, Chap. 2). Since E is not pluripolar, it follows, then, that there exists a non-pluripolar subset $E' \subset E$ such that the set $\{'z = 'z^0\} \setminus G$ is discrete for all $'z^0 \in E'$.

Using the class R^0 (cf. 1.3), it is easy to show that the envelope of holomorphy $\hat{G}$ is single sheeted (lies in $\mathbb{C}^n$). Then, the set $('U \times \mathbb{C}) \setminus \hat{G}$ is pseudoconcave and $\{'z = 'z^0\} \setminus \hat{G}$ is discrete for all $'z^0 \in E'$. By Oka's Theorem (1.4) $('U \times \mathbb{C}) \setminus \hat{G}$ is an analytic subset of U.

Remark. With the help of the rational functions

$$g(z) = z^N \cdot \frac{1 - \bar{a}_1 z}{z - a_1} \cdots \cdots \frac{1 - \bar{a}_m z}{z - a_m},$$

$|a_j| \geq r, j = 1, \ldots, m; N = 1, 2, \ldots$, we can show the following result.

Theorem 2. *Suppose f is holomorphic in the polydisc $'U \times \{|z_n| < r\}$, $r > 0$, and for each fixed $'z \in 'U$ the function $f('z, z_n)$ has a finite number of singularities in the disc $|z_n| < R$, $R > r$. Then the singularity set of f in the polydisc $'U \times \{|z_n| < R\}$ is analytic.*

2.2. Pseudoconcave Sets and Analytic Tubes. The following theorem is due to Alexander and Wermer (1983).

Theorem. *Let $U_1 : |z_1| < 1$ be the unit disc and S a non-empty pseudoconcave subset of the domain $U_1 \times \mathbb{C}$ such that $S \subset \{|z_2 - a(z_1)| < r\}$, where $a(z_1)$ is a function continuous in U_1. Then there exists a function $f(z_1)$ holomorphic in the disc U such that $|f(z_1) - a(z_1)| < 4r$.*

In other words, if a pseudoconcave set lies in some tube of constant radius, then it lies in a slightly larger tube with analytic center. A number of questions arise in connection with this theorem:

1. It is true that in any neighborhood U of a pseudoconcave set S, there exists an analytic set of positive dimension $A \subset U$ (Alexander-Wermer (1983))?

2. In the proof of the theorem, essential use is made of the fact that U_1 is a plane disc. Is it possible to replace U_1 by an arbitrary domain in $\mathbb{C}^n$?

3. Is it possible to replace the disc-tube $|z_2 - a(z_1)| < r$ in the theorem by a lemniscate-tube $|P(z_1, z_2)| < r$, where P is a polynomial in z_2?

It would be interesting also to study these questions with variable radius $r('z)$. The solutions to these problems might turn out to be useful, for example, in studying the capacitary properties of arbitrary pseudoconcave sets.

§ 3. Multidimensional Analog of the Bernstein-Walsh Theorem and Separately Analytic Functions

The classical theorem of Bernstein-Walsh establishes a connection between the speed of approximation by polynomials, of a function f given on a compact set $K \subset \mathbb{C}$, and the holomorphic extension of f to certain standard neighborhoods of K. In 3.2 of this section, an analogous theorem will be proved in $\mathbb{C}^n$.

3.1. The Generalized Green Function. For a compact set $K \subset \mathbb{C}^n$, the generalized Green function $V(z, K)$ is most easily defined with the help of the class L, consisting of all functions $u \in \text{Psh}(\mathbb{C}^n)$ such that $u(z) \le \alpha + \ln(1 + |z|)$, $z \in \mathbb{C}^n$, where $\alpha = \alpha(u)$ is constant.

We set $V(z, K) = \sup\{u(z) \in L : u|_K \le 0\}$; the regularization $V^*(z, K)$ is called the *generalized Green function* (or simply, Green function) for the compact set K.

For a non-pluripolar compact set K, the function V^* exists ($V^* \not\equiv +\infty$) and belongs to the class L; $V^* \equiv +\infty$ if and only if K is pluripolar.

As for $\mathscr{P}$-measures (2.1, Chap. 2), we have

Theorem 1. *If a compact set K is such that $V^*|_K \equiv 0,$*[8] *then the function V^* is continuous in $\mathbb{C}^n$ and $V^*(z, K) \equiv V(z, K)$.*

If P is a polynomial of degree s, then the function $\frac{1}{s} \ln |P|$ belongs to L. Thus, we have the *Bernstein-Walsh inequality*

$$\frac{1}{s} \ln |P(z)| \leq \frac{1}{s} \ln \|P\|_K + V(z, K), \quad z \in \mathbb{C}^n. \tag{6}$$

In the classical case when $n = 1$, the proof of the above mentioned Bernstein-Walsh theorem is based on the construction of a sequence of polynomials P_s such that $\|P_s\|_K \leq 1$ and $\frac{1}{s} \ln |P_s|$ converges to $V(z, K)$ as $s \to \infty$. There is a similar connection between the function $V(z, K)$ and polynomials for $n > 1$ also.

Theorem 2. *For any compact set $K \subset \mathbb{C}^n$, we have the inequality*

$$V(z, K) = \sup \left\{ \frac{1}{\deg P} \ln |P(z)| : \|P\|_K \leq 1 \right\}. \tag{7}$$

Proof. Any compact set K in $\mathbb{C}^n$ can be approximated by pluriregular compacta

$$K_\delta : \mathrm{dist}(z, K) \leq \delta, \quad \delta > 0.$$

Thus, it is sufficient to prove the theorem for pluriregular compacta.

Consider the auxiliary psh function

$$h(z, w) = |w| \exp V\left(\frac{z}{w}, K\right),$$

which is homogeneous and continuous in $\mathbb{C}^{n+1}$.

Fix a point $(z^0, \omega^0) \neq 0$ and a number ε, $0 < \varepsilon < h(z^0, \omega^0)$. From Bremermann's Theorem (cf. 2.4, Chap. 1), it follows that the circled compact set $K = \{(z, \omega) : h(z, \omega) \leq h(z^0, \omega^0) - \varepsilon\}$ is polynomially convex. Consequently, it is convex with respect to homogeneous polynomials (cf. 3.3, Chap. 3). It follows that there exists a homogeneous polynomial $Q_s(z, \omega)$ such that $|Q_s(z^0, \omega^0)| > \|Q_s\|_K$. Let

$$P_s = \frac{Q_s}{\|Q_s\|_K} h(z^0, \omega^0) - \varepsilon.$$

Then $|P_s|^{1/s} \leq h$ on the compact set K and from the homogeneity of the functions $|P_s|^{1/s}$ and h we obtain the inequality

$$|P_s|^{1/s} \leq h \quad \text{everywhere in } \mathbb{C}^{n+1}.$$

---

[8] Such compact sets are said to be pluriregular in $\mathbb{C}^n$.

On the other hand, at the point (z^0, ω^0), we have

$$|P_s(z^0, \omega^0)|^{1/s} > h(z^0, \omega^0) - \varepsilon.$$

Since $\varepsilon > 0$ is arbitrary and also the point (z^0, ω^0), we obtain that

$$h(z, \omega) = \sup\{|P_s(z, \omega)|^{1/s} : |P_s|^{1/s} \le h\}.$$

Setting, here, $\omega = 1$, we have (7). $\square$

3.2. The Main Result. Let $K \subset \mathbb{C}^n$ be a pluriregular compact set and f a continuous function on K. The following connection holds between the speed at which the error $e_m(f, K)$ of best polynomial approximation decreases and the holomorphy of f in domains of the form

$$G_R = \{z \in \mathbb{C}^n : V^*(z, K) < \ln R\}, \quad R > 0.$$

Theorem. *The function f extends holomorphically to G_R if and only if*

$$\overline{\lim_{m \to \infty}} \; e_m^{1/m}(f, K) \le \frac{1}{R}.$$

Proof. a) If this inequality holds, then there exists a sequence of polynomials $P_m(z)$ such that

$$\lim_{m \to \infty} \|f - P_m\|_K^{1/m} \le \frac{1}{R}$$

Thus $\|P_{m+1} - P_m\|_K \le \dfrac{2}{R^m}$ and by (6)

$$|P_{m+1}(z) - P_m(z)| \le \frac{2}{R^m} \exp\{(m + 1)V(z, K)\}, \quad z \in \mathbb{C}^n.$$

From this it follows that the series $P_0 + \sum\limits_{m=1}^{\infty} (P_{m+1} - P_m)$ converges uniformly on compact subsets of G_R, to a holomorphic function which, on K, coincides with f.

b) To prove the other direction of the theorem, we use (7) which, together with Theorem 1 of 3.1, allow us, for arbitrary fixed $\varepsilon > 0$, to select a finite set of polynomials P_j, $\deg P_j = s, j = 1, 2, \ldots, N$, such that

$$\sup\left\{\frac{1}{s} \ln |P_j|, j = 1, 2, \ldots, n\right\} > \ln R - \varepsilon,$$

for all $z \in \partial G_R$. It follows that if we set $\mathscr{P}_j = \dfrac{P_j}{e^{-\varepsilon s} R^s}$, then the *Weil polyhedron*

$\Pi = \{|\mathscr{P}_j| < 1 : j = 1, 2, \ldots, N\}$ is a compact subset of G_R, and

$$K \subset \left\{|\mathscr{P}_j|^{1/s} \le \frac{1}{e^{-\varepsilon} R}, j = 1, 2, \ldots, N\right\}. \tag{8}$$

The function f which is holomorphic in G_R can be expanded in Π in a series via the polynomials $\mathscr{P}_j$ (cf., e.g., Shabat (1976)):

$$f(z) = \sum_{|k|=0}^{\infty} A_k \mathscr{P}_1^{k_1} \dots \mathscr{P}_N^{k_N},$$

where $k = (k_1, k_2, \dots, k_N)$ are multi-indices, $|k| = k_1, k_2, \dots, k_N$, and A_K are polynomials of a fixed degree which depends on $\mathscr{P}_1, \mathscr{P}_2, \dots, \mathscr{P}_N$. The A_k satisfy a Cauchy type inequality: $\|A_k\|_{\Pi} \leq C(f)$, $|k| = 0, 1, \dots$, where $C(f)$ is a constant independent of k.

Let

$$Q_j = \sum_{|k|=0}^{q} A_k \mathscr{P}_1^{k_1} \dots \mathscr{P}_N^{k_N},$$

be a polynomial of degree $j = t + sq$. Then according to (8)

$$\|f - Q_j\|_K \leq \sum_{|k|=q+1}^{\infty} C(f) \cdot \frac{1}{(e^{-\varepsilon} R)^{s|k|}} \leq C(f, N) \cdot \frac{1}{(e^{-\varepsilon} R)^{sq}},$$

where $C(f, N)$ is a constant independent of f. From this it follows that

$$\varlimsup_{\substack{j \to \infty \\ j = t + sq}} \|f - Q_j\|_K^{1/j} \leq \frac{1}{e^{-\varepsilon} R} \quad \text{and, since } \varepsilon > 0 \text{ is arbitrary, } \varlimsup_{m \to \infty} e_m^{1/m}(f, K) \leq \frac{1}{R}. \quad \square$$

In case the compact set K is not pluriregular, the second part of the theorem, in general, does not hold on account of the discontinuity of $V^*(z, K)$: one can construct a compact set K and a function f, holomorphic in the domain $G_R = \{V^*(z, K) < \ln R\}$, and continuous on K, such that $\varlimsup_{m \to \infty} e_m^{1/m}(f, K) > \frac{1}{R}$. However, approximating the compact set K by pluriregular compacta K_δ, one can show part b) of the theorem in the following weaker form: if f is holomorphic in a neighborhood of the set $\{V(z, K) \leq \ln R\}$, then

$$\varlimsup_{m \to \infty} e_m^{1/m}(f, K) \leq \frac{1}{R}.$$

3.3. Green Functions for Circled Sets. Projective Capacity. Let K be a *circled compact set* in $\mathbb{C}^n$, i.e. for each point $z^0 \in K$ the set K contains all points of the form $e^{i\varphi} z^0$, $\varphi \in R$. We show that the polynomially convex hull $\hat{K}$ of such a compact set coincides with $\tilde{K}$, the convex hull of K with respect to homogeneous polynomials. For this it is sufficient to show that if $z^0 \in \tilde{K}$, i.e. if $|Q(z^0)| \leq \|Q\|_K$ for all homogeneous polynomials Q, then this inequality holds also for all other polynomials.

Fix a number $0 < \sigma < 1$ and consider a polynomial $P(z) = \sum_{s=0}^{N} Q_s(z)$ with norm $\|P\|_k = 1$. By Cauchy's inequality on the slices $\{z = \lambda \alpha, \lambda \in \mathbb{C}\}$ the norms $\|Q_s\|_K$ of the homogeneous polynomials are also bounded by one and, consequently, $|Q_s(z^0)| \leq 1$. Hence $|P_s(\sigma z^0)| \leq \sum\limits_{s=0}^{N} |Q_s(z^0)| \sigma^s \leq \dfrac{1}{1-\sigma}$. Thus, $\sigma z^0 \in \hat{K}$, since this inequality is also satisfied by the polynomials P^j for arbitrary j. Letting σ tend to 1, we obtain that $z^0 \in \hat{K}$.

Making use of this remark, we prove the following curious result.

Theorem 1. *If K is a circled compact subset of the closed unit ball $\overline{B(0,1)}$, then*

$$\hat{K} = \left\{ z : |z| \cdot \exp V\left(\frac{z}{|z|}, K \right) \leq 1 \right\}.$$

Indeed, if Q_s is a homogeneous polynomial with $\|Q_s\|_K = 1$, then according to (6), we have

$$\left| Q_s\left(\frac{z}{|z|} \right) \right|^{1/s} \leq \exp V\left(\frac{z}{|z|}, K \right).$$

Consequently,

$$|Q_s(z)| = |z|^s \cdot \left| Q_s\left(\frac{z}{|z|} \right) \right| \leq \left[|z| \cdot \exp V\left(\frac{z}{|z|}, K \right) \right]^s,$$

and, hence,

$$\hat{K} \supset \left\{ z : |z| \cdot \exp V\left(\frac{z}{|z|}, K \right) \leq 1 \right\}.$$

In fact, here, instead of the inclusion $\supset$, we have equality, since the set on the right hand side is polynomially convex.

Corollary (compare Alexander (1981)). *If a circled compact set K is contained in the unit sphere $S(0,1)$, then its polynomially convex hull $\hat{K}$ contains the ball*

$$|z| \leq \exp \left\{ -\sup_{|\xi|=1} V(\xi, K) \right\}.$$

The expression on the right hand side of this inequality is related to the so called *projective capacity* $\mathscr{C}(K)$, introduced by Alexander (1981). A quantity close to $\mathscr{C}(K)$, namely, τ-capacity, was considered by the author (1982b).

In closing this section, we introduce another application of the estimate (6), this time to questions of holomorphic continuation.

Proposition 1. *Let $\sum\limits_{s=0}^{\infty} Q_s(z)$ be a formal series of homogenous polynomials Q_s, and let $\mathscr{L}$ be a family of complex lines $l \ni 0$. If for each line $l \in \mathscr{L}$ this series converges in the disc $l \cap B(0,1)$, then it converges normally in the domain*

$$G = \left\{ z \in \mathbb{C}^n : |z| \cdot \exp V^*\left(\frac{z}{|z|}, E \right) < 1 \right\},$$

where $E = \bigcup\limits_{l \in \mathscr{L}} l \cap S(0,1)$.

Indeed, fix $\varepsilon > 0$ and set

$$E_N = \left\{ w \in S(0, 1) : \left| \sum_{s=0}^{\infty} Q_s(w)\omega^s \right| \le N, |\omega| \le 1 - \varepsilon \right\}.$$

According to the Cauchy inequality,

$$|Q_s(w)| \le \frac{N}{(1 - \varepsilon)^s}, \quad w \in E_N.$$

Thus, by the Bernstein-Walsh inequality

$$|Q_s(w)| \le \frac{N}{(1 - \varepsilon)^s} [\exp V^*(w, E_N)]^s, \quad w \in S(0, 1),$$

and, hence, the series $\sum_{s=0}^{\infty} Q_s(z)$ converges uniformly in the interior of the set

$$G_N = \left\{ z \in \mathbb{C}^n : |z| \cdot \exp V^*\left(\frac{z}{|z|}, E_N\right) < 1 - \varepsilon \right\}.$$

Letting N tend to ∞ and ε to zero, we obtain the convergence of the series inside G. $\square$

Proposition 2. *Let*

$$f = \sum_{s=0}^{\infty} P_s(z, \bar{z})$$

be a formal power series such that the restriction $f|_l$ represents a holomorphic function in the disc $l \cap B(0, 1)$, for each complex line $l \ni 0$. Then the series converges in $B(0, 1)$ and its sum is holomorphic there.

Indeed from the holomorphicity of the restrictions of the series to complex lines l, it follows easily that the terms of the homogenous polynomials P_s, which contain $\bar{z}$ are zero. The rest of the proof follows from Proposition 1. $\square$

From Proposition 2 we easily have the following.

Theorem 2 (Forelli, cf. Rudin (1980)). *If f is infinitely differentiable at the point 0 and the restriction $f|_l$ is holomorphic in the disc $l \cap B(0, 1)$ for each complex line $l \ni 0$, then f is holomorphic in the ball $B(0, 1)$.*

The function

$$f(z_1, z_2) = \frac{z_1^{k+1} \bar{z}_2}{z_1 \bar{z}_1 + z_2 \bar{z}_2},$$

which is k times continuously differentiable in $\mathbb{C}^2$, is not covered by Theorem 2 $\left(f \text{ does not determine a formal power series } \sum_{s=0}^{\infty} P_s(z, \bar{z}) \right)$.

The function $f(z_1, z_2) = |z_1|^2 - |z_2|^2$ is identically zero on the complex lines $z_2 = e^{i\theta} z_1$, $\theta \in [0, 2\pi]$, which form a non pluripolar set. Nevertheless, f is not holomorphic at any point. This example shows that formal power series, containing $\bar{z}$, also are not covered by Proposition 1.

3.4. Separately Analytic Functions. Consider two domains $D \subset \mathbb{C}^n(z)$, $G \subset \mathbb{C}^n(w)$, two sets $E \subset D$, $F \subset G$ and set $X = (D \times F) \cup (E \times G)$. A function $f(z, w)$, defined in X is said to be *separately-analytic* if it is holomorphic as a function of w in G for each fixed $z \in E$ and holomorphic as a function of z in D for each fixed $w \in F$.

If $E = D$ and $F = G$, we obtain a function which is holomorphic in each variable in $D \times G$; in this case f is jointly holomorphic in $D \times G$, according to Hartogs Theorem. There is a deep connection between the notion of separate analyticity and the well known "edge of the wedge" theorem of N.N. Bogolyubov (cf. e.g. Vladimirov (1964), Zakharyuta (1976)), which for $n = m = 1$ corresponds to the situation when E and F are intervals of the real axis.

Consideration of this particular case naturally gives rise to the general question of holomorphic continuation of f to a neighborhood of X. In this section we show the possibility of holomorphic continuation to a neighborhood which is defined in terms of $\mathscr{P}$-measures of E and F with respect to D and G. In the sequel, for the sake of simplicity, we shall suppose that D and G are strictly pseudoconvex domains. We shall also suppose that E and F are compact, although from the C-measurability property of capacity (3.4, Chap. 2), it will follow that the theorem stated below holds also for arbitrary Borel subsets F and E.

Theorem. *If a function $f(z, w)$ is separately analytic on the set $X = (D \times F) \cup (E \times G)$, then it extends holomorphically to a neighborhood of the set*

$$\hat{X} = \{(z, w) \in D \times G : \omega^*(z, E, D) + \omega^*(w, F, G) + 1 < 0\}.$$

In the case of pluriregular compact sets K and F, the set $\hat{X}$ is itself a neighborhood of X.

The theorem has content ($\hat{X} \neq \varnothing$) if neither of the sets E and F is pluripolar, which we shall assume below.

An important part in the proof of the theorem is the construction of a special orthogonal basis for the pair $E \subset D$, however, here, we will only list the properties of this basis, refering, for further details, to the works of Zakharyuta (1974, 1976).

Let $D \subset \mathbb{C}^n$ be a strictly pseudoconvex domain and $E \subset D$ a non pluripolar compact set. We shall associate to the domain D any Hilbert space H_0 such that

$$\mathcal{O}(\bar{D}) \subset H_0 \subset C\mathcal{O}(\bar{D}) = C(\bar{D}) \cap \mathcal{O}(D). \tag{9}$$

We define also a Hilbert space H_1 associated to the compact set E. It is the closure of the space $\mathcal{O}(D)$ with respect to the norm

$$\|f\|_{H_1} = \left(\int_E |f(z)|^2 \, d\sigma(z) \right)^{1/2},$$

where the measure $\sigma = (dd^c \omega^*(z, E, D))^n$ is concentrated on E (cf. Chap. 2).

Let, now, $\{h_k(z)\}$ be a common orthogonal basis in the spaces H_0, H_1, satisfying the conditions:

$$\|h_k\|_{H_0} = \mu_k, \quad \|h_k\|_{H_1} = 1, \quad k = 1, 2, \ldots .$$

Such a basis exists since, by construction, H_0 is densely and completely continuously embedded in the space H_1. In addition (cf., e.g. Zakharyuta (1974))

$$\frac{1}{L} k^{1/n} \leq \ln \mu_k \leq L \cdot k^{1/n}, \tag{10}$$

where L is a constant.

From the continuous embedding (9), it follows that

$$|h_k(z)| \leq C \cdot \|h_k\|_{H_0} = C \cdot \mu_k, \tag{11}$$

where C is a constant.

Consider the set $A_k = \{z \in E : |h_k(z)| > k\}$. According to the Chebyshev inequality,

$$\sigma(A_k) \leq \frac{1}{k^2} \int_E |h_k(z)|^2 \, d\sigma = \frac{1}{k^2} \|h_k\|_{H_1} = \frac{1}{k^2}, \quad k = 1, 2, \ldots .$$

Consequently, $\sum_{k=1}^{\infty} \sigma(A_k) < \infty$ and $\lim_{s \to \infty} \sigma\left(\bigcup_{k=s}^{\infty} A_k \right) = 0$. Set $E_s = E \setminus \bigcup_{k=s}^{\infty} A_k$, $E' = \bigcup_{k=1}^{\infty} E_s$. Then $\sigma(E \setminus E') = 0$, and from the majorization principle (cf. 1.3, Chap. 2), we have the identity $\omega^*(z, E, D) \equiv \omega^*(z, E', D)$. Consequently, if $s \to \infty$

$$\omega^*(z, E_s, D) \downarrow \omega^*(z, E, D), \quad z \in D. \tag{12}$$

Since, by construction, $|h_k(z)| \leq k$, $z \in E_s$, $k \geq s$, then, using (11) and with the help of the two constants theorem (2.1, Chap. 2), we obtain

$$|h_k(z)| \leq C(s) \cdot k \cdot \mu_k^{1 + \omega^*(z, E_s, D)}, \quad k = 1, 2, \ldots, \tag{13}$$

where $C(s)$ is a constant independent of k.

We pass to the proof of the theorem.

We fix domains D' and G' such that $E \subset D' \Subset D$, $F \subset G' \Subset G$ and we denote

$$E_N = \{z \in E : \|f(z, w)\|_{\overline{G}'} \leq N\}$$
$$F_N = \{w \in F : \|f(z, w)\|_{\overline{D}'} \leq N\}.$$

Then $E = \bigcup\limits_{N=1}^{\infty} E_N$, $F = \bigcup\limits_{N=1}^{\infty} F_N$ and thus, after some N_0, the sets E_N and F_N will be non pluripolar. Let us show that the function $f(z, w)$ is continuous on $(D' \times F_N) \cup (E_N \times G')$, $N \geq N_0$. Indeed, as z varies in E_N the functions $f(z, w)$ form a compact family of functions (in w) on G'. In particlar, it is equicontinuous. The same is true with respect to $z \in D'$ as w varies in F_N. Moreover, since the sets E_N and F_N are not pluripolar, and hence are sets of uniqueness for holomorphic functions,

$$\lim_{z \to z^0} f(z, w) = f(z^0, w), \quad z, z^0 \in E_N, w \in G',$$

$$\lim_{w \to w^0} f(z, w) = f(z, w^0), \quad z \in D', w, w^0 \in F_N.$$

From this follows the continuity of f on the designated set. In particular, f is continuous on the compact set $E_N \times F_N$.

We construct, for $E_N \subset D'$ the Hilbert spaces H_0, H_1 and their common orthogonal basis $\{h_k\}$, satisfying conditions (10) and (13) with $E = E_N$, $D = D'$; consider also the analogous basis $\{e_k\}$ for $F_N \subset G'$.

We associate to f its formal double Fourier series

$$f \sim \sum_{k,j}^{\infty} a_{kj} h_k(z) e_j(w), \tag{14}$$

whose coefficients are defined by

$$a_{kj} = \int_{E_N \times F_N} f(z, w) \, d\sigma_z(h_k) \, d\mu_w(e_j),$$

with

$$d\sigma_z(h_k) = h_k(z) \cdot (dd^c \omega(z, E_N, D'))^n,$$

$$d\mu_w(e_j) = e_j(w) \cdot (dd^c \omega(w, F_N, G'))^n,$$

Borel measures on E_N, and F_N.

Let us show that the series (14) converges uniformly in the open set

$$\hat{X}_N = \{(z, w) \in D' \times G' : \omega^*(z, E_N, D') + \omega^*(w, F_N, G') < -1\}.$$

Since, for fixed $w \in F_N$, the function $f(z, w)$ belongs to H_0, we have

$$\int_{E_N} f(z, w) \, d\sigma_z(h_k) = (f, h_k)_{H_1} = \mu_k^{-2} \cdot (f, h_k)_{H_0} \leq \frac{c(f)}{\mu_k}.$$

Analogously, we get:

$$|a_{kj}| \leq \frac{c(f)}{v_j},$$

where the sequence $\{v_j\}$ is to the basis $\{e_j\}$ as $\{\mu_k\}$ is to $\{h_k\}$. From this and by

(13), we obtain, for $v_j \leq \mu_k$, the estimate

$$|a_{kj}h_k(z)e_j(w)| \leq c(s,f) \cdot k \cdot j \cdot \mu_k^{-1} \cdot \mu_k^{1+\omega^*(z,E_{N,s},D')} \cdot v_j^{1+\omega^*(w,F_{N,s},G')}$$

$$\leq c(s,f) \cdot k \cdot j \cdot \mu_k^{1+\omega^*(z,E_{N,s},D')+\omega^*(w,F_{N,s},G')}$$

and an analogous estimate for $\mu_k \leq v_j$. Thus, for any compact set $S \subset \hat{X}_N$, choosing s sufficiently large and using (12), we obtain the estimate

$$|a_{kj}h_k(z)e_j(w)| \leq c(s,f) \cdot k \cdot j(\max\{\mu_k, v_j\})^{-\delta},$$

where $\delta = \delta(S) > 0$. By (10), this estimate ensures the uniform convergence in $\hat{X}_N$ of the series (14).

Clearly, the sum, $f_N(z,w)$ of this series is a holomorphic extension of the function $f(z,w)$ to $\hat{X}_N$. Approximating the domains D and G by a non-decreasing sequence of domains D'_m and G'_m, we construct a holomorphic extension of the function f to the set $\hat{X}$. $\square$

Remarks. The results of §1 are due to the author (1982b, 1984); the main result of this section (1.3, Theorem 2) gives a positive answer to a question posed by Gonchar. The use of convergence in measure in the definition of the class R^0 was needed because the question of rapid uniform approximation of functions having a pluripolar set of singularities is open for $n > 1$. In the particular case when f has an analytic set of singularities, the possibility of such approximation by rational functions was shown by Chirka (1974).

Theorems 1 and 2 of 2.1 are due to Chirka and Kazaryan (cf. Kazaryan (1983) where a particular case is analyzed).

An analogue of the Bernstein-Walsh Theorem in $\mathbb{C}^n$ was shown by Siciak (1962) and later, using a different method, by Zakharyuta. See also Korevaar (1986).

Theorem 2 of 3.1 is due to Zakharyuta. The proof presented here is due to Siciak. The theorem is 3.4 was proved by Zakharyuta (1976). It can be extended without difficulty to the case of several, rather than two, groups of variables; more precisely, to the case where f is a separately analytic function on the set

$$X = \bigcup_{j=1}^{\infty} \{E_1 \times \cdots \times E_{j-1} \times D_j \times E_{j+1} \times \cdots \times E_k\},$$

where

$$E_j \subset D_j \subset \mathbb{C}^{n_j}, \quad n_j \in N, \quad j = 1, 2, \ldots, k.$$

In Chapter 3 we presented a series of applications of complex potential theory. Of course, we were not able to cover all such applications in all fields of function theory. For example, in recent years, the basic objects of this theory ($\mathscr{P}$-measures, Green functions, capacities, etc.) have begun to be used in Nevanlinna theory (cf. Shabat (1982)). In particular, with the help of $\mathscr{P}$-measures, it has been shown that for any holomorphic mapping $f : \mathbb{C}^n \to \mathbb{P}^m$, the union of its defective divisors, in the sense of Valiron, is a pluripolar set.

References*

Alexander, H. (1981): Projective capacity. Ann. Math. Stud. *100*, No. 1, 3–27, Zbl.494.32001

Alexander, H., Wermer, J. (1983): On the approximation of singularity sets by analytic varieties. Pac. J. Math. *104*, No. 2, 263–268, Zbl.543.32005

Bedford, E., Taylor, B.A. (1976): The Dirichlet problem for a complex Monge-Ampère equation. Invent. Math. *37*, No. 1., 1–44, Zbl.315.31007

Bedford, E., Taylor, B.A. (1982): A new capacity for plurisubharmonic functions. Acta Math. *149*, No. 1–2, 1–40, Zb.547.32012

Brelot, M. (1959): Éléments de la théorie classique du potential. Paris: Centre de Documentation Universitaire. 198 pp., Zbl.84,309

Brelot, M. (1972): Les étapes et les aspects multiples de la théorie du potentiel. Enseign. Math., II. Ser. *18*, 1–36, Zbl.235.31002

Bremermann, H.J. (1956): On the conjecture of the equivalence of the plurisubharmonic functions and the Hartogs functions. Math. Ann. *131*, No. 1, 76–86, Zbl.70,76

Bremermann, H.J. (1959): On a generalized Dirichlet problem for plurisubharmonic functions and pseudoconvex domains. Characterization of Shilov boundaries. Trans. Am. Math. Soc. *91*, No. 2, 246–276, Zbl.91,75

Chirka, E.M. (1974): Expansion in series and speed of rational approximation for holomorphic functions with analytic singularities. Mat. Sb., Nov. Ser. *93*, No. 2, 314–324. Engl. trans.: Math. USSR, Sb. *22*, 323–332 (1975), Zbl.286.32002

Chirka, E.M. (1976): Rational approximation of holomorphic functions with singularities of finite order. Mat. Sb., Nov. Ser. *100*, No. 1, 137–155. Engl. transl.: Math. USSR, Sb. *29*, 123–138 (1978), Zbl.328.30032

Goluzin, G.M. (1966): Geometric Theory of Functions of a Complex Variable. Moscow: Nauka. 628 pp. English transl.: Providence 1969, Zbl.148,306

Gonchar, A.A. (1972): A local condition of single valuedness for analytic functions. Mat. Sb., Nov. Ser. *89*, No. 1, 148–164. Engl. transl.: Math. USSR, Sb. *18*, 151–167 (1973), Zbl.247,30033

Gonchar, A.A. (1974): A local condition of single valuedness for analytic functions of several variables. Mat. Sb., Nov. Ser. *93*, No. 2, 296–313. Engl. transl.: Math. USSR, Sb. *22*, 305–322 (1975), Zbl.286.32001

Hartogs, F. (1906): Zur Theorie der analytischen Funktionen mehrerer unabhängiger Veränderlichen, insbesondere über die Darstellung derselben durch Reihen welche nach Potenzen einer Veränderlichen fortschreizen. Math. Ann. *62*, 1–88, Jbuch37,444

Harvey, R. (1977): Holomorphic chains and their boundaries. Proc. Symp. Pure Math. *30*, No. 1, 309–382, Zbl.374.32002

Hayman, W.K., Kennedy, P.B. (1976): Subharmonic Functions. London, New York, San Francisco: Academic Press. 284 pp., Zbl.419.31001

Josefson, B. (1978): On the equivalence between locally polar and globally polar sets for plurisubharmonic functions of $\mathbb{C}^n$. Ark. Mat. *16*, No. 1, 109–115, Zbl.383.31003

Kazaryan, M.V. (1983): On the holomorphic continuation of functions with special singularities in $\mathbb{C}^n$. Dokl., Akad. Nauk Arm. SSR 76, No. 1, 13–17 (Russian), Zbl.549.32007

Keldysh, M.V., Lavrent'ev, M.A. (1937): On the stability of the solution to the Dirichlet problem. Izv. Akad. Nauk SSSR, Ser. Mat. *1*, No. 4, 551–595 (Russian), Jbuch63,1040

Landkof, N.S. (1966): Foundations of Modern Potential Theory. Moscow: Nauka. 515 pp. Engl. transl.: New York, Berlin, Heidelberg: Springer-Verlag 1972, Zbl.148,103

Lelong, P. (1941): Sur quelques problèmes de la théorie des fonctions de deux variables complexes. Ann. Sci. Ecole Norm., III. Ser. *58*, 83–177, Zbl.26,15

---

* For the convenience of the reader, references to reviews in Zentralblatt für Mathematik (Zbl.), compiled using the MATH database, and Jahrbuch über die Fortschritte der Mathematik (Jbuch) have, as far as possible, been included in this bibliography.

Lelong, P. (1957): Ensembles singuliers impropres des fonctions plurisousharmoniques. J. Math. Pures Appl., IX. Ser. *36*, No. 3, 263–303, Zbl.122,319

Lelong, P. (1966): Fonctions entières (*n* variables) et fonctions plurisousharmoniques de type exponentiel. Applications à l'analyse fonctionnelle. Sovrem. Probl. Teor. Anal. Funkts., Konf. Erevan 1965, 188–209, Zbl.173,327

Nishino, T. (1962): Sur les ensembles pseudoconcaves. J. Math. Kyoto Univ. *1*, 225–245, Zbl.109,55

Oka, K. (1942): Sur les fonctions analytiques de plusieurs variables. VI. Domaines pseudoconvexes. Tôhoku Math. J. *49*, No. 1, 15–52, Zbl.60,240

Oka, K. (1953): Sur les fonctions analytiques de plusieurs variables. IX. Domaines finis sans point critique interieur. Jap. J. Math. *23*, No. 1, 97–155, Zbl.53,243

Privalov, I.I. (1937): Subharmonic Functions. Moscow: Gostekhizdat. 200 pp. (Russian)

Rudin, W. (1980): Function Theory in the Unit Ball of $\mathbb{C}^n$. New York, Berlin, Heidelberg: Springer-Verlag. 436 pp., Zbl.495.32001

Sadullaev, A. (1976): A boundary uniqueness theorem in $\mathbb{C}^n$. Mat. Sb., Nov. Ser. *101*, No. 4, 568–583. Engl. transl.: Math. USSR, Sb. *30*, 501–514 (1978), Zbl.346.32024

Sadullaev, A. (1980): The operator $(dd^c u)^n$ and condenser capacities. Dokl. Akad. Nauk SSSR *251*, No. 1, 44–47. Engl. transl.: Sov. Math., Dokl. *21*, 387–391 (1980), Zbl.488.31005

Sadullaev, A. (1981): Plurisubharmonic measures and capacities on complex manifolds. Usp. Mat. Nauk *36*, No. 4, 53–105. Engl. transl.: Russ. Math. Surv. *36*, No. 4, 61–119 (1981), Zbl.475.31006

Sadullaev, A. (1982a): Continuation of plurisubharmonic functions from a submanifold. Dokl. Akad Nauk UzSSR *5*, No. 1, 3–4 (Russian)

Sadullaev, A. (1982b): Rational approximation and pluripolar sets. Mat. Sb., Nov. Ser. *119*, No. 1, 96–118. Engl. transl.: Math. USSR, Sb. *47*, 91–113 (1984), Zbl.511.32011

Sadullaev, A. (1984): Criteria for rapid rational approximation in $\mathbb{C}^n$. Mat. Sb., Nov. Ser. *125*, No. 2, 269–279. Engl. transl.: Math. USSR, Sb. *53*, 271–281 (1986), Zbl.592.32013

Shabat, B.V. (1976): Introduction to Complex Analysis, Vol. 2. Moscow: Nauka. 400 pp. French transl.: Moscow: MIR 1990, Zbl.578.32001

Shabat, B.V. (1982): Distribution of the Value of Holomorphic Mappings. Moscow: Nauka. 288 pp. English transl.: Transl. Math. Monogr. *61*, Providence (1985), Zbl.537.32008

Siciak, J. (1962): On some extremal functions and their applications in the theory of analytic functions of several complex variables. Trans. Am. Math. Soc. *105*, No. 2, 322–357, Zbl.111,81

Siciak, J. (1969): Separately analytic functions and envelopes of holomorphy of some lower dimensional subsets of $\mathbb{C}^n$. Ann. Pol. Math. *22*, No. 1, 145–171, Zbl.185,152

Slodkowski, Z. (1990): Polynomial hulls with convex fibers and complex geodesics. J. Funct. Anal. *94*, 156–176, Zbl.717.32009

Vladimirov, V.S. (1964): Methods of the Theory of Functions of Several Complex Variables. Moscow: Nauka. 411 pp. French transl.: Les fouctions de plusieurs variables complexes et leurs application. Paris: Dunod 1967, 338 pp., Zbl.125,319

Zakharyuta, V.P. (1974): Extremal plurisubharmonic functions, Hilbert scales and isomorphisms of spaces of analytic functions of several variables, I, II. Teor. Funk. Funkts. Anal. Priloz. *19*, 133–157, Zbl.336.46031; *21*, 65–83, Zbl.336.46032. (Russian)

Zakharyuta, V.P. (1976): Separately analytic functions, generalized Hartogs theorem, and envelopes of holomorphy. Mat. Sb., Nov. Ser. *101*, No. 1, 57–76. Engl. transl.: Math. USSR, Sb. *30*, 51–67 (1978), Zbl.357.32002

References Added (by Author) in Translation

Alexander, H., Wermer, J. (1985): Polynomial hulls with convex fibers. Math. Ann. *271*, 99–109, Zbl.538.32011

Alexander, H., Wermer, J. (1989): Polynomial hulls of sets with intervals as fibers. Complex Variables, Theory Appl. *11*, 11–19, Zbl.673.32017

Aupetit, B. (1984): Geometry of pseudoconvex open sets and distribution of values of analytic multivalued functions. Contemp. Math. *32*, 15–34, Zbl.595.32027

Berndtsson, B., Ransford, T.J. (1986): Analytic multifunctions, the $\bar{\partial}$-equation, and a proof of the Corona theorem. Pac. J. Math. *124*, 57–72, Zbl.602.32002

Cegrell, U. (1978): Construction of capacities on $\mathbb{C}^n$. Upps. Univ. Dep. of Math. *1*, 1–18

Cegrell, U. (1986): Sums of continuous plurisubharmonic functions and the Complex Monge-Ampère operator in $\mathbb{C}^n$. Math. Z. *193*, 373–380, Zbl.624.31004

Cegrell, U. (1990): The Dirichlet problem for the Complex Monge-Ampère operator: Stability in L^2. Univ. of Umea, Dep. of Math., 9, 1–7, appeared in Mich. Math. J. *39*, 145–151 (1992)

Demailly, J.P. (1987): Mesures de Monge-Ampère et mesures pluriharmoniques. Math. Z. *194*, 519–564, Zbl.595.32006

Fornaess, J.E., Wiegerinck, J. (1989): Approximation of plurisubharmonic functions. Ark. Mat. *27*, No. 2, 257–272, Zbl.693.32009

Korevaar, J. (1986): Polynomial approximation numbers, capacities and extended Green functions for $\mathbb{C}$ and $\mathbb{C}^n$. Approximation Theory V, Proc. Int. Symp. College Station 1986, 97–127, Zbl.612.41011

Lempert, L. (1982): Holomorphic retracts and intrinsic metrics in convex domains. Anal. Math. *8*, 257–261, Zbl.509.32015

Levenberg, N. (1985): Monge-Ampère measures associated to extremal plurisubharmonic functions in $\mathbb{C}^n$. Trans. Am. Math. Soc. *289*, 333–343, Zbl.541.31009

Nguyen Thanh Van; Zeriahi, A. (1983): Familles de polynômes presque partout bornées. Bull. Sci. Math., II. Ser. *107*, 81–91, Zbl.523.32011

Nguyen Thanh Van (1989): Bases polynomiales et approximation des fonctions séparément harmoniques dans $\mathbb{C}^v$. Bull. Sci. Math., II. Sér. *113*, No. 3, 349–361, Zbl.677.41006

Slodkowski, Z. (1986): An analytic set-valued selection and its applications to the Corona theorem, to polynomial hulls and joint spectra. Trans. Am. Math. Soc. *294*, 367–377, Zbl.594.32008

Slodkowski, Z. (1990): Polynomial hulls with convex fibers and complex geodesics. J. Funct. Anal. *94*, 156–176, Zbl.717.32009

Zeriahi, A. (1985): Capacité, constante de Chebyshev et polynômes orthogonaux associés a un compact de $\mathbb{C}^n$. Bull. Sci. Math. II. Ser. *109*, 325–335, Zbl.583.31006

Zeriahi, A. (1990): Bases de Schauder et isomorphismes d'espaces de fonctions holomorphes, C.R. Acad. Sci., Paris, Sér. I *310*, 691–694, Zbl.721.32002

III. Function Theory in the Ball

A.B. Aleksandrov

Translated from the Russian
by P.M. Gauthier

Contents

Preface

In the theory of functions of several complex variables, the ball occupies a crucial position: on the one hand, the ball is the simplest example of a strictly pseudoconvex domain with smooth boundary; on the other hand, it is the simplest bounded classical domain.

The theory of functions in the ball is very well and rather completely set forth in the book of Rudin (1980). Rudin's book has had a great influence both on the choice of material for the present paper as well as on its presentation. In the present paper, several results[1] are also exposed which were clarified in the four years following the appearance of Rudin's book. Moreover, the BMO[2] space and the atomic technique, which were completely absent in Rudin (1980), are herein presented. Almost all results of this paper (in which the group of automorphisms of the ball, harmonic analysis on the sphere and related aspects of the "pure" theory of functions on classical domains play no role) hold for strictly pseudoconvex domains with C^2-boundary and sometimes, even with weaker conditions on the boundary.

We remark that such important topics of the theory of functions as the $\bar{\partial}$-problem and related questions, analysis on the Heisenberg group, proper holomorphic mappings, and the maximal ideal space of the algebra $H^\infty(B)$ will not be treated at all in this paper.

For the $\bar{\partial}$-problem, we refer the interested reader to the survey of G.M. Khenkin and E.M. Chirka (1975), the book of Khenkin and Leiterer (1984), and Rudin (1980).

We remark also that the proofs of many of the results herein make use of appropriate theorems on the solution of the $\bar{\partial}$-equation and variants thereof.

For proper holomorphic mappings, we refer the reader to the paper of S.I. Pinchuk (1986). Concerning the maximal ideal space of $H^\infty(B)$, a certain amount of information can be gathered from Axler-Shapiro (1983), McDonald (1979), and Rudin (1983).

The corona problem, which to this day has not been solved in the multidimensional case, is discussed in Khenkin-Leiterer (1984).

The author is very grateful to G.M. Khenkin for many helpful suggestions. I also extend my sincere gratitude to S.A. Vinogradov for consultations on atomic theory and to V.V. Peller for consultations on the theory of Hankel and Töplitz operators.

---

[1] For the convenience of the reader, at the end of this paper we append a list of those problems presented in Rudin (1980) which have since been solved.

[2] BMO-Bounded Mean Oscillation.

Chapter 1
Introduction

§ 1. Preliminary Information

1.1. Notation. The symbols $\langle \, , \, \rangle$ and $| \ |$ will denote the scalar product and the norm in $\mathbb{C}^n$.

$$\langle z, w \rangle \overset{\text{def}}{=} \sum_{j=1}^{n} z_j \bar{w}_j, \quad |z| \overset{\text{def}}{=} \sqrt{\langle z, z \rangle}.$$

Let $\{e_1, e_2, \ldots, e_n\}$ be the standard basis in $\mathbb{C}^n$, i.e. $\langle z, e_j \rangle = z_j$ for all $z \in \mathbb{C}^n$ and all $j \in \{1, 2, \ldots, n\}$. We shall denote the open unit ball in $\mathbb{C}^n$ by $B_n \overset{\text{def}}{=} \{z \in \mathbb{C}^n : |z| < 1\}$. The boundary of this ball is the unit sphere $S_n \overset{\text{def}}{=} \{z \in \mathbb{C}^n : |z| = 1\}$. We set $\mathbb{D} = B_1$ and $\mathbb{T} = S_1$. The number n will (generally) be fixed and so we will usually write B and S rather than B_n and S_n. We denote by $\mathbb{P}^{n-1}$ the complex projective space of dimension $n - 1$ (i.e. the collection of all one-dimensional linear subspaces of $\mathbb{C}^n$). The symbol $\pi = \pi_n$ will denote the canonical projection from $\mathbb{C}^n \setminus \{0\}$ (or S) onto $\mathbb{P}^{n-1}$.

To each multi-index $\alpha \in \mathbb{Z}_+^n \overset{\text{def}}{=} \{\alpha \in \mathbb{Z}^n = \alpha_j \geq 0\}$, we associate two numbers:

$$|\alpha| \overset{\text{def}}{=} \sum_{j=1}^{n} \alpha_j, \quad \alpha! \overset{\text{def}}{=} \prod_{j=1}^{n} \alpha_j!.$$

The first of these notions will not give rise to ambiguity for in any given situation it will be clear whether we are dealing with a vector in $\mathbb{C}^n$ or a multi-index. We set

$$z^\alpha \overset{\text{def}}{=} \prod_{j=1}^{n} z_j^{\alpha_j}, \quad (z \in \mathbb{C}^n, \alpha \in \mathbb{Z}^n).$$

1.2. Integration on the Sphere. To every positive measure μ we associate the space $L^0(\mu)$ of all μ-measurable μ-almost everywhere finite functons (functions equal μ-almost everywhere are identified). As usual

$$L^p(\mu) \overset{\text{def}}{=} \left\{ f \in L^0(\mu) : \|f\|_{L^p} \overset{\text{def}}{=} \left(\int |f|^p \, d\mu \right)^{1/p} < +\infty \right\}, \quad (0 < p < +\infty),$$

$$L^\infty(\mu) \overset{\text{def}}{=} \{ f \in L^0(\mu) : \|f\|_{L_\infty} \overset{\text{def}}{=} \text{ess sup } |f| < +\infty \}.$$

The symbol $\sigma = \sigma_n$ (respectively $\hat{\sigma}$) will denote the unique probability Borel measure on S (respectively on $\mathbb{P}^{n-1}$), invariant with respect to all unitary transformations of $\mathbb{C}^n$, $m \overset{\text{def}}{=} \sigma_1$. Let $\nu = \nu_n$ denote Lebesgue measure in $\mathbb{C}^n$ normalized by the condition $\nu(B) = 1$.

We denote by $C(K)$ the space of all continous functions on a compact set K. The dual space of $C(K)$ is identified in a natural way with the space $M(K)$ of all regular Borel measures on K.

The space $L^1(\sigma)$ is identified in a natural way with the set of all absolutely continous (with respect to σ) measures in $M(S)$. In the future we will often write $L^p(S)$ instead of $L^p(\sigma)$.

We present several convenient integral formulae, which will hold if either the left member or the right member has a meaning. The letter f will denote a Borel function (whose domain of definition will be defined by the formula itself).

Let $z = (u, w)$, where $u \in \mathbb{C}^k$, $w \in \mathbb{C}^{n-k}$, i.e.

$$z_j = \begin{cases} u_j(1 \leq j \leq k), \\ w_{j-k}(k+1 \leq j \leq n), \end{cases}$$

Then

$$\int_{S_n} f \, d\sigma_n = \binom{n-1}{k-1} \int_{B_{n-k}} (1 - |w|^2)^{k-1}$$
$$\times \left(\int_{S_k} f((1 - |w|^2)^{1/2}\zeta, w) \, d\sigma_k(\zeta) \right) dv_{n-k}(w). \tag{1}$$

Here and in the sequel $\binom{p}{m} \overset{\text{def}}{=} \dfrac{p!}{m!(p-m)!}$. We note two important cases of (1).

$$\int_{S_n} f(z_1, z_2, \ldots, z_{n-1}) \, d\sigma_n(z_1, z_2, \ldots, z_n)$$
$$= \int_{B_{n-1}} f(z_1, z_2, \ldots, z_{n-1}) \, dv_{n-1}(z_1, z_2, \ldots, z_{n-1}), \tag{2}$$

$$\int_{S_n} f(\langle z, \zeta \rangle) \, d\sigma_n(z) = (n-1) \int_D f(\xi)(1 - |\xi|^2)^{n-2} \, dv_1(\xi) \tag{3}$$

for all $\zeta \in S$.

We present two more formulae.

$$\int_{\mathbb{C}^n} f \, dv = 2n \int_0^\infty r^{2n-1} \left(\int_S f(r\zeta) \, d\sigma(\zeta) \right) dr, \tag{4}$$

$$\int_{\mathbb{C}^n} f(z)|z|^{-2n} \, dv_n(z) = n \int_{S^n} \left(\int_C f(\zeta z)|z|^{-2} \, dv_1(z) \right) d\sigma_n(\zeta). \tag{5}$$

It is easy to see that the family $\{z^\alpha\}_{\alpha \in \mathbb{Z}_+^n}$ forms an orthogonal system in $L^2(B, v)$ and in $L^2(S, \sigma)$.

The following formulae hold:

$$\int_S |z^\alpha|^2 \, d\sigma(z) = \frac{(n-1)! \prod_{j=1}^{n} \Gamma(\alpha_j + 1)}{\Gamma\left(n + \sum_{j=1}^{n} \alpha_j\right)}, \tag{6}$$

$$\int_B |z^\alpha|^2 \, dv(z) = \frac{n! \prod_{j=1}^{n} \Gamma(\alpha_j + 1)}{\Gamma\left(n + 1 + \sum_{j=1}^{n} \alpha_j\right)}. \tag{7}$$

Formula (7) follows from formulae (6) and (2). In the right members of formulae (6) and (7), we have intentionally used the gamma function (instead of the factorial), in order for these formulae to make sense not only for integers α_j, but also for complex numbers α_j (Re $\alpha_j > -1$), if in the left members instead of $|z^\alpha|^2$ we write $\prod_{j=1}^{n} |z_j|^{2\alpha_j}$.

1.3. Differentiation Operators. We set

$$D_j = \frac{1}{2}\left(\frac{\partial}{\partial x_j} - i\frac{\partial}{\partial y_i}\right), \quad \bar{D}_j = \frac{1}{2}\left(\frac{\partial}{\partial x_j} + i\frac{\partial}{\partial y_i}\right),$$

where $x_j = \operatorname{Re} z_j$, $y_j = \operatorname{Im} z_j$. Let Ω be an open subset of $\mathbb{C}^n$. A function $f \in C^1(\Omega)$ is said to be *holomorphic* if $\bar{D}_j f = 0$ for all $j \in \{1, 2, \ldots, n\}$. Every holomorphic function is infinitely differentiable and has a power series expansion:

$$f(z) = \sum_{\alpha \in z_+^n} \frac{(D^\alpha f)(a)}{\alpha!}(z - a)^\alpha,$$

in the neighbourhood of any point $a \in \Omega$, where $D^\alpha f = D_1^{\alpha_1} D_2^{\alpha_2} \ldots D_n^{\alpha_n} f$. A function $f \in C^2(\Omega)$ is called *pluriharmonic* if $D_k \bar{D}_j f = 0$ for all $k, j \in \{1, 2, \ldots, n\}$. A real pluriharmonic function is locally the real part of a holomorphic function.

To each holomorphic function f we associate the function $\mathcal{R}_0 f = \sum_{j=1}^{n} z_j D_j f$.

If f and g are holomorphic in a neighbourhood of the ball B and $f(0)g(0) = 0$, then

$$\int_S f\bar{g}\, d\sigma = \frac{2^{k+l-1}}{n(k + l - 1)!} \times \int_B (\mathcal{R}_0^k f)(\overline{\mathcal{R}_0^l g})\left(\log \frac{1}{|z|}\right)^{k+l-1} |z|^{-2n}\, dv(z), \quad (8)$$

where $k, l \in Z_+$, and $k + l > 0$. This formula easily reduces to the one-dimensional case with the help of (5).

1.4. Manifolds. Let M be a smooth (real) manifold (with boundary). We shall denote the tangent space to the manifold M at the point $p \in M$ by $T_p(M)$. We shall consider only manifolds which are submanifolds of $\mathbb{C}^n$. In this case the tangent space $T_p(M)$ can be thought of, in a natural way, as a $\mathbb{R}$-linear subspace of $\mathbb{C}^n$.

A smooth manifold $M \subset \mathbb{C}^n$ is said to be *generic* at a point $p \in M$ if $T_p(M) + iT_p(M) = \mathbb{C}^n$. Along with the tangent space $T_p(M)$, we may associate to each point $p \in M$ its *complex tangent space* $T_p^C(M) \overset{\text{def}}{=} T_p(M) \cap iT_p(M)$. When $M = S$, we have:

$$T_p(S) = \{z \in \mathbb{C}^n : \operatorname{Re}\langle z, p\rangle = 0\}, \quad T_p^C(S) = \{z \in \mathbb{C}^n : \langle z, p\rangle = 0\}.$$

A smooth manifold $M \subset \mathbb{C}^n$ is said to be *totally real* at a point $p \in M$ if $T_p^C(M) = \{0\}$. If a manifold $M \subset \mathbb{C}^n$ is generic (respectively totally real) at one of its points, then dim $M \geq n$ (respectively dim $M \leq n$). We remark also that the dimension of a manifold $M \subset S$, which is totally real at one of its points, is at most $n - 1$. A smooth manifold $M \subset S$ is said to be *integral* (or *complex tangential*) at a point $p \in M$ if $T_p(M) \subset T_p^C(S)$. We shall say that a manifold is *generic* (respectively *totally real, integral*) if it is such at each of its points.

1.4.1. Theorem (see Khenkin (1976), Rudin (1980)). *Let M be a C^1-manifold, $M \subset S$. If M is integral, then $T_p(M)$ is orthogonal to $iT_p(M)$ (in $\mathbb{R}^{2n} \simeq \mathbb{C}^n$) for each $p \in M$, i.e. $\langle \xi, \eta \rangle \in \mathbb{R}$ for all $\xi, \eta \in T_p(M)$ and all $p \in M$.*

1.4.2. Corollary. *Under the conditions of Theorem 1.4.1, the manifold M is totally real, and consequently, dim $M \leq n - 1$.*

It is clear that a manifold $M \subset S$ of dimension n (and even $2n - 2$) can be integral at some (but not all!) of its points ($n \geq 2$).

§2. Automorphisms of the Ball

An *automorphism* of a domain $\Omega \subset \mathbb{C}^n$ is a biholomorphic mapping of the domain onto itself. We shall denote by $\mathrm{Aut}(\Omega)$ the group of all automorphisms of a domain Ω.

2.1. Description of the Automorphisms of the Ball. The group $\mathrm{Aut}(B_n)$ for $n = 2$ was described by Poincaré (see Rudin (1980)). The case of arbitrary n is completely analogous to this particular case (cf., for example, Shabat (1976) and Rudin (1980)). Each automorphism $\mathscr{A}$ of the ball B is a fractional linear transformation, i.e. is of the form:

$$w_\mu = \frac{a_{\mu 0} + \sum_{v=1}^{n} a_{\mu v} z_v}{a_{00} + \sum_{v=1}^{n} a_{0v} z_v} \quad (w = \mathscr{A}z).$$

Each automorphism $\mathscr{A}$ is represented by a unique such matrix $A = \{a_{\mu v}\}, 0 \leq \mu \leq n, 0 \leq v \leq n$, up to non-zero multiple factors. In order to characterize all matrices which correspond to automorphisms of the ball, we denote by J the matrix $\begin{pmatrix} -I_1 & 0 \\ 0 & I_n \end{pmatrix}$, where I_k is the identity $(k \times k)$-matrix. A matrix A is said to be *J-unitary* if $A^* J A = J$. In other words, the matrix A induces an operator in $\mathbb{C}^{n+1}$ which preserves the sesquilinear form $\langle J \cdot, \cdot \rangle$, also called an indefinite metric.

A matrix A determines an automorphism of the ball B if and only if $A = cA_0$, where $c \in \mathbb{C} \setminus \{0\}$, and A_0 is a J-unitary matrix. Thus, we may identify the group $\mathrm{Aut}(B_n)$ with the group of all J-unitary matrices factored by its centre, the subgroup of scalar matrices $\mathbb{T} \cdot I_{n+1}$.

The group $\mathrm{Aut}(B_n)$ is a connected Lie group of dimension $n^2 + 2n$.

To each point $a \in B$ we may associate a unique automorphism φ_a, enjoying the following properties:

1. $\varphi_a(0) = a$, $\varphi_a(a) = 0$;
2. $\varphi_a = \varphi_a^{-1}$;
3. φ_a has a unique fixed point.

Such an automorphism can be described by an explicit formula

$$\varphi_a(z) = \begin{cases} -z, & a = 0, \\ \dfrac{a - P_a z - (1 - |a|^2)^{1/2} Q_a z}{1 - \langle z, a \rangle}, & a \in B \backslash \{0\}. \end{cases}$$

Here, P_a denotes the orthogonal projection onto the one-dimensional subspace $\mathbb{C} \cdot a$, and Q_a the projection onto its orthogonal complement, i.e.

$$P_a z = \frac{\langle z, a \rangle}{\langle a, a \rangle} a,$$

$$Q_a z = z - \frac{\langle z, a \rangle}{\langle a, a \rangle} a.$$

The automorphism $\varphi_b \circ \varphi_a$ carries the point $a \in B$ to the point $b \in B$. Thus the group $\mathrm{Aut}(B)$ is transitive on B. Any automorphism of the ball B which fixes the origin is a unitary operator. Consequently, any automorphism $\psi \in \mathrm{Aut}(B)$ has a unique representation of the form:

$$\psi = U \circ \varphi_a \quad (\text{respectively } \psi = \varphi_a \circ U),$$

where U is a unitary operator. A point $a \in B$ is uniquely determined by the equation $\psi(a) = 0$ (correspondingly $\psi(0) = a$).

We present two convenient formulae, in which $\psi \in \mathrm{Aut}(B)$ and $a = \psi^{-1}(0)$,

$$1 - \langle \psi(z), \psi(w) \rangle = \frac{(1 - \langle a, a \rangle)(1 - \langle z, w \rangle)}{(1 - \langle z, a \rangle)(1 - \langle a, w \rangle)} \quad (z, w \in \bar{B}), \tag{9}$$

$$\det_{\mathbb{R}} \psi'(z) = \left(\frac{1 - |a|^2}{|1 - \langle z, a \rangle|^2} \right)^{n+1}. \tag{10}$$

In the ball B there exists a unique (up to a multiplicative constant) regular Borel measure which is invariant with respect to automorphisms of the ball. From formulas (9) and (10) it follows that such a measure is τ,

$$d\tau(z) = K(z) \, dv(z),$$

where $K(z) = (1 - |z|^2)^{-n-1}$.

2.2. The Bergman Metric. There exists, in the ball B, a Riemannian metric, unique up to a multiplicative constant, which is invariant with respect to the group $\mathrm{Aut}(B)$:

$$ds^2 = (n + 1) \left(\sum_{j=1}^{n} \frac{dz_i \, \overline{dz_j}}{1 - |z|^2} + \sum_{k=1}^{n} \sum_{j=1}^{n} \frac{\bar{z}_k z_j \, dz_k \, \overline{dz_j}}{(1 - |z|^2)^2} \right). \tag{11}$$

This metric is called the *Bergmann metric*. Formula (11) shows that this metric is a Hermitian metric. Moreover, this metric is Kähler (cf. Wells (1973)) since its associated differential form ω (of bidegree $(1, 1)$) is exact. The exactness of this form is most easily established by noting that $\omega = (1/2)i\partial\bar{\partial}\log K$.

In view of its invariance, the Bergman distance between points a and b $(a, b \in B)$ is equal to the distance between the points 0 and $|\varphi_a(b)|e_1$. Thus,

$$\rho(a, b) = \int_0^{|\varphi_a(b)|} \frac{\sqrt{(n + 1)}\, dt}{1 - t^2} = \sqrt{(n + 1)} \log \frac{1 + |\varphi_a(b)|}{1 - |\varphi_a(b)|}.$$

2.3. The Cayley Transform. In order to define the Cayley transform in several variables, consider the domain $\Omega = \Omega_n$, given by

$$\Omega_n \stackrel{\text{def}}{=} \left\{ z \in \mathbb{C}^n : \operatorname{Im} z_1 > \sum_{j=2}^n |z_j|^2 \right\}.$$

The mapping $\Phi : B \to \Omega$,

$$\Phi(z) = i\frac{e_1 + z}{1 - z_1},$$

is called the *Cayley Transform*. Its inverse transform ψ can be given in the following form:

$$\psi(z) = \frac{2z}{i + z_1} - e_1.$$

The mapping Φ induces an isomorphism between the groups $\operatorname{Aut}(B)$ and $\operatorname{Aut}(\Omega)$. A subgroup of $\operatorname{Aut}(\Omega)$ which naturally arises is the group of "non-isotropic" dilatations $\{\delta_t\}_{t>0}$, where

$$\delta_t z = (t^2 z_1, t z_2, \ldots, t z_n).$$

If $t \neq 1$, then δ_t fixes only two points: 0 and ∞. Consequently, for $t \neq 1$, the composition $\psi \circ \delta_t \circ \Phi$ is an automorphism of the ball B which fixes only the points e_1 and $-e_1$.

Another important subgroup of the group $\operatorname{Aut}(\Omega)$ is the group of "shifts" h_a $(a \in \partial\Omega)$, defined in the following way:

$$h_a(w) = \left(w_1 + a_1 2i \sum_{j=2}^n w_j \bar{a}_j, w_2 + a_2, \ldots, w_n + a_n \right).$$

The set $\partial\Omega$ can be endowed with a group structure by defining an operation $a \# b$ such that $h_{a \# b} = h_a \circ h_b$. This last equation is equivalent to $a \# b = h_a(b)$. The topoligical group $(\partial\Omega, \#)$ is called the *Heisenberg group*. We remark that for $a \neq 0$, the automorphism h_a fixes only the point at infinity. Consequently, the automorphism $\psi \circ h_a \circ \Phi$ $(a \neq 0)$ of the ball has the unique fixed point e_1.

Just as in the one dimensional situation, one can construct two "parallel" function theories, in B or in Ω (and on the sphere S or on the Heisenberg group $\partial\Omega$, if we have boundary behavior of functtuions in mind). Certain facts from one theory automatically carry over to the other. In the present paper we shall not dwell, to any extent, on the theory of functions in the domain Ω nor on the Heisenberg group $\partial\Omega$. For a treatment of analysis on the Heisenberg group, we refer the reader to Greiner and Stein (1977) and Rothschild and Stein (1976).

§3. $\mathcal{U}$-Invariant Subspaces

Let $\mathcal{U} = \mathcal{U}(n)$ denote the group of all unitary transformations of $\mathbb{C}^n$. Consider the representation of the group $\mathcal{U}$ on the space $L^2(S)$ given by $f^U(\zeta) = f(U\zeta)$ ($\zeta \in S$), where $f \in L^2(S)$, $U \in \mathcal{U}(n)$. This representation (as any representation of a compact group) can be decomposed as a direct sum of finite-dimensional irreducible representations. In the present situation, these irreducible representations turn out to be pairwise non-equivalent. In order to give this decomposition explicitly, we introduce some notations.

3.1. The Spaces $H(p, q)$. Let $\mathbb{C}[u_1, u_2, \ldots, u_r]$ denote the ring of all complex polynomials in the variables $u_1, u_2, \ldots, u_r$. Let $H(p, q)$ denote the space of homogenous harmonic polynomials

$$f \in \mathbb{C}[z_1, z_2, \ldots z_n, \bar{z}_1, \bar{z}_2, \ldots \bar{z}_n]$$

of degree p in the variables $z_1, z_2, \ldots z_n$ and of degree q in the variables $\bar{z}_1, \bar{z}_2, \ldots \bar{z}_n$. In the one-dimensional case, $H(p, 0) = \mathbb{C}z^p$, $H(0, q) = \mathbb{C}\bar{z}^q$, and all remaining spaces $H(p, q)$ ($pq \neq 0$) are null. We denote the dimension of the space $H(p, q)$ by $D(p, q, n)$. It is not difficult to verify that

$$D(p, q, n) = \binom{p + n - 2}{p}\binom{q + n - 2}{q}\frac{p + q + n - 1}{n - 1} \quad (n \geq 2).$$

3.1.1. Theorem (cf. Rudin (1980), Sect. 12.2). *All the spaces $H(p, q)$ ($p, q \geq 0$) are $\mathcal{U}$-invariant and*

$$L^2(S) = \bigoplus_{p,q \geq 0} H(p, q).$$

Any closed $\mathcal{U}$-invariant subspace $E \subset L^2(S)$ is of the form:

$$E = \bigoplus_{(p,q) \in A} H(p, q),$$

where $A \subset \mathbb{Z}_+^2$. If T is a non-zero $\mathcal{U}$-invariant operator from $H(p, q)$ to $H(r, s)$, then $p = r, q = s$ and $T = cI$, where I is the identity operator, and $c \in \mathbb{C}$.

Let $K_{pq}(z, \zeta)$ denote the reproducing kernel for the space $H(p, q)$, $H(p, q) \subset L^2(S)$. This means that $K_{pq}(\cdot, z) \in H(p, q)$ and

$$f(z) = \langle f, K_{pq}(\cdot, z)\rangle_{L^2(S)}$$

for all $f \in H(p, q)$ and all $z \in \mathbb{C}^n$, i.e.

$$f(z) = \int_S f(\zeta) \overline{K_{pq}(\zeta, z)} \, d\sigma(\zeta)$$

or (since a reproducing kernel is always symmetric: $K_{pq}(z, \zeta) = \overline{K_{pq}(\zeta, z)}$)

$$f(z) = \int_S K_{pq}(z, \zeta) f(\zeta) \, d\sigma(\zeta).$$

From the $\mathscr{U}$-invariance of $H(p, q)$ it follows that $K_{pq}(Uz, U\zeta) = K_{pq}(z, \zeta)$ for all $z, \zeta \in \mathbb{C}^n$ and all $U \in \mathscr{U}$. The integral operator with kernel $K_{pq}(z, \zeta)$ defines a $\mathscr{U}$-invariant orthogonal projection from $L^2(S)$ onto $H(p, q)$. Since the square of the Hilbert-Schmidt norm of this projection (as well as its trace) is equal to the dimension of the space $H(p, q)$, we have

$$D(p, q, n) = K_{pq}(z, z) = \int_S |K_{pq}(z, \zeta)|^2 \, d\sigma(\zeta)$$

for all $z \in S$ and all $p, q \geq 0$.

We remark also that

$$|K_{pq}(z, \zeta)| \leq |z|^{p+q} |\zeta|^{p+q} \cdot D(p, q, n), \tag{12}$$

$$K_{pq}(z, \zeta) = \overline{K_{pq}(\zeta, z)} = K_{pq}(z, \zeta) \tag{13}$$

for all $z, \zeta \in \mathbb{C}^n$ and all $p, q \geq 0$.

3.2. Explicit Formulae for the Kernel $K_{pq}(z, \zeta)$. We shall express the kernel $K_{pq}(z, \zeta)$ via special functions. To this end we recall at first the definition of the *hypergeometric function*

$$F(\alpha, \beta, \gamma; x) \overset{\text{def}}{=} \sum_{k \geq 0} \left(\prod_{j=0}^{k-1} \frac{(\alpha + j)(\beta + j)}{(\gamma + j)} \right) \frac{x^k}{k!}.$$

The following formula is essentially contained in the proof of Proposition 12.2.6 in Rudin (1980):

$$K_{pq}(z, \zeta) = D(p, q, n) \langle z, \zeta \rangle^p \langle \zeta, z \rangle^q F\left(-p, -q, n - 1; 1 - \frac{|z|^2 |\zeta|^2}{|\langle z, \zeta \rangle|^2} \right).$$

We may also express the kernel $K_{pq}(z, \zeta)$ via Jacobi polynomials. By $\mathscr{P}_m^{\alpha, \beta}(\alpha, \beta > -1)$ we denote the polynomial of degree m, orthogonal in $L^2([0, 1], x^\alpha (1 - x)^\beta \, dx)$ to all such polynomials of lesser degree, and normalized by the condition $\mathscr{P}_m^{\alpha, \beta}(1) = 1$. The following equation holds (cf. Szegö (1959)):

$$\mathscr{P}_m^{\alpha, \beta}(x) = (-1)^m \left(\prod_{j=1}^{m} \frac{\alpha + j}{\beta + j} \right) F(-m, \alpha + \beta + 1 + m,; x).$$

It is not difficult to show that for $n \geq 2$

$$K_{pq}(z, \zeta) = D(p, q, n)(\langle z, \zeta \rangle)^{p-q} \mathscr{P}_q^{p-q, n-2} \left(\frac{|\langle z, \zeta \rangle|^2}{|z|^2 |\zeta|^2} \right) |z|^{2q} |\zeta|^{2q},$$

if $p \geq q$.

3.3. Generalized Functions on the Sphere S.
To each distribution $f \in \mathscr{D}'(S)$ we may associate a family of harmonic polynomials $\{K_{pq} f\}_{p,q \geq 0}$:

$$(K_{pq} f)(z) \overset{\text{def}}{=} \int_S K_{pq}(z, \zeta) f(\zeta) \, d\sigma(\zeta)$$

(the integral is understood in the sense of the theory of distributions). We have:

$$f = \sum_{p,q \geq 0} K_{pq} f \quad (\text{in } \mathscr{D}'(S)).$$

Moreover, the series $\sum_{p,q \geq 0} K_{pq} f$ converges normally in the interior of the ball B and represents therein a harmonic function. Thus, the space $\mathscr{D}'(S)$ can be identified with the class of harmonic[3] functions in the ball B such that

$$\sup_{z \in B} |u(z)| (1 - |z|)^N < +\infty$$

for some $N \in \mathbb{N}$.

We remark that a series $\sum_{p,q \geq 0} f_{pq}$, where $f_{pq} \in H(p, q)$, converges in $\mathscr{D}'(S)$ if and only if $\|f_{pq}\|_{L^2(S)} = 0((p + q + 1)^{-N})$, for some $N \in \mathbb{N}$ (instead of the L^2 norm here one can substitute any L^r-norm ($1 \leq r \leq +\infty$). The series $\sum_{p,q \geq 0} f_{pq}$ converges in $C^\infty(S)$ if and only if $\|f_{pq}\|_{L^r} = 0((p + q + 1)^{-N})$ for all $N \in \mathbb{N}$ ($1 \leq r \leq +\infty$).

$$\sigma(f) \overset{\text{def}}{=} \{(p, q) \in \mathbb{Z}_+^2 : K_{pq} f \not\equiv 0\} \quad (n \geq 2).$$

A distribution $f \in \mathscr{D}'(S)$ is said to be $\mathbb{C}$-*invariant* if $f = f^U$ for all scalar operators $U \in \mathscr{U}$. In other words f can be represented in the form $f = g \circ \pi$, where $g \in \mathscr{D}'(\mathbb{P}^{n-1})$. It is easy to see that a function $f \in \mathscr{D}'(S)$ is $\mathbb{C}$-invariant if and only if

$$\sigma(f) \subset \{(p, q) \in \mathbb{Z}_+^2 : p = q\}.$$

An analogous notion to T-invariance may also be introduced for measurable functions on the sphere S. A measurable function $f : S \to \mathbb{C}$ is said to be $\mathbb{C}$-*invariant* if $f = f^U$ almost everywhere on S for all scalar operators $U \in \mathscr{U}$, or (which is the same) if f can be represented in the form $f = g \circ \pi$, almost everywhere on S, where g is a measurable function on $\mathbb{P}^{n-1}$.

---

[3] In some situations it is preferable to identify the space $\mathscr{D}'(S)$ with the analogous class of M-harmonic functions. For the definition of M-harmonic functions cf. Chap. 2, Sect. 2.5.

3.4. The Tangential Cauchy-Riemann Equations. We shall say that a distribution $f \in \mathcal{D}'(S)$ ($n \geq 2$) satisfies the *tangential Cauchy-Riemann equations*, if $(z_j \overline{D}_k - z_k \overline{D}_j)(f) = 0$ for all k and j (the operator $z_j \overline{D}_k - z_k \overline{D}_j$ may be correctly defined in an obvious manner on $\mathcal{D}'(S)$).

3.4.1. Theorem. *Let $f \in \mathcal{D}'(S)$ ($n \geq 2$). Then $\sigma(f) \subset \{(p, q) \in \mathbb{Z}_+^2 : q = 0\}$ if and only if f satisfies the tangential Cauchy-Riemann equations.*

For further results in this direction, see Khenkin and Chirka (1975) and Rudin (1980).

3.5. Multiplicative Properties of the Space $H(p, q)$. Let $H(p, q)H(r, s)$ denote the set of all funtions $f \in C(S)$, which can be represented in the form $f = \sum\limits_{j=1}^{N} g_j h_j$, $g_j \in H(p, q)$, $h_j \in H(r, s)$. Clearly, $H(p, q)H(r, s)$ is a $\mathcal{U}$-invariant subspace of $L^2(S)$.

3.5.1. Theorem (see Theorem 12.4.4 in Rudin (1980)). *If $n \geq 3$, then*

$$H(p, q)H(r, s) = \sum_{j=0}^{\mu} H(p + r - j, q + s - j), \tag{14}$$

where $\mu = \min(p, s) + \min(q, r)$.

It is interesting to note that for $n = 2$ the situation is essentially more complicated (see Rudin (1980)): the left side of (14), in general, is a proper subset of the right side. Thus for $n = 2$ there are "more" $\mathcal{U}$-invariant subalgebras of $C(S)$ than for $n > 2$ (see Rudin (1980)).

3.6. Ryll-Wojtaszsczyk Polynomials

3.6.1. Theorem (Ryll-Wojtaszsczyk (1983)). *There exists a positive number $C(n)$ having the following property: for any $p \in Z_+$, there is a polynomial $f_p \in H(p, 0)$ such that $|f_p| \leq 1$ everywhere in B and $\|f_p\|_{L^2(S)} \geq C(n)$.*

This theorem has already found several applications[4] in the theory of functions holomorphic in the ball: functions with Carleman singularities (Wojtaszczyk (1982)), functions with a "large set of roots" (Alexander (1982)), and inner functions[5] (Aleksandrov (1984)). Moreover, we may find several appli-

---

[4] These applications are discussed in more detail at the end of this paper (cf. "Update on problems ...", 2).

[5] Recently, with the help of Theorem 3.6.1, the author has constructed a proper holomorphic mapping of the ball B into a polydisc of sufficiently high dimension (A.B. Aleksandrov, Proper holomorphic mappings from the ball into a polydisc. Dokl. Akad. Nauk SSSR, 1985. Analogous results were obtained in a somewhat different way by Löw, E.; The ball in $\mathbb{C}^n$ is a closed complex submanifold of a polydisc. Invent. Math. 83, 405–410 (1986).)

The ball here can be replaced by an arbitrary strictly pseudoconvex domain Ω with C^2-boundary. Löw also showed the existence of a proper holomorphic mapping f from Ω to a ball B of sufficiently high dimension. Moreover, in this case, one may additionally require the continuity of f up to the boundary (cf. Löw, E.; Embedding and proper holomorphic maps of strictly pseudoconvex domains into polydiscs and balls. Math. Z. 190, 401–410 (1985).)

cations in the work of Ryll-Wojtaszsczyk (1983) itself. Most likely, Theorem 3.6.1 may turn out to be very useful in other questions of function theory in the ball also. It is unknown to the author whether one can find such polynomials $f_{pq} \in H(p, q)$ such that $|f_{pq}| \leq 1$ everywhere in B and $\inf_{p,q \geq 0} \|f_{pq}\|_{L^2(S)} > 0$. We remark that in Ryll-Wojtaszsczyk (1983) the following inequality is essentially proven

$$\sup \left\{ \frac{\|f\|_{L^2(S)}}{\|f\|_{L^\infty(S)}} : f \in H(p, q) \setminus \{0\} \right\} > \frac{1}{2} \sqrt{\pi} \|K_{pq}\|_{L^1(S \times S)}^{-1}.$$

It is easy to see that for $p \geq q$

$$\|K_{pq}\|_{L^1(S \times S)} = 2(n - 1) D(p, q, n) \times \int_0^1 x^{(p-q)/2} (1 - x)^{n-2} |\mathscr{P}_q^{p-q,n-2}(x)| \, dx.$$

One easily verifies that $\sup_{p \geq 0} \|K_{pq}\|_{L^1(S \times S)} = C(n, q) < +\infty$ for all n and q. On the other hand well known estimates for weighted L^1-norms of Jacobi polynomials (Szegö (1959)) yield the following inequality:

$$C_1(n) p^{n(3/2)} \leq \|K_{pq}\|_{L^1(S \times S)} \leq C_2(n) p^{n(3/2)}$$

for all $n \geq 2$ and all $p \geq 1$. Consequently,

$$\sup_{p,q \geq 0} \|K_{pq}\|_{L^1(S \times S)} = +\infty$$

for all $n \geq 2$.

§4. Nonisotropic Quasimetrics on the Sphere S

4.1. Elementary Properties of Nonisotropic Quasimetrics. For many questions in the theory of functions in the ball, different tangential directions on the sphere S turn out to have differing roles. Among these directions, a particular role is played by the complex tangential directions, i.e. those directions which are defined by vectors from the complex tangent space $T_\zeta^C(S)$ ($\zeta \in S$).

We now introduce the nonisotropic quasimetric d on the sphere S, which in a quantitative way "pins down" this particularity of complex tangential directions. Set

$$d(\zeta, \xi) = |1 - \langle \zeta, \xi \rangle| (\zeta, \xi \in S).$$

The function $d^{1/2}$ is a metric on S, and hence

$$d(\zeta, \xi) \leq 2(d(\zeta, \eta) + d(\eta, \xi))$$

for all $\zeta, \xi, \eta \in S$.

For $n = 1$ this quasimetric d coincides with the usual Euclidean metric: $d(\zeta, \eta) = |\zeta - \eta|$. For $n \geq 2$ we have the inequality:

$$\tfrac{1}{2} |\zeta - \xi|^2 \leq d(\zeta, \xi) \leq |\zeta - \xi| (\zeta, \xi \in S).$$

The first inequality turns into an equality when $\langle \zeta, \xi \rangle \in \mathbb{R}$; and the second, when $\langle \zeta, \xi \rangle \in \mathbb{C}$.

Quite often (and we see this repeatedly) the quasimetric d turns out to be significantly more convenient than the usual Euclidean metric. For example, the quasimetric d is very practical in studying the Cauchy kernel C (see Chap. 2, 2). This stems from the fact that the set

$$\{\xi \in S : |C(\zeta, \xi)| > t\}$$

is a ball for the quasimetric d.

Let $Q(\zeta, r)$ denote the open d-ball with center at the point $\zeta \in S$ and radius r, i.e.

$$Q(\zeta, r) \stackrel{\text{def}}{=} \{\xi \in S : |1 - \langle \zeta, \xi \rangle| < r\}.$$

Set

$$B(\zeta, r) \stackrel{\text{def}}{=} \{\xi \in S : |\xi - \zeta| < r\}.$$

Simple calculations show that the projection of the ball $Q(\zeta, r)$ onto the tangent space $T_\zeta(S)$ is contained in the set

$$\tilde{Q}(\zeta, r) = \{(w, x) \in T_\zeta^C(S) \oplus i\zeta\mathbb{R} : |w| < 2\sqrt{r}, |x| < r\}$$

and contains the set $\tilde{Q}\left(\zeta, \dfrac{r}{4}\right)$. Thus, the ball $Q(\zeta, r)$, roughly speaking, represents a "curvilinear ellipsoid" obtained by dilating the usual Euclidean ball $B(\zeta, r)$ by a factor of approximately $\approx r^{-1/2}$ in the complex tangential direction. From the above it follows that

$$\sigma(Q(\xi, r)) \asymp r^n \quad (0 < r < 2).$$

The symbol $a \asymp b$ denotes "$a < c_1 b$ and $b < c_2 a$, for some positive c_1 and c_2".

We remark also that the quasimetric d admits a natural extension to a quasimetric d_1 on the ball $\bar{B}$:

$$d_1(\xi, \zeta) = \frac{||\xi|^2 - \langle \xi, \zeta \rangle|}{|\xi|} \quad (\xi, \zeta \in \bar{B}; |\xi| \geq |\zeta|)$$

The function $\sqrt{d_1}$ is a metric on the ball $\bar{B}$.

4.2. Hausdorff Measure and Dimension.

On each quasimetric space (X, ρ) we may introduce, in a natural manner, the α-dimensional *Hausdorff measure* h_α and *Hausdorff dimension*. We recall the respective definitions:

$$h_\alpha(A) \stackrel{\text{def}}{=} \lim_{\varepsilon \to 0_+} \inf \{\Sigma (\operatorname{diam} A_j)^\alpha : \operatorname{diam} A_j < \varepsilon, \cup A_j = A\} \quad (\alpha > 0),$$

where $\operatorname{diam} A = \sup\{\rho(x, y) : x, y \in A\}$. The *Hausdorff dimension* of a set $A \subset X$ is the number

$$\sup\{\alpha : h_\alpha(A) = +\infty\} = \inf\{\alpha : h_\alpha(A) = 0\}.$$

To each subset E of the sphere S, we may associate its usual α-dimensional Hausdorff measure, $h_\alpha(E) = h_\alpha(E, \mathbb{C}^n)$, considering E as a subset of $\mathbb{C}^n$. We shall denote the corresponding Hausdorff dimension by $\dim(E, \mathbb{C}^n)$. However, in the theory of functions in the ball, sometimes it is natural to consider the α-dimensional Hausdorff measure $h_\alpha(E, S)$ viewing E as a subset of the sphere S endowed with the quasimetric d.

Let $\dim E$ denote the topological dimension of a set E (Hurewicz and Wallman (1941)). It is well known (op. cit.) that

$$\dim E \leq \dim(E, \mathbb{C}^n) \text{ and } \dim E \leq 2 \dim(E, S).$$

The number 2 in the second inequality is explained by the fact that 2 is the smallest among all numbers α such that the function $d^{1/\alpha}$ is a metric. It is easy to verify the following inequalities:

$$h_\alpha(E, S) \leq C(\alpha, n) \begin{cases} h_\alpha(E), & 0 < \alpha \leq 1, \\ h_{2\alpha-1}(E), & \alpha \geq 1. \end{cases}$$

$$h_\alpha(E) \leq C(\alpha, n) \begin{cases} h_{\alpha/2}(E, S), & 0 < \alpha \leq 2n - 2, \\ h_{\alpha+1}(E, S), & \alpha \geq 2n - 2. \end{cases}$$

From these, one obtains corresponding inequalities for $\dim(E, S)$ and $\dim(E, \mathbb{C}^n)$, from which, in particular, it follows that

$$\dim(E, \mathbb{C}^n) = 0 \Leftrightarrow \dim(E, S) = 0,$$

$$\dim(E, \mathbb{C}^n) = 2n - 1 \Leftrightarrow \dim(E, S) = n.$$

4.2.1. Proposition. *Let M be a compact manifold of class C^1 and dimension m, $M \subset S$. Then*
a) *if M is integral, then*

$$0 < h_{m/2}(M, S) < +\infty;$$

b) *if M is not integral, then*

$$0 < h_{(m+1)/2}(M, S) < +\infty;$$

Chapter 2
Fundamental Integral Representations

§1. Fundamental Spaces of Functions Holomorphic in the Ball

We denote by $H(B)$ the space of all functions holomorphic in the ball. We introduce on $H(B)$ the topology of uniform convergence on compact subsets of

B. We define differential operators $\mathscr{R}_0$, $\mathscr{R} : H(B) \to H(B)$,

$$\mathscr{R}_0 f \overset{\text{def}}{=} \sum_{j=1}^{n} z_j D_j f,$$

$$\mathscr{R} f \overset{\text{def}}{=} f + \mathscr{R}_0 f.$$

It is easy to see that if

$$f(z) = \sum_{k \in z_+^n} a_k z^k \quad (z \in B)$$

is the Maclaurin series expansion of f, then

$$(\mathscr{R}_0 f)(z) = \sum_{k \in z_+^n} |k| a_k z^k \quad (z \in B).$$

$$(\mathscr{R} f)(z) = \sum_{k \in z_+^n} (|k| + 1) a_k z^k \quad (z \in B).$$

To every $\alpha \in \mathbb{R}$ we may associate the power of order α of $\mathscr{R}$,

$$(\mathscr{R}^\alpha f)(z) = \sum_{k \in z_+^n} (|k| + 1)^\alpha a_k z^k \quad (z \in B).$$

Everything which will be said below concerning the operator $\mathscr{R}$ and its powers $\mathscr{R}^\alpha$ will have a natural analog for the operator $\mathscr{R}_0$ and its powers. However, one should keep in mind that in order to investigate the operator $\mathscr{R}_0^\alpha$ (and in particular in order to define its fractional powers), it is natural to consider the space $H(B)$ factored by the constants, i.e. $H(B)/\mathbb{C}$.

We remark that

$$\mathscr{R}^\alpha(f_\zeta) = (\mathscr{R}^\alpha f)_\zeta \tag{1}$$

for all $\zeta \in \bar{B}$, where f_ζ is the slice function, i.e. $f_\zeta(\lambda) = f(\lambda\zeta)$ $(\lambda \in \mathbb{D})$. This formula allows us to reduce many multidimensional results concerning the operator $\mathscr{R}^\alpha$ to one-dimensional ones. For $\alpha > 0$, the operator $\mathscr{R}^\alpha$ will be called an *operator of fractional differentiation order* α, and for $\alpha < 0$, an *operator of fractional integration order* $(-\alpha)$.

1.1. Notation. We denote by $A^m(B)$ $(0 \leq m \leq +\infty)$ the set of all functions $f \in C^m(\bar{B})$, which are holomorphic in B. Set $A(B) = A^0(B)$. We remark that $A^m(B) = \mathscr{R}^{-m}(A(B))$ $(0 \leq m < +\infty)$. This can be derived from (16) in Chap. 3.

Let $\Lambda_A^{-\alpha}(B)$ $(\alpha > 0)$ denote the space of all functions $f \in H(B)$ such that

$$\|f\|_{\Lambda_A^{-\alpha}} \overset{\text{def}}{=} \sup_{z \in B} |f(z)| (1 - |z|)^\alpha < +\infty.$$

The closure, in this space, of all polynomials is denoted by $\lambda_A^{-\alpha}(B)$. In other words, $\lambda_A^{-\alpha}(B)$ is the set of all $f \in H(B)$ such that

$$f(z) = o((1 - |z|)^{-\alpha})(|z| \to 1).$$

From the corresponding one-dimensional result (see Duren (1970)), and from (1) we have:

$$\mathscr{R}^{\alpha - \beta}(\Lambda_A^{-\beta}(B)) = \Lambda_A^{-\alpha}(B), \tag{2}$$

$$\mathscr{R}^{\alpha - \beta}(\lambda_A^{-\beta}(B)) = \lambda_A^{-\alpha}(B), \tag{3}$$

where $\alpha, \beta \in (0, +\infty)$. Formulas (2) and (3) allow us to define the spaces $\Lambda^\alpha_A(B)$ and $\lambda^\alpha_A(B)$, for $\alpha \geq 0$, so that these formulae hold for all $\alpha, \beta \in \mathbb{R}$. It is well known (op. cit.) that for $n = 1$ and $\alpha \in (0, 1)$, the class $\Lambda^\alpha_A(B) \overset{\text{def}}{=} \Lambda^\alpha_A$ is the space of all holomorphic functions in the disc $\mathbb{D}$ which satisfy a Hölder condition of order α. The case $\alpha = 1$ corresponds to the Zygmund class. In the general case

$$\Lambda^\alpha_A = \Lambda^\alpha_A(\mathbb{D}) \quad (p < \alpha \leq p + 1)$$

is the space of all holomorphic functions f in the disc $\mathbb{D}$ such that $f^{(p)} \in \Lambda^{\alpha-p}_A$. In the multidimensional case, (1) shows that a function f holomorphic in B belongs to the class $\Lambda^\alpha_A(B)$ if and only if $\{f_\zeta\}_{\zeta \in S}$ is a bounded family in the space Λ^α_A. We shall speak more of the degree of smoothness of the functions in $\Lambda^\alpha_A(B)$ in Chapter 3. The space $\Lambda^0_A(B)$ is called the *Bloch space*. As in the one-dimensional case, functions from the class $\Lambda^0_A(B)$ may fail almost everywhere to have radial boundary values on S (see Ryll-Wojtaszczyk (1983)). It is easy to see that $\bigcap_{\alpha \in R} \Lambda^\alpha_A(B) = A^\infty(B)$. Set $\mathscr{D}'_A(B) \overset{\text{def}}{=} \bigcup_{\alpha \in R} \Lambda^\alpha_A(B)$. The space $\mathscr{D}'_A(B)$ is the set of all functions $f \in H(B)$ for which $\lim_{r \to 1-} f_r$, exists in the space of distributions $\mathscr{D}'(S)$. Here, and in the sequel, $f_r(z) = f(rz)$, $0 < r < 1$, $z \in B$. The space $\mathscr{D}'_A(B)$ can be identified in a natural manner with the space of all distributions in $\mathscr{D}'(S)$ which satisfy the tangential Cauchy-Riemann equations ($n \geq 2$).

1.2. The Nevanlinna and Smirnov Classes. The *Nevanlinna class* $N(B)$ is the set of all functions f holomorphic in the ball B such that

$$\sup_{0 < r < 1} \int_S \log^+ |f_r| \, d\sigma < +\infty.$$

It is easy to see that the class $N(B)$ is an algebra. If $f \in N(B)$, then for almost all $\zeta \in S$, the slice-funcion f_ζ belongs to classical Nevanlinna class $N \overset{\text{def}}{=} N(\mathbb{D})$. Thus each function $f \in N(B)$ has radial boundary values almost everywhere on the sphere S, and we shall denote this boundary function by the same letter f. If $f \not\equiv 0$ in B, then $\log |f| \in L^1(S)$. Thus the class $N(B)$ can be identified (mod sets of measure 0) with a subalgebra of the space $L^0(S)$ of all measurable functions on the sphere S.

The *Smirnov class* $N_*(B)$ is the space of all functions $f \in N(B)$ such that

$$\lim_{r \to 1-} \int_S \log^+ |f_r| \, d\sigma = \int_S \log^+ |f| \, d\sigma. \tag{4}$$

Formula (4) is equivalent to the following condition: the family $\{\log^+ |f_r|\}_{0 < r < 1}$ has uniformly absolutely continuous integrals. For functions $f \in N(B)$ we have $f \in N_*(B)$ if and only if $f_\zeta \in N_*$ for almost all $\zeta \in S$. Here, N_* denotes the classical Smirnov class $N_*(\mathbb{D})$. In the class $N_*(B)$, we may introduce the metric

$$\rho(f, g) = \int_S \log(1 + |f - g|) \, d\sigma. \tag{5}$$

The space $N_*(B)$ endowed with the metric ρ is a complete topological algebra continuously imbedded in the algebra $H(B)$.

1.3. The Hardy Classes. The *Hardy class* $H^p(B)$ is the space of all functions $f \in H(B)$ such that

$$\|f\|_{H^p}^p \overset{\text{def}}{=} \sup_{0 < r < 1} \int_S |f_r|^p \, d\sigma < +\infty \quad (0 < p < +\infty),$$

$$\|f\|_{H^\infty} \overset{\text{def}}{=} \sup_{z \in B} |f(z)|.$$

It easy to see that $H^p(B) \subset N_*(B)$ for all $p \in (0, +\infty)$. The spaces $H^p(B)$ are Banach spaces for $p \in [1, +\infty]$, and are p-Banach[6] spaces for $p \in (0, 1)$. The class $H^p(B)$ has a canonical imbedding as a subspace of $L^p(S)$ ($0 < p \le +\infty$);

$$H^p(B) = N_*(B) \cap L^p(S).$$

To each $p \in (0, +\infty)$ and $\alpha > 0$, we associate the space $\mathcal{H}_\alpha^p(B)$, consisting of all functions $f \in H(B)$ such that

$$\|f\|_{\mathcal{H}_\alpha^p}^p \overset{\text{def}}{=} \int_B |f(z)|^p \, d\mu_{\alpha p}(z) < +\infty,$$

where

$$d\mu_s(z) = \frac{\Gamma(n+s)}{n! \, \Gamma(s)} (1 - |z|^2)^{s-1} \, dv(z).$$

The multiplicative constant is so chosen that the measure μ_s is a probability measure.

The space $\mathcal{H}_\alpha^p(B)$ is a Banach space for $p \ge 1$ and a p-Banach space for $p \in (0, 1)$. Moreover, the spaces $H^2(B)$ and $\mathcal{H}_\alpha^2(B)$ are Hilbert spaces.

To each pair of natural numbers $n, m, n > m$, we associate two operators: ($\mathcal{F}_m^n : H(B_n) \to H(B_m)$ and $\mathcal{K}_m^n : H(B_m) \to H(B_n)$.

$$(\mathcal{F}_m^n F)(z_1, z_2; \ldots, z_m) = F(z_1, z_2, \ldots, z_m, 0, \ldots, 0),$$

$$(\mathcal{K}_m^n f)(z_1, z_2, \ldots, z_n) = f(z_1, z_2, \ldots, z_m).$$

1.3.1. Theorem. *Let m and n be natural numbers $m < n$, and let $\alpha, p \in (0, +\infty)$. Then*

$$\mathcal{F}_m^n H^p(B_n) = \mathcal{H}_{(n-m)/p}^p(B_m), \tag{6}$$

$$\mathcal{F}_m^n(\mathcal{H}_\alpha^p(B_n)) = \mathcal{H}_{\alpha+(n-m)/p}^p(B_m). \tag{7}$$

Moreover,

$$\|\mathcal{F}_m^n F\|_{\mathcal{H}_{(n-m)/p}^p} \le \|F\|_{H^p}, \quad \|\mathcal{F}_m^n F\|_{\mathcal{H}_{\alpha+(n-m)/p}^p} \le \|F\|_{\mathcal{H}_\alpha^p},$$

$$\|\mathcal{F}_m^n F\|_{H^p} = \|f\|_{\mathcal{H}_{(n-m)/p}^p}, \quad \|\mathcal{K}_m^n f\|_{\mathcal{H}_\alpha^p} = \|f\|_{\mathcal{H}_{\alpha+(n-m)/p}^p}.$$

This theorem follows easily from (1) in Chapter 1 and Fubini's Theorem.

---

[6] A p-Banach space differs from a Banach space in that instead of the triangle inequality, one requires the weaker inequality $\|x + y\|^p \le \|x\|^p + \|y\|^p$ ($0 < p < 1$).

The following assertion follows from Theorem 1.3.1 for $m = 1$ and from the corresponding one-dimensional result.

1.3.2. Theorem. *Let* $\alpha, p \in (0, +\infty)$. *Then*

$$H^p(B) \subset \lambda_A^{-n/p}(B),$$

$$\mathcal{H}_\alpha^p(B) \subset \lambda_A^{-(n/p)-\alpha}(B).$$

We introduce several more results concerning the spaces $H^p(B)$ and $\mathcal{H}_\alpha^p(B)$.

1.3.3. Theorem. *Let* $\alpha, p, q \in (0, +\infty)$, *where* $\dfrac{n}{p} - \dfrac{n}{q} = \alpha$. *Then*

$$\mathcal{R}^{-\alpha}(H^p(B)) \subset H^q(B),$$

$$H^p(B) \subset \mathcal{H}_\alpha^q(B).$$

It is not difficult to give a simple proof of this theorem (see Aleksandrov (1981) p. 25) using Theorem 1.3.2 and a theorem on the maximal function:

$$\sup_{0 < r < 1} |f_r| \in L^p(S), \tag{8}$$

when $f \in H^p(B)$. The inclusion (8) easily reduces to the corresponding one-dimensional result.

1.3.4. Theorem. *Let* $\alpha, p \in (0, +\infty)$. *Then*

$$\mathcal{R}^{-\alpha}(\mathcal{H}_\alpha^q(B)) \subset H^p(B) \quad (p \le 2),$$

$$\mathcal{R}^\alpha(H^p(B)) \subset \mathcal{H}_\alpha^p(B) \quad (p \ge 2).$$

This theorem reduces to the one-dimensional case with the help of (5) in Chapter 1.

§2. Fundamental Integral Representations

Various integral representations play a very important role in the theory of holomorphic functions (see Aizenberg-Yuzhakov (1979) and Khenkin-Leiterer (1984)). In this section we briefly pause to consider only a few of these.

2.1. The Cauchy Kernel. *The Cauchy (or Cauchy-Szegö) kernel is the* reproducing kernel $C(z, \zeta)$ for the space $H^2(B)$. This means that

$$f(z) = \langle f, C(\cdot, z) \rangle_{H^2}$$

for all $f \in H^2(B)$ and all $z \in B$, i.e.

$$f(z) = \int_S f(\zeta)\overline{C(\zeta, z)} \, d\sigma(\zeta)$$

or (on account of the symmetry property $C(z, \zeta) = \overline{C(\zeta, z)}$)

$$f(z) = \int_S C(z, \zeta) \, d\sigma(\zeta) \tag{9}$$

for all $f \in H^2(B)$ and all $z \in B$.

The following formula holds

$$C(z, \zeta) = (1 - \langle z, \zeta \rangle)^{-n}(z, \zeta \in \bar{B}).$$

Formula (9) automatically carries over to functions of Hardy class $H^1(B)$. Moreover, it is also valid for all $f \in \mathscr{D}'_A(B)$, provided the integral therein is interpreted in the sense of distributions. The operator

$$(Cf)(z) = \int_S C(z, \zeta) f(\zeta) \, d\sigma(\zeta) \qquad (10)$$

is the orthogonal projection from $L^2(S)$ onto $H^2(B)$ and is called the *Riesz projection*. We remark that the right member of (10) is meaningful for $f \in L^1(S)$, and even for $f \in \mathscr{D}'(S)$, provided the integral is understood in the sense of distributions. In the latter case (i.e. when $f \in \mathscr{D}'(S)$), we shall also use the notation Cf. In particular, if $\mu \in M(S)$, then

$$(C\mu)(z) = \int_S C(z, \zeta) \, d\mu(\zeta).$$

The operator C plays a very important role in the theory of functons in the ball and we shall speak of C again many times (in particular, in Chap. 3).

2.2. The Bergman Kernel. *The Bergman kernel* is the reproducing kernel $K(z, \zeta)$ for the space $\mathscr{H}^2_{1/2}(B)$. It has the following form:

$$K(z, \zeta) = (1 - \langle z, \zeta \rangle)^{-n-1} \quad (z, \zeta \in \bar{B}).$$

Consequently, the orthogonal projection T from $L^2(B)$ onto $\mathscr{H}^2_{1/2}(B)$ can be given in the following form:

$$(Tf)(z) = \int_B K(z, \zeta) f(\zeta) \, d\nu(\zeta).$$

In particular,

$$f(z) = \int_B K(z, \zeta) f(\zeta) \, d\nu(\zeta).$$

for all $z \in B$ and all $f \in \mathscr{H}^2_{1/2}(B)$, and therefore, for all $f \in \mathscr{H}^1_1(B)$.

The reproducing kernel for the space $\mathscr{H}^2_{\alpha/2}(B)$ has the following form:

$$K_\alpha(z, \zeta) = (1 - \langle z, \zeta \rangle)^{-n-\alpha} \quad (z, \zeta \in \bar{B}).$$

The corresponding orthogonal projections are studied in Forelli-Rudin (1974) and Rudin (1980). These authors also investigated integral operators with kernel K_α, for complex α.

2.3. The Invariant Poisson Kernel. *The invariant Poisson kernel* is the kernel

$$P(z, \zeta) \overset{\text{def}}{=} \frac{C(z, \zeta) C(\zeta, z)}{C(z, z)} = \left(\frac{1 - |z|^2}{|1 - \langle z, \zeta \rangle|^2} \right)^n \quad (z \in B, \zeta \in S).$$

Let $f \in H^1(B)$. Then $g = f \cdot \dfrac{C(\cdot, z)}{C(z, z)} \in H^1(B)$. Applying (9) to the functon g we obtain

$$f(z) = \int_S P(z, \zeta) f(\zeta)\, d\sigma(\zeta) \tag{14}$$

for all $z \in B$. Formula (14) holds also for all $f \in \overline{H^1(B)}$ since $P(z, \zeta) \geq 0$.

The measure $P(z, \cdot)\, d\sigma$ is invariant with respect to the group of all automorphisms of the ball which fix the point $z \in B$.

2.4. The "Harmonic" Poisson Kernel. In the ball B, there exists one more important kernel, the usual Poisson kernel $P_h(z, \zeta)$, which solves the Dirichlet problem

$$P_h(z, \zeta) = \frac{1 - |z|^2}{|\zeta - z|^{2n}}.$$

Since every function $f \in H^1(B)$ is harmonic in B, we have

$$f(z) = \int_S P_h(z, \zeta) f(\zeta)\, d\sigma(\zeta).$$

However, for $n \geq 2$, the kernel P_h does not reflect the "complexness" of the ball B (the analog of this kernel can be considered also in $\mathbb{R}^{2n+1}$) and has no specific connection with the Cauchy kernel C nor with the group of all automorphisms of the ball B. Since the equality $P_h = P$ holds only for $n = 1$, the invariant Poisson kernel does not solve the Dirichlet problem for $n \geq 2$.

2.5. Which Problem does the Invariant Poisson Integral Solve? By the *invariant Laplacian* $\widetilde{\Delta}$, we mean the Laplace-Beltrami operator (see de Rham (1955)) on the ball B, associated with the Bergman metric, i.e.

$$\widetilde{\Delta} \stackrel{\text{def}}{=} 4(1 - |z|^2) \sum_{j=1}^{n} \sum_{k=1}^{n} (\delta_{jk} - z_j \bar{z}_k) D_j \bar{D}_k,$$

where δ_{jk}. is the Kronecker symbol. This operator is invariant with respect to all automorphisms of the ball. From the obvious equality $(\widetilde{\Delta} f)(0) = (\Delta f)(0)$, it thus follows that

$$(\widetilde{\Delta} f)(a) = (\Delta(f \circ \varphi_a))(0) \tag{15}$$

for all $a \in B$. We introduce two more formulae for the invariant Laplacian: where $f_a : \dfrac{1}{|a|} \mathbb{D} \to \mathbb{C}$ is the slice function $f_a(t) = f(ta)$;

$$(\widetilde{\Delta} f)(a) = (1 - |a|^2)((\Delta f)(a) - (\Delta f_a)(1)),$$

$$(\widetilde{\Delta} f)(a) = \lim_{r \to 0+} \frac{4n}{r^2} \int_S (f(\varphi_a(r\zeta)) - f(a))\, d\sigma(\zeta).$$

A function $f \in C^2(B)$ is called *M-harmonic* if $\tilde{\Delta}f \equiv 0$. A function $f \in C(B)$ is M-harmonic if and only if

$$f(\psi(0)) = \int_S (f(\psi(r\zeta))\, d\sigma(\zeta)$$

for all $\psi \in \text{Aut}(B)$ and all $r \in (0, 1)$.

Many properties of M-harmonic functions are analogous to properties of harmonic functions. For example, the maximum principle holds for M-harmonic functions. Moreover, each M-harmonic function is real analytic.

The following theorem follows from the fact that the invariant Poisson kernel is an approximate identity.

2.5.1. Theorem (see Rudin (1980)). *Let u be a function M-harmonic in the ball* B.

1. *If* $1 < p \le +\infty$, *then*

$$u(z) = \int_S P(z, \zeta)f(\zeta)\, d\sigma(\zeta) \tag{16}$$

for some function $f \in L^p(S)$ *if and only if* $\sup\limits_{0<r<1} \|u_r\|_{L^p(s)} < +\infty$. *Moreover* $\lim\limits_{r \to 1-} u_r = f$ *in the space* $L^p(S)$ *(for* $p = +\infty$ *one should use the weak topology* $\sigma(L^\infty, L^1)$).

2. *Formula* (16) *holds for some function* $f \in L^1(S)$ *if and only if* $\lim\limits_{r \to 1-} u_r$ *exists in the space* $L^1(S)$, *and then* $f = \lim\limits_{r \to 1-} u_r$.

3. *Formula* (16) *holds for some function* $f \in C(S)$ *if and only if the function u is uniformly continuous in B, and* $f = \lim\limits_{r \to 1-} u_r$ *(in* $C(S)$).

4. *A function u can be represented in the form*

$$u(z) = \int_S P(z, \zeta)\, d\mu(\zeta), \tag{17}$$

where $\mu \in M(S)$ *if and only if* $\sup\limits_{0<r<1} \|u_r\|_{L^1} < +\infty$ *and in the weak topology* $\sigma(M(S), C(S))$ *on the space* $M(S)$ $\mu = \lim\limits_{r \to 1-} u_r$.

5. *Formula* (17) *holds for some positive measure* μ *if and only if* $u \ge 0$.

Every pluriharmonic function u is M-harmonic since u is annihilated by all operators $D_j\bar{D}_k$. However, the class of M-harmonic functions is far from being exhausted by the pluriharmonic functions.

We remark that the operator $\tilde{\Delta}$ "interacts" very poorly with dilatations. The following assertion speaks to this.

2.5.2. Proposition (Rudin (1980)). *If the function u is M-harmonic in the ball* B *and if* $r \in (0, 1)$, *then the M-harmonicity of the functions* u_r *is equivalent to the pluriharmonicity of the function* u.

We remark further that a function $u \in C(B)$ is *pluriharmonic if and only if u is* both M-harmonic and harmonic (see Rudin (1980)).

2.6. $H(p, q)$-Expansion of the Cauchy and Poisson Kernels. The following expansions hold:

$$\frac{1}{1 - \langle z, \zeta \rangle^n} = \sum_{p \geq 0} K_{p0}(z, \zeta) = \sum_{p \geq 0} \binom{p + n - 1}{p} \langle z, \zeta \rangle^p \quad (\langle z, \zeta \rangle \in \mathbb{D}),$$

$$\frac{1 - |z|^2}{|\zeta - z|^{2n}} = \sum_{p, q \geq 0} K_{pq}(z, \zeta) \quad (z \in B, \zeta \in S),$$

$$\left(\frac{1 - |z|^2}{|1 - \langle z, \zeta \rangle|^2} \right)^n = \sum_{p, q \geq 0} K_{pq}(z, \zeta) \cdot \frac{F(p, q, p + q + n; |z|^2)}{F(p, q, p + q + n; 1)}.$$

Here, $F(\alpha, \beta, \gamma; \chi)$ denotes the hypergeometric function (see Chap. 1, 3.2).

Chapter 3
Boundary Properties of the Cauchy Integral and the Invariant Poisson Integral

§ 1. The Maximal Function

1.1. Properties of the Maximal Function. To each measure $\mu \in M(S)$ we may associate the maximal function $M\mu : S \to [0, +\infty]$,

$$(M\mu)(\zeta) \stackrel{\text{def}}{=} \sup_{r > 0} \frac{|\mu|(Q(\zeta, r))}{\sigma(Q(\zeta, r))}.$$

For such a maximal function (associated with the nonisotropic quasimetric d), the usual properties of the "isotropic" maximal function hold. We present some of these.

1.1.1. Theorem. *The operator M is of weak type $(1, 1)$, i.e.*

$$\sigma\{M\mu > t\} \leq \frac{C(n) \|\mu\|}{t}$$

for all $t > 0$.

Moreover, it is clear that $M(L^\infty(S)) \subset L^\infty(S)$. Invoking, now, the Marcinkiewicz Interpolation Theorem (see Rudin (1980)), we obtain the following theorem.

1.1.2. Theorem. *For each $p \in (1, +\infty]$, there is a constant $A = A(p, n)$ such that*

$$\int_S |Mf|^p \, d\sigma \le A \int_S |f|^p \, d\sigma.$$

Each measure μ has a Lebesgue decomposition $\mu = f\sigma + \mu^s$. The function f can be found by the formula

$$f(\zeta) = \lim_{r \to 0+} \frac{\mu(Q(\zeta, r))}{\sigma(Q(\zeta, r))},$$

valid for almost all $\zeta \in S$. We remark also that the singular part μ^s of the measure μ is concentrated on the set $\{M\mu = +\infty\}$. Almost every point $\zeta \in S$ is a *Lebesgue point* for the summable function f, i.e.

$$\lim_{r \to 0+} \frac{1}{\sigma(Q(\zeta, r))} \int_{Q(\zeta, r)} |f - f(\zeta)| \, d\sigma = 0$$

for almost all $\zeta \in S$.

1.2. K-Limits. To each point $\zeta \in S$ and $\alpha > 1$, we associate a domain $D_\alpha(\zeta)$,

$$D_\alpha(\zeta) \overset{\text{def}}{=} \left\{ z \in B : |\langle z - \zeta, \zeta \rangle| < \frac{\alpha}{2}(1 - |z|^2) \right\}.$$

We remark that for each unitary operator $U \in \mathcal{U}(n)$, we have

$$D_\alpha(U\zeta) = UD_\alpha(\zeta).$$

Hence, in order to try to better imagine the "structure" of the domain $D_\alpha(\zeta)$, it is sufficient to restrict one's attention to the case $\zeta = e_1$. The intersection $D_\alpha(e_1) \cap \mathbb{C} \cdot e_1$ is the usual angular domain

$$\left\{ z \in \mathbb{D} : |1 - z| < \frac{\alpha}{2}(1 - |z|^2) \right\}.$$

However, the intersection of this domain with the $(2n - 1)$-dimensional space

$$\mathbb{R}_{2n-1} = \{ z \in \mathbb{C}^n : \operatorname{Im} z_1 = 0 \}$$

is the ball

$$\left\{ z \in \mathbb{R}_{2n-1} : \left(z_1 - \frac{1}{\alpha} \right)^2 + \sum_{j=2}^n |z_j|^2 < \left(1 - \frac{1}{\alpha} \right)^2 \right\},$$

which is tangent to the sphere S at the point e_1.

Thus, the boundary of the domain $D_\alpha(\zeta)$ is tangent to the sphere in the complex tangential direction and the tangency is of order 1 (i.e. as for a circle or a line).

To each function F, given on the ball B, and to each $\alpha > 1$, we may associate the *maximal function* $M_\alpha F : S \to [0, +\infty]$,

$$(M_\alpha F)(\zeta) \stackrel{\text{def}}{=} \sup\{|F(\zeta)| : z \in D_\alpha(\zeta)\}.$$

The function $M_\alpha F$ is measurable since it is lower semicontinuous.

1.2.1. Theorem (Korányi, see Rudin (1980)). *For each $\alpha > 1$ there is a constant $A = A(\alpha, n)$, such that*

$$M_\alpha(P\mu) \le AM\mu.$$

1.2.2. Definition. We shall say that a function $F : B \to \mathbb{C}$ has *K-limit* $\lambda \in \mathbb{C}$ at $\zeta \in S$ if

$$\lambda = \lim_\zeta F|_{D_\alpha(\zeta)}$$

for each $\alpha > 1$. In this case, we write $\lambda = K\text{-}\lim_\zeta F$.

From Theorem 1.2.1, the following result of Korányi (see Rudin (1980)) easily follows.

1.2.3. Theorem. *Let $\mu \in M(S)$ have Lebesgue decomposition; $\mu = f \cdot \sigma + \mu^s$. Then*

$$f(\zeta) = K\text{-}\lim_\zeta P\mu$$

for almost all $\zeta \in S$.

We present also the multidimensional analog of Plessner's Theorem (see Garnett (1981)).

1.2.4. Theorem (see Rudin (1980)). *Let $f \in H(B)$. Then the sphere S can be written as the union of three measurable sets E_k, E_C, and E_N, such that*
1) $\sigma(E_N) = 0$;
2) *the function f has a finite K-limit at each point of E_k;*
3) *the image $f(D_\alpha(\zeta))$ is everywhere dense in $\mathbb{C}$, for all $\zeta \in E_C$ and all $\alpha > 1$.*

1.2.5. Theorem (Korányi-Stein, see Rudin (1980)). *Every function f in Nevanlinna class $N(B)$ has finite K-limits almost everywhere on S.*

1.3. The Lindelöf-Chirka Theorem. We pause now briefly to consider a multidimensional generalization of a theorem of Lindelöf discovered by E.M. Chirka. A mapping $\Gamma : [0, 1) \to B$ will be called a ζ-curve ($\zeta \in S$), if $\lim_{t \to 1} \Gamma(t) = \zeta$. By a *special* ζ-curve we mean a ζ-curve Γ, such that

$$\lim_{t \to 1} \frac{|\Gamma(t) - \gamma(t)|^2}{1 - |\gamma(t)|^2} = 0,$$

where γ is the orthogonal projection of the curve Γ into the complex line $\mathbb{C}\zeta$, i.e.

$$\gamma = \langle \Gamma, \zeta \rangle \zeta.$$

A special ζ-curve will be called a (B, ζ)-*curve*, if there are numbers $\alpha > 1$ and $t_0 < 1$ such that $\Gamma(t) \in D_\alpha(\zeta)$ for all $t \in (t_0, 1)$. Finally, we shall say that $\lambda \in \mathbb{C}$ is the B-limit of a function f at a point $\zeta \in S$, if

$$\lambda = \lim_{t \to 1} f(\Gamma(t))$$

for each (B, ζ)-curve Γ. In this case, we shall write $\lambda = B\text{-}\lim_\zeta f$.

It is clear that the existence of a K-limit follows from that of a B-limit. The converse is not true. For example, the function

$$f(z_1, z_2) = \frac{z_2^2}{1 - z_1^2}$$

bounded and holomorphic in the ball B^2 has B-limit at the point $(1, 0)$ (equal to zero), but does not have a K-limit at $(1, 0)$ since

$$f(t, c\sqrt{1 - t^2}) = c^2.$$

The function f is also interesting in that its power series

$$f(z_1, z_2) = \sum_{k \geq 0} z_1^{2k} z_2^2$$

converges absolutely in the closed ball $\overline{B^2}$, although the function f in not continuous on the closed ball (In the one-dimensional case, this is impossible.).

1.3.1. Theorem (E.M. Chirka (1973)). *Let* $f \in H^\infty(B)$, $\zeta \in S$, *and suppose* Γ *is a special* ζ-curve *such that*

$$\lim_{t \to 1-} f(\Gamma(t)) = L.$$

Then $B\text{-}\lim_\zeta f = L$.

The example of the function $f(z_1, z_2) = \dfrac{z_2^2}{(1 - z_1^2)^{1+\varepsilon}}$ (not having a B-limit at the point $(1, 0)$) shows that the condition $f \in H^\infty(B)$ is essential.

Further results and examples relating to multidimensional generalizations of the Lindelöf Theorem can be found in the paper of Chirka (1973), the survey of Khenkin and Chirka (1975), and Rudin's book (1980).

1.4. Carleson Measures. Stein (1970, Chap. 7, 4.4) remarked that the angular maximal function is convenient not only for studying the boundary behavior of functions, but also for obtaining imbedding theorems in the spirit of Carleson (see Garnett (1981)). We shall need an analog of Stein's result in the ball; for this we require the following definition.

1.4.1. Definition. A positive Borel measure μ on the ball B is called a *Carleson measure* if

$$\mathscr{E}(\mu) \stackrel{\text{def}}{=} \sup\{r^n \mu(B(\zeta, r)) : \zeta \in S, r > 0\} < +\infty, \tag{1}$$

where

$$B(\zeta, r) = \{z \in B : |\langle z - \zeta, \zeta \rangle| < r\}.$$

S.A. Vinogradov (see Nikol'skij (1985)) suggested another convenient form of condition (1):

$$\sup_{z \in B} \int_B P(z, \zeta) \, d\mu(\zeta) = \sup_{z \in B} \frac{\displaystyle\int_B |C(z, \zeta)|^2 \, d\mu(\zeta)}{\displaystyle\int_S |C(z, \zeta)|^2 \, d\sigma(\zeta)} < +\infty.$$

1.4.2. Theorem. *Let f be a continuous function in the ball B and let μ be a Carleson measure on B. Then*

$$\int_B |f|^p \, d\mu \leq A(n, \alpha)\mathscr{E}(\mu) \int_S |M_\alpha f|^p \, d\sigma \quad (p > 0, \alpha > 2).$$

Theorem 1.4.2 was implicitly proved by Power (1985).

1.4.3. Definition. *A positive Borel measure μ on the ball B is called a Carleson measure of order t, if*

$$\mathscr{E}(\mu, t) \overset{\text{def}}{=} \sup\{r^{-n}(\mu(B(\zeta, r)))^{1/t} : \zeta \in S, r > 0\} < +\infty,$$

1.4.4. Theorem *Let f be a continuous function on the ball B; $0 < p \leq q < +\infty$; and μ a Carleson measure of order $\dfrac{q}{p}$. Then*

$$\left(\int_B |f|^q \, d\mu\right)^{p/q} \leq A(n, \alpha)\mathscr{E}\left(\mu, \frac{q}{p}\right) \int_S |M_\alpha f|^p \, d\sigma \quad (\alpha > 2).$$

In order to deduce Theorem 1.4.4 from Theorem 1.4.2, it is sufficient to invoke the following equality

$$\left(\int_B |f|^q \, d\mu\right)^{p/q} = \sup\left\{\int_B |f|^p |g| \, d\mu : g \in L^{q/(q-p)}(\mu), \int |g|^{q/(q-p)} \, d\mu \leq 1\right\}.$$

§2. The "Real" Hardy Class

2.1. The Carleson-Duren-Hörmander Theorem. Let $\mathscr{H}^p(S)$ $(0 < p < +\infty)$ denote the set of all (complex) functions u M-harmonic in the ball B and such that $M_\alpha u \in L^p(S)$. One can show (see Fefferman-Stein (1972)) that this definition is independent of $\alpha > 1$. Every function $u \in \mathscr{H}^p(S)$ is the invariant Poisson integral of a distribution $f \in \mathscr{D}'(S)$; we shall identify functions in the class $\mathscr{H}^p(S)$ with (generalized) functions in $\mathscr{D}'(S)$ ($u \mapsto f$), and we shall not introduce a special notation for such functions. It is easy to see that $\mathscr{H}^p(S) \subset L^p(S)$, for $p \geq 1$. In fact, $\mathscr{H}^p(S) = L^p(S)$, if $1 < p < +\infty$.

From Theorem 1.4.4 we have the following:

2.1.1. Theorem (Carleson-Duren-Hörmander, (see Hörmander (1967), Duren (1970))). *Suppose $0 < p \leq q < +\infty$ and μ is a Carleson measure of order $\dfrac{q}{p}$. Then*

$$\mathcal{H}^p(S) \subset L^q(\mu).$$

The space $\mathcal{H}^p(B)$ is a subspace of $\mathcal{H}^p(S)$. For $n \geq 2$, $\mathcal{H}^p(B)$ consists of all functions $f \in \mathcal{H}^p(S) \subset \mathcal{D}'(S)$ which satisfy the tangential Cauchy-Riemann equations.

Each function $f \in \mathcal{H}^p(S)$ $(0 < p < +\infty)$ has finite K-limits almost everywhere on S. However, for $p < 1$, a function $f \in \mathcal{H}^p(S)$ is no longer uniquely determined by these boundary values. Indeed, it easy to see that $M(S) \subset \mathcal{H}^p(S)$, for all $p < 1$. However, the invariant Poisson integral of a singular measure has zero boundary values almost everywhere on S.

2.2. Atomic Decomposition in Hardy Spaces

2.2.1. Theorem. *The Riesz projection C projects the space $\mathcal{H}^p(S)$ onto the space $H^p(B)$ $(0 < p < +\infty)$.*

For $p \in (1, +\infty)$ this result was obtained by Korányi and Vagi (1971). In this case, Theorem 2.2.1 is equivalent to the continuity in $L^p(S)$ of the singular operator determined by the Cauchy kernel $C(z, \zeta) = (1 - \langle z, \zeta \rangle)^{-n}$ $(z, \zeta \in S)$. As in the one-dimensional case, this operator is of weak type $(1, 1)$, i.e.

$$\sigma\{|C\mu| > t\} \leq \frac{A(n)}{t} \|\mu\|$$

for all $\mu \in M(S)$.

The case $p \in (0, 1]$ was considered by Garnett and Latter (1978). They used their own atomic decomposition of the space $\mathcal{H}^p(S)$, for $0 < p \leq 1$. We now pass to this decomposition.

2.2.2. Definition. A function $b \in L^\infty(S)$ is said to be *p-atomic* $(0 < p \leq 1)$ if one of the following two conditions is satisfied:

1) $\|b\|_{L^\infty} \leq 1$;
2) there exists a point $\zeta \in S$ and an $r > 0$ such that simultaneously
 i) supp $b \subset Q(\zeta, r)$;
 ii) $\|b\|_{L^\infty} \leq (\sigma(Q(\zeta, r)))^{-1/p}$;
 iii) $\displaystyle\int_S bP \, d\sigma = 0$ for all polynomials $P \in C[z_1, z_2, \ldots, z_n, \bar{z}_1, \bar{z}_2, \ldots, \bar{z}_n]$ of

degree no greater than $2n\left(\dfrac{1}{p} - 1\right)$.

In particular, if $p > \dfrac{2n}{2n+1}$, then condition iii) means that $\displaystyle\int_S b \, d\sigma = 0$.

From the following two assertions, it follows that $\|b\|_{\mathcal{H}^p(S)} \leq C(p, n)$ for any p-atom b $(0 < p \leq 1)$.

2.2.3. Proposition. *Let* $\zeta \in S$, $r > 0$, *and* $0 < p < 1$. *Suppose a measure* $\mu \in M(S)$ *satisfies the following properties:*

1) $\mathrm{supp}\, \mu \subset Q(\zeta, r)$;

2) $\displaystyle\int_S P\, d\mu = 0$ *for all polynomials* $P \in \mathbb{C}[z_1, z_2, z_n, \ldots, \bar{z}_1, \bar{z}_2, \ldots, \bar{z}_n]$ *of degree no greater than* $2n\left(\dfrac{1}{p} - 1\right)$. *Then* $\mu \in \mathscr{H}^p(S)$ *and* $\|\mu\|_{\mathscr{H}^p} \leq C(p, n)\|\mu\|_M \cdot r^{(n/p)-n}$.

2.2.4. Proposition. *Let* $\zeta \in S$, *and* $r > 0$. *Suppose that* $f \in L \log L$ *(i.e.* $f \in L^1(S)$ *and* $\displaystyle\int_S |f| \log(1 + |f|)\, d\sigma < +\infty$) *and satisfies the following conditions:*

1) $\mathrm{supp}\, f \subset Q(\zeta, r)$;

2) $\displaystyle\int_S f\, d\sigma = 0$. *Then* $f \in \mathscr{H}^1(S)$ *and* $\|f\|_{\mathscr{H}^1} \leq C(n)\|f\|_{L \log L}$.

Here, $\| \; \|_{L \log L}$. *denotes the norm in the Orlicz space* $L \log L$, *i.e.*

$$\|f\|_{L \log L} \overset{\mathrm{def}}{=} \inf\left\{A > 0 : \int \frac{|f|}{A} \log\left(1 + \frac{|f|}{A}\right) \leq 1\right\}.$$

It is not possible to weaken the condition $f \in L \log L$ in Proposition 2.2.4, for if $f \in \mathscr{H}^1(S)$ and $f \geq 0$, then $f \in L \log L$ (see Rudin (1980)).

From Propositions 2.2.3 and 2.2.4 it follows that

$$\sum_{k \geq 1} \alpha_k b_k \in \mathscr{H}^p(S) \tag{2}$$

for any sequence $\{b_k\}_{k \geq 1}$, of p-atoms provided that

$$f = \sum_{k \geq 1} |\alpha_k|^p < +\infty \quad (\alpha_k \in \mathbb{C}).$$

Garnett and Latter (1978) showed that the functions of type (2) exhaust the space $\mathscr{H}^p(S)$ $(0 < p \leq 1)$.

2.2.5. Theorem (op. cit.). *Let* $0 < p \leq 1$. *Any function* $f \in \mathscr{H}^p(S)$ *can be represented in the form*

$$f = \sum_{k \geq 1} \alpha_k b_k$$

(the series converges in $\mathscr{H}^p(S)$ *and hence in* $\mathscr{D}'(S)$, *where the* b_k *are* p-*atoms and* $\alpha_k \in \mathbb{C}$.

$$A(p, n)\|f\|_{\mathscr{H}^p} \leq \left(\sum_{k \geq 1} |\alpha_k|^p\right)^{1/p} \leq B(p, n)\|f\|_{\mathscr{H}^p}. \tag{3}$$

2.2.6. Remark. Suppose $N \in \mathbb{N}\left(N > 2n\left(\dfrac{1}{p} - 1\right)\right)$. We may also require that all the atoms b_k are orthogonal to all polynomials of degree at most N (of course, in this case the constants A and B in (3) will depend also on N).

We remark that, for $p \leq \dfrac{2n-1}{2n}$, Garnett and Latter considered somewhat different atoms, requiring in iii) that $P \in \mathbb{C}[\xi_1, \xi_2, \ldots, \xi_{2n-1}]$, where $\xi_1, \xi_2, \ldots,$ ξ_{2n-1} are specially chosen local coordinates in $Q(\zeta, r) \subset S$. Their approach allows one also to diminish the degree of the polynomials P with respect to the coordinates which are "orthogonal" to the complex tangent space (at the point ζ). However, this approach has its own drawback: the orthogonality condition depends on the ball $Q(\zeta, r)$ (more precisely, on its center). For $n = 1$, Theorem 2.2.5 was obtained by Coifman (1974).

Theorem 2.2.1 for $p \in (0, 1]$ follows easily from Theorem 2.2.5.

2.2.7. Definition. A function f, which can be represented in the form $f = Cb$, where b is a p-atom, is called a *holomorphic p-atom*.

The following theorem on holomorphic p-atoms follows from Theorems 2.2.1 and 2.2.5.

2.2.8. Theorem (Garnett-Latter (1978)). *Each function $f \in H^p(B)$ $(0 < p \leq 1)$ can be represented in the form $f = \sum\limits_{k \geq 1} \alpha_k b_k$, where the b_k are holomorphic p-atoms and $\alpha_k \in \mathbb{C}$; moreover*

$$A(p, n) \|f\|_{H^p} \leq \left(\sum_{k \geq 1} |\alpha_k|^p \right)^{1/p} \leq B(p, n) \|f\|_{H^p}.$$

For $p = 1$, this theorem was proved by Coifman, Rochberg, and Weiss (1976) by-passing Theorems 2.2.1 and 2.2.5.

Theorems on atomic decompositions are very strong tools in the theory of Hardy spaces (see Coifman-Weiss (1977)).

The following theorem is deduced in Coifman-Rochberg-Weiss (1976) and Garnett-Latter (1978) from Theorem 2.2.8.

2.2.9. Theorem. *Let $f \in H^p(B)$ $(0 < p \leq 1)$. Then there are two sequences of functions $\{F_k\}_{k \geq 1}$ and $\{G_k\}_{k \geq 1}$ from $H^{2p}(B)$ such that*
1)

$$\sum_{k \geq 1} \|F_k\|_{H^{2p}}^p \|G_k\|_{H^{2p}}^p \leq C(p, n) \|f\|_{H^p}^p; \tag{4}$$

2) $f = \sum\limits_{k \geq 1} F_k G_k$ (the series converges in $H^p(B)$ on account of (4)).

In the one-dimensional case, thanks to the availability of the inner-outer factorization (see Hoifman (1962)), a stronger assertion holds:

$$H^p = H^p \cdot H^r \quad \left(\frac{1}{p} = \frac{1}{q} + \frac{1}{r} \right).$$

Horowitz (1977) proved an analogous assertion for the one-dimensional spaces $\mathcal{H}_\alpha^p$:

$$\mathcal{H}_{\alpha/p}^p = \mathcal{H}_{\alpha/q}^q \cdot \mathcal{H}_{\alpha/r}^r.$$

Gowda (1983) showed that in the multidimensional case the set $H^q(B) \cdot H^r(B)$ (respectively the set $\mathscr{H}^q_{\alpha/q}(B) \cdot \mathscr{H}^r_{\alpha/r}(B)$ is a set of first category in $H^p(B)$ (respectively in $\mathscr{H}^p_{\alpha/p}(B)$).

At the present time, it is unknown whether in Theorem 2.2.9 one may require that the sequences $\{F_k\}_{k\geq 1}$ and $\{G_k\}_{k\geq 1}$ be finite.

Theorems on atomic decompositions (with different atoms) were obtained by Coifman and Rochberg (1980) for the spaces $\mathscr{H}^p_\alpha(B)$ $(0 < p, \alpha < +\infty)$. These results yield the following theorem of Lindenstrauss and Pelcziński (1971), even for $p \leq 1$.

2.2.10. Theorem. *Let* $\alpha, p \in (0, +\infty)$. *Then the space* $\mathscr{H}^p_\alpha(B)$ *is isomorphic to* l^p.

Recall that

$$l^p \overset{\text{def}}{=} \left\{ \{x_j\}_{j\geq 1} : \sum_{j\geq 1} |x_j|^p < +\infty \right\}.$$

We remark that the space $H^p(B)$ is isomorphic to l^p only for $p = 2$.

For further results concerning isomorphisms and non-isomorphisms of spaces of holomorphic functions, see Chap. 5, 1.3.

§3. Dual Spaces for Hardy Spaces $\mathscr{H}^p(S)$ and Spaces of Smooth Functions

3.1. Dual Spaces and Spaces of Multipliers

3.1.1. Definition. A topological vector space X will be called a space of generalized functions on the sphere S, if the following two conditions are satisfied:
1) $C^\infty(S) \subset X \subset \mathscr{D}'(S)$ (here, $\subset$ means continuously imbedded);
2) the set $C^\infty(S)$ is dense in X.

To each such space X, we may associate its dual space X' as well as the space of multipliers $\mathscr{M}X$. The space X' consists of all distributions $\varphi \in \mathscr{D}'(S)$ for which the functional

$$f \mapsto (f, \varphi),$$

defined at first on the space $C^\infty(S)$, can be extended to a continuous linear functional on the entire space X. The space $\mathscr{M}X$ is the set of all distributions $\varphi \in \mathscr{D}'(S)$ such that the operator

$$f \mapsto f \cdot \varphi,$$

at first defined on $C^\infty(S)$, can be extended as a continuous operator from X into X.

It is clear that $(\mathscr{H}^p(S))' = \mathscr{H}^q(S)$, for $1 < p < +\infty$, where q is the conjugate of p, i.e. $\dfrac{1}{p} + \dfrac{1}{q} = 1$. Moreover, $\mathscr{M}(\mathscr{H}^p(S)) = L^\infty(S)$, for $1 < p < +\infty$.

In order to describe the spaces $(\mathscr{H}^p(S))'$ and $\mathscr{M}(\mathscr{H}^p(S))$ for $p \leq 1$, we require several definitions.

We define, first of all, the degree of smoothness (with respect to the quasi-metric d) of functions in $\Lambda^\alpha(S)$.

3.1.2. Definition. A function $f \in C(S)$ is said to belong to the space $\Lambda^\alpha(S)$ if for each $\zeta \in S$ and each $r > 0$, there is a polynomial $P \in \mathbb{C}[z_1, z_2, \ldots, z_n, \bar{z}_1, \bar{z}_2, \ldots, \bar{z}_n]$ such that $\deg P \leq 2\alpha$ and

$$|f - P| \leq C_f \cdot r^\alpha \qquad (5)$$

everywhere on $Q(\zeta, r)$.

We set $\|f\|_{\Lambda^\alpha} \overset{\text{def}}{=} \|f\|_C + C_f^0$ where C_f^0, is the smallest quantity C_f, for which (5) holds for all $\zeta \in S$ and all $r > 0$.

3.1.3. Definition. We denote by $\text{BMO}(S)$ the space of all functions $f \in L^1(S)$ such that

$$\|f\|_{\text{BMO}} \overset{\text{def}}{=} \|f\|_{L^1} + \sup_{\zeta \in S, 0 < r} \inf_{c \in \mathbb{C}} \frac{1}{\sigma(Q(\zeta, r))} \int_{Q(\zeta,r)} |f - c|\, d\sigma < +\infty.$$

3.1.4. Definition. We denote by $\text{BMO}_{\log}(S)$ the space of all functions $f \in L^1(S)$ such that

$$\|f\|_{\text{BMO}} \overset{\text{def}}{=} \|f\|_{L^1} + \sup_{\zeta \in S, 0 < r < 2} \inf_{c \in \mathbb{C}} \frac{\log \dfrac{10}{r}}{\sigma(Q(\zeta, r))} \int_{Q(\zeta,r)} |f - c|\, d\sigma < +\infty.$$

None of the spaces $\Lambda^\alpha(S)$, $\text{BMO}(S)$, and $\text{BMO}_{\log}(S)$ are separable. The closure of $C^\infty(S)$ in these spaces are denoted respectively by $\lambda^\alpha(S)$, $\text{VMO}(S)$, and $\text{VMO}_{\log}(S)$. It is not difficult to verify that

$$\lambda^\alpha(S) = \left\{ f \in \Lambda^\alpha(S) : \lim_{\varepsilon \to 0+} C_f(\varepsilon) = 0 \right\},$$

where $C_f(\varepsilon)$ is the smallest quantity C_f, for which (5) holds for all $\zeta \in S$ and all $r \in (0, \varepsilon]$. Moreover (see Coifman-Rochberg-Weiss (1976))

$$\text{VMO}(S) = \left\{ f \in \text{BMO}(S) : \lim_{\varepsilon \to 0+} \sup_{\zeta \in S, 0 < r \leq \varepsilon} \inf_{c \in \mathbb{C}} \frac{1}{\sigma(Q(\zeta, r))} \int_{Q(\zeta,r)} |f - c|\, d\sigma = 0 \right\}.$$

We shall not require an analogous description of the space $\text{VMO}_{\log}(S)$.

3.1.5. Theorem (Garnett-Latter (1978)). *Let* $0 < p < 1$. *Then*

$$\mathscr{M}(\mathscr{H}^p(S)) = (\mathscr{H}^p(S))' = \Lambda^{(n/p)-n}(S).$$

3.1.6. Theorem (Coifman-Rochberg-Weiss (1976) and Janson (1976)).

$$(\mathscr{H}^1(S))' = \text{BMO}(S), \qquad (6)$$

$$\mathscr{M}\mathscr{H}^1(S) = L^\infty(S) \cap \text{BMO}_{\log}(S). \qquad (7)$$

3.1.7. Corollary.

$$\mathcal{M}(\text{BMO}(S)) = L^\infty(S) \cap \text{BMO}_{\log}(S).$$

Theorems 3.1.5 and 3.1.6 can be deduced from Theorem 2.2.5. Formula (7) in the one-dimensional case was obtained by Janson (1976).

From the inclusions $\mathcal{H}^1(S) \supset L^p(S) = \mathcal{H}^p(S)$ for all $p > 1$, we obtain the inclusions $\text{BMO}(S) \subset L^p(S)$ for all $p < +\infty$. Moreover, the following result of John and Nirenberg (see Garnett (1981)) holds.

3.1.8. Theorem. *There exist positive constants $A(n)$ and $B(n)$ such that*

$$\int_S e^{|f|} \, d\sigma \le B(n)$$

for all $f \in \text{BMO}(S)$ such that $\| f \|_{\text{BMO}} \le A(n)$.

3.2. The Cauchy Integral in Spaces of Smooth Functions.

Let us denote by $\{\tilde{\Lambda}^\alpha(S)\}_{\alpha > 0}$ the classes of smooth functions on S with respect to the usual Euclidean metric on S. The space $\tilde{\Lambda}^\alpha(S)$ is the set of all functions $f \in C(S)$ such that for each point $\zeta \in S$ and each $r > 0$, there is a polynomial $P \in \mathbb{C}[z_1, z_2, \ldots, z_n, \bar{z}_1, \bar{z}_2, \ldots, \bar{z}_n]$ such that $\deg P \le \alpha$ and

$$|f(z) - P(z)| \le C_f \cdot r^\alpha$$

for all $z \in S$ such that $|z - \zeta| \le r$. It is easily seen that $\tilde{\Lambda}^\alpha(S) \supset \Lambda^\alpha(S)$ for all $\alpha > 0$. Analogously, one can define the space $\widehat{\text{BMO}}(S)$ (the BMO-space associated with the Euclidean metric on S).

One can show that $\Lambda_A^\alpha(B) = \Lambda^\alpha(S) \cap A(B)$. Let us set $\text{BMO}_A(B) = H^1(B) \cap \text{BMO}(S)$, and $\text{VMO}_A(B) = H^1(B) \cap \text{VMO}(S)$.

The following theorem follows from Theorem 2.2.1 and duality.

3.2.1. Theorem. *The operator C projects the space $\Lambda^\alpha(S)$ onto $\Lambda_A^\alpha(B)$ $(\alpha > 0)$ and the space $\text{BMO}(S)$ onto $\text{BMO}_A(B)$.*

We remark also that $\text{BMO}_A(B) = C(L^\infty(S))$ and $\text{VMO}_A(B) = C(C(S))$.

3.2.2. Theorem (Ahern-Schneider (1980)). *Let $f \in L^1(S)$. Suppose that for almost every $\zeta \in S$, the slice function f_ζ belongs to $\Lambda^\alpha(\mathbb{T})$ $(\alpha > 0)$, and*

$$\operatorname*{ess\,sup}_{\zeta \in S} \| f_\zeta \|_{\Lambda^\alpha} < +\infty.$$

Then $Cf \in \Lambda_A^\alpha(B)$.

3.2.3. Corollary. *The opertor C projects $\tilde{\Lambda}^\alpha(S)$ onto $\Lambda_A^\alpha(B)$.*

From Theorem 3.2.2 (and no less from Corollary 3.2.3), the "doubling of smoothness" phenomenon, first detected by Stein (1973), appears.

3.2.4. Corollary. $A(B) \cap \tilde{\Lambda}^\alpha(S) = \Lambda_A^\alpha(B)$.

This corollary shows that a function of class $A(B) \cap \tilde{A}^\alpha(S)$ is "twice smoother" in the complex tangential direction. We remark also that an analogous phenomenon holds also for functions satisfying a Hölder condition with respect to the L^p-metric.

We now present several results of Krantz (1980) on the inter-relation between the spaces $\mathrm{BMO}(S)$ and $\widetilde{\mathrm{BMO}}(S)$.

An analog to Corollary 3.2.3 for $\widetilde{\mathrm{BMO}}(S)$ holds only partially:

$$C(\widetilde{\mathrm{BMO}}(S)) = \mathrm{BMO}_A(B), \tag{8}$$

however,

$$\mathrm{BMO}_A(B) \not\subset \widetilde{\mathrm{BMO}}(S).$$

Thus, the operator C does not act on the space $\widetilde{\mathrm{BMO}}(S)$. From (8) it follows that $\widetilde{\mathrm{BMO}}(S) \cap H^1(B) \subset \mathrm{BMO}_A(B)$. However, the last inclusion is strict, i.e. there is no analog of Corollary 3.2.4 for $\widetilde{\mathrm{BMO}}(S)$.

§4. Dual Spaces of Some Spaces of Holomorphic Functions

4.1. The Duren-Romberg-Shields Theorem. A topological vector space X will be called a space of holomorphic functions if

1) $A^\infty(B) \subset X \subset \mathscr{D}'_A(B)$ (continuous imbeddings);
2) $A^\infty(B)$ is dense in X.

We define the dual space as the set of all functions $\varphi \in \mathscr{D}'_A(B)$ such that the functional $f \mapsto (f, \varphi)$, defined at first on $A^\infty(B)$, extends as a continuous linear functional on all of X. Just as in §4 we may define the multiplier space $\mathscr{M}X$.

It is easily seen that $\mathscr{M}H^p(B) = H^\infty(B)$ for all $p > 0$. Moreover, $(H^p(B))^* = H^q(B)$, for $1 < p < +\infty$, $\frac{1}{p} + \frac{1}{q} = 1$.

4.1.1. Theorem. *If $p \in (0, 1)$, then*

$$(H^p(B))^* = \Lambda_A^{(n/p)-n}(B).$$

This theorem follows easily from the theorem on atomic decompositions in $H^p(B)$ (see Garnett-Latter (1978)). We remark that in the one-dimensional case, Theorem 4.1.1 was proved by Duren, Romberg, and Shields (1969) five years before the appearance of the theorem of Coifman (1974) on the atomic characterization of H^p. A proof of Theorem 4.1.1 analogous to the proof of the "one-dimensional" theorem of Duren-Romberg-Shields can be given without using the rather non-trivial Theorme 2.2.5.

4.1.2 Theorem. *If $0 < p \leq 1$, then*

$$(\mathscr{H}_\alpha^p(B))^* = \Lambda_A^{\alpha+(n/p)-n}(B), \tag{9}$$

$$(\lambda_A^\alpha(B))^* = \mathscr{H}_\alpha^1(B). \tag{10}$$

One essentially derives (9) for $p = 1$ and (10) in the process of proving Theorem 4.1.1. The case of arbitrary $p < 1$ reduces to Theorem 4.1.1 and the case $p = 1$ with the help of the inclusion

$$H^q(B) \subset \mathcal{H}_\alpha^p(B) \subset \mathcal{H}_{\alpha+(n/p)-n}^1(B),$$

where $\dfrac{n}{q} = \dfrac{n}{p} + \alpha$.

One can show that, for $p > 1$ and $\alpha > 0$, one has

$$(\mathcal{H}_\alpha^p(B))^* = \mathcal{R}^{-\alpha p}(\mathcal{H}_{\alpha p/q}^q(B));$$

where q is the conjugate exponent. The spaces $\mathcal{R}^{-\beta}(\mathcal{H}_\alpha^q(B))$ $(\beta > \alpha)$ are Besov spaces and have a more direct description. At this point, we shall not dwell further on these spaces.

4.2. The Dual Space of $H^1(B)$

4.2.1. Theorem (Coifman-Rochberg-Weiss (1976)).

$$(H^1(B))^* = \text{BMO}_A(B),$$

$$(\text{VMO}_A(B))^* = H^1(B).$$

4.2.2. Theorem. *Let f be a function holomorphic in B. Then the following assertions are equivalent:*

1) $f \in \text{BMO}_A(B)$;

2) *The measure* $\left| \sum\limits_{j=1}^n z_j D_j f \right|^2 (1 - |z|)\, dv(z)$ *is a Carleson measure*;

3)

$$\|f\|_G \overset{\text{def}}{=} \sup_{a \in B} \left\| \frac{f - f(a)}{(1 - \langle z, a \rangle)^n} \right\|_{H^2} (1 - |a|^2)^{n/2} < +\infty. \tag{11}$$

The equivalence of 1) and 2) was proved in Coifman-Rochberg-Weiss (1976). The functional $\|f\|_G$ is a seminorm. The equivalence of 1) and 3) in the one-dimensional case was proved by Garsia (see Garnett (1981)). For the multidimensional case, see Axler-Shapiro (1983). An important property of the seminorm of Garcia is its invariance with respect to automorphisms of the ball, i.e.

$$\|f \circ \varphi\|_G = \|f\|_G$$

for all $\varphi \in \text{Aut}(B)$. From this it follows that if $f \in \text{BMO}_A(B)$, then

$$\sup_{\varphi \in \text{Aut}(B)} \inf_{C \in \mathbb{C}} \|f \circ \varphi - c\|_{\text{BMO}} < +\infty.$$

The following theorem follows easily from the Hahn-Banach Theorem.

4.2.3. Theorem.

$$(C(L^1(S))^* = H^\infty(B),$$

$$(A(B))^* = C(M(S)).$$

§ 5. The Töplitz and Hankel Operators

5.1. The Töplitz and Hankel Operators on the Space $H^2(B)$. The theory of Hankel and Töplitz operators is very rich, already in the one-dimensional case (see, for example, Nikol'skij (1985)). We remark also that this theory is very useful in many questions in function theory, in operator theory, and in probability theory (see Nikol'skij (1985) and Khrushchev-Peller (1982)). It will not be possible for us to dwell on this theory and its applications in detail. We shall restrict ourselves merely to some definitions and isolated results.

To each distribution $\varphi \in \mathscr{D}'(S)$, we may associate the *Töplitz operator* T_φ and the two *Hankel operators* H_φ and V_φ (defined at first on the set $A^\infty(B)$):

$$T_\varphi f \overset{\text{def}}{=} C(\varphi f),$$

$$H_\varphi f \overset{\text{def}}{=} C_-(\varphi f),$$

$$V_\varphi f \overset{\text{def}}{=} \varphi f - C(\varphi f),$$

where

$$(C_- h)(z) = \int_S \frac{h(\zeta)}{(1 - \langle \zeta, z \rangle)^n} \, d\sigma(\zeta) - \int_S h(\zeta) \, d\sigma(\zeta) \quad (z \in B).$$

The operator C_- projects the space L^2 onto the space[7] $\{\bar{f} : f \in H^2(B) \ominus \mathbb{C}\}$. It is clear that $H_\varphi = V_\varphi$ in the one-dimensional case.

We remark also that

$$T_{\varphi_1} = T_{\varphi_2} \Leftrightarrow \varphi_1 = \varphi_2,$$

$$H_{\varphi_1} = H_{\varphi_2} \Leftrightarrow C_- \varphi_1 = C_- \varphi_2,$$

$$V_{\varphi_1} = V_{\varphi_2} \Leftrightarrow \varphi_1 - C\varphi_1 = \varphi_2 - C\varphi_2.$$

A nonzero operator T_φ cannot act compactly on any space $H^p(B)$ and a nonzero operator V_φ cannot have a finite-dimensional image ($n \geq 2$).

5.1.1. Theorem (Davie-Jewell (1977)). *The operator T_φ acts continuously on $H^2(B)$ if and only if $\varphi \in L^\infty(S)$; moreover, $\|T_\varphi\| = \|\varphi\|_{L^\infty}$.*

Let us denote by $\sigma_{L^\infty}(\varphi)$ the essential image of a function φ, i.e. the spectrum of φ, with respect to the algebra $L^\infty(S)$. We denote by $\text{conv}(K)$ the convex hull of a set K.

5.1.2. Theorem (Davie-Jewell (1977)). *Let $\varphi \in L^\infty(S)$. Then*

$$\sigma_{L^\infty}(\varphi) \subset \sigma(T_\varphi) \subset \text{conv}\, \sigma_{L^\infty}(\varphi),$$

where $\sigma(T_\varphi)$ is the spectrum of the operator T_φ.

---

[7] The symbol $L \ominus M$ denotes the orthogonal complement of a subspace M of a Hilbert space L.

We remark that in the one-dimensional case the spectrum of a Töplitz operator is connected (Theorem of Widom (1964)). In the multidimensional case, however, the spectrum of the operator T_φ need not be connected even for $\varphi \in C(S)$ (see Davie-Jewell (1977)). An example can be constructed (op. cit.) of the form $\varphi = h(|z_1|)$. In general, it should be remarked that if the function φ is $\mathbb{T}$-invariant, then $T_\varphi(H(n, 0)) \subset H(n, 0)$ $(n \geq 0)$, i.e. the space $H^2(B)$ can be decomposed as a sum of finite-dimensional T_φ-invariant subspaces.

Töplitz operators can be characterized by the following identity (Davie-Jewell (1977)):

$$\sum_{j=1}^{n} T_{\bar{z}_j} T_\varphi T_{z_j} = T_\varphi.$$

We set $H^p_-(B) = \{f \in L^p(S) : \bar{f} \in H^p(B), \bar{f}(0) = 0\}$.

5.1.3. Theorem (Coifman-Rochberg-Weiss (1976)). *The operator H_φ acts continuously from $H^2(B)$ to $H^2_-(B)$ if and only if $C_-\varphi \in \mathrm{BMO}(S)$ or (which is the same) there exists a function $\psi \in L^\infty(S)$ such that $C_-\varphi = C_-\psi$. The compactness of the operator H_φ is equivalent to the condition $C_-\varphi \in \mathrm{VMO}(S)$, i.e. $C_-\varphi = C_-\psi$ for some function $\psi \in C(S)$.*

We remark further that

$$\|H_\varphi\| \leq \inf\{\|\psi\|_{L^\infty} : C_-\psi = C_-\varphi\} \leq C(n)\|H_\varphi\|. \tag{12}$$

A theorem of Nehari (see Nikol'skij (1985)) asserts that in the one-dimensional case, we may take 1 for $C(1)$ and both inequalities in (12) become equalities. It is not difficult to show that this is not the case in the multidimensional situation:

$$C(n) \geq \frac{\|z_1\|_{H^2}}{\|z_1\|_{H^1}} > 1.$$

From Theorem 5.1.3 and the equality $H_\varphi = C_- \circ V_\varphi$ we obtain the following

5.1.4. Proposition. *If the operator V_φ acts continuously from $H^2(B)$ into $L^2(S) \ominus H^2(B)$, then $C_-\varphi \in \mathrm{BMO}(S)$.*

In addition the following proposition is obvious:

5.1.5. Proposition. *If $\varphi - C\varphi = \psi - C\psi$ for some function $\psi \in L^\infty(S)$, then the operator V_φ acts from $H^2(B)$ into $L^2(S) \ominus H^2(B)$.*

Necessary and sufficient conditions for the continuity of an operator V_φ are unknown.

5.2. The Töplitz and Hankel Operators on the Spaces $H^p(B)$ $(0 < p \leq +\infty)$. If $1 < p < +\infty$, then $T_\varphi(H^p(B)) \subset H^p(B)$ if and only if $\varphi \in L^\infty(S)$, moreover, if $p \neq 2$, equality of norms (as in Theorem 5.1.1) fails. The inclusion $\sigma_{L^\infty}(\varphi) \subset \sigma(T_\varphi)$ holds in any space $H^p(B)$ $(0 < p \leq +\infty)$ regardless of the Töplitz operator T_φ. However, the inclusion $\sigma(T_\varphi) \subset \mathrm{conv}\, \sigma_{L^\infty}(\varphi)$ for $p \neq 2$ is false already in the one-dimensional case (see Gokhberg-Krupnik (1973)). If $1 < p < +\infty$, then the inclusion $H_\varphi(H^p(B)) \subset H^p_-(B)$ is equivalent to the condition $C_-\varphi \in \mathrm{BMO}(S)$.

5.2.1. Theorem. *Let* $0 < p < 1$; $\varphi \in \Lambda^{(n/p)-n}(S)$. *Then*

$$T_\varphi(H^p(B)) \subset H^p(B), \quad V_\varphi(H^p(B)) \subset \mathscr{H}^p(S), \quad H_\varphi(H^p(B)) \subset H^p_-(B).$$

5.2.2. Theorem. *Let* $\varphi \in L^\infty(S) \cap (\mathrm{BMO}_{\log}(S)$. *Then*

$$T_\varphi(H^1(B)) \subset H^1(B), \quad V_\varphi(H^1(B)) \subset \mathscr{H}^1(S), \quad H_\varphi(H^1(B)) \subset H^1_-(B).$$

Theorems 5.2.1 and 5.2.2 follow easily from Theorems 3.1.5 and 3.1.6 respectively.

In all theorems of this section (on the continuity of the operators H_φ and V_φ), H_φ and V_φ turn out to be compact provided we assume that the function φ belongs to the closure of $C^\infty(S)$ in the corresponding space.

5.2.3. Theorem (Rudin (1980)). *If the modulus of continuity* ω_φ *of the function* φ *satisfies the Dini condition, i.e.*

$$\int_0^2 \frac{\omega_\varphi(t)}{t}\, dt < +\infty,$$

then

$$T_\varphi(H^\infty(B)) \subset H^\infty(B), \quad T_\varphi(A(B)) \subset A(B), \quad V_\varphi(H^\infty(B)) \subset C(S).$$

Moreover, the operator $V_\varphi : H^\infty(B) \to C(S)$ *is compact.*

5.3. Töplitz Operators and Multipliers. Let X be a space of holomorphic functions (see § 4). We define

$$\mathscr{T}(X) \overset{\mathrm{def}}{=} \{\varphi \in \mathscr{D}'(S) : T_\varphi \in \mathscr{L}(X)\},$$

where $\mathscr{L}(X)$ is the set of continuous operators on the space X. It is clear that $\mathscr{F}(X) \cap \mathscr{D}'_A(B) = \mathscr{M}X$.

We further define

$$\mathscr{M}_*X \overset{\mathrm{def}}{=} \{\varphi \in \mathscr{T}(X) : \overline{\varphi} \in \mathscr{D}'_A(B)\}.$$

It is not difficult to show that $\mathscr{M}X = \mathscr{M}_*X^*$ and $\mathscr{M}X^* = \mathscr{M}_*X$. As an application, we have the following theorem.

5.3.1. Theorem. *Let* $\varphi \in H^\infty(B)$. *Then*

$$T_{\overline{\varphi}}(\Lambda^\alpha_A(B)) \subset \Lambda^\alpha_A(B), \quad T_{\overline{\varphi}}(\lambda^\alpha_A(B)) \subset \lambda^\alpha_A(B) \quad (a > 0);$$

$$T_{\overline{\varphi}}(\mathrm{BMO}_A(B)) \subset \mathrm{BMO}_A(B), \quad T_{\overline{\varphi}}(\mathrm{VMO}_A(B)) \subset \mathrm{VMO}_A(B);$$

$$T_{\overline{\varphi}}(C(L^1(S))) \subset C(L^1(S)).$$

We remark also that $T_{\overline{\varphi}}(C(M(S))) \subset C(M(S))$, if $\varphi \in A(B)$.

5.4. Applications of Töplitz Operators to a Problem of Gleason. Let $f \in X \subset$ $H(B)$, $a \in B$. Are there functions $g_1, g_2, \ldots g_n \in X$ such that

$$f - f(a) = \sum_{j=1}^{n} (z_j - a_j)g_j?$$

This question was posed by Gleason for $X = A(B)$, $a = 0$. An affirmative answer was obtained by Lejbenzon (see Khenkin (1971)). At the present time there are several ways to solve this problem (see Rudin (1980)). We shall consider the method of Ahern and Schneider (see Rudin (1980)).

To each point $a \in B$ we associate an operator $\mathscr{R}_a : H(B) \to H(B)$,

$$\mathscr{R}_a f \overset{\text{def}}{=} \sum_{j=1}^{n} (z_j - a_j)D_j f.$$

Set

$$g_j(a, z) \overset{\text{def}}{=} \int_0^1 (D_j f)(a + t(z - a))\, dt.$$

It is easy to verify the following equality

$$f(z) - f(a) = \sum_{j=1}^{n} (z_j - a_j)g_j(a, z). \tag{13}$$

Set

$$(\mathscr{R}_a^{-1} f)(z) \overset{\text{def}}{=} \int_0^1 (f(a + t(z - a)) - f(a)) \frac{dt}{t}.$$

Then $g_j(a, \cdot) = D_j \mathscr{R}_a^{-1} f$, and (13) signifies that $\mathscr{R}_a \mathscr{R}_a^{-1} f = f - f(a)$. We remark that the power series expansion of the function $g_j(a, \cdot)$ in the vicinity of the point a is easily expressed via the corresponding expansion of the function f:

$$(z_j - a_j)g_j(a, z) = \sum_{\substack{\alpha \in z_+^n \\ \alpha \neq 0}} \frac{\alpha_j}{|\alpha|} \frac{(D^\alpha f)(a)}{\alpha!}(z - a)^\alpha, \tag{14}$$

for $|z - a| < 1 - |a|$.

We shall say that the Gleason problem is canonically solvable in the space X at the point $a \in B$ if $(D_j \mathscr{R}_a^{-1})X \subset X$ for all $j \in \{1, 2, \ldots, n\}$

A direct calculation shows that if $f \in \mathscr{D}_A'(B)$, then

$$g_j(a, z) = \int_S \frac{C(z, \zeta) - C(a, \zeta)}{\langle z - a, \zeta \rangle} f(\zeta) \bar{\zeta}_j \, d\sigma(\zeta)$$

(the integral is to be understood in the sense of distributions). The identity

$$\frac{C(z, \zeta) - C(a, \zeta)}{\langle z - a, \zeta \rangle} = C(z, \zeta) \sum_{j=0}^{n-1} \frac{(1 - \langle z, \zeta \rangle)^j}{(1 - \langle a, \zeta \rangle)^{j+1}} \tag{15}$$

shows that

$$g_k(a, z) = \sum_{|\alpha| \leq n-1} z^\alpha T_{\varphi_{k,\alpha}} f,$$

where $\varphi_{k,\alpha}$ are functions antiholomorphic in some neighborhood of the closed ball $\bar{B}$. Consequently, the following theorem holds.

5.4.1. Theorem. *Let X be a space of holomorphic functions. Suppose*
1) $z_j X \subset X$ *for all* $j \in \{1, 2, \dots, n\}$;
2) $T_{\bar{\varphi}} X \subset X$ *for all functions* φ *holomorphic in some neighbourhood of the closed ball* $\bar{B}$.
Then the Gleason problem is canonically solvable for the space X and for all $a \in B$.

The results of the previous section allow us to apply Theorem 5.4.1 to the following spaces X:

$$A(B), \quad H^p(B) \quad (p > 0), \quad \Lambda_A^\alpha(B) \quad (\alpha > 0), \quad \lambda_A^\alpha(B) \quad (\alpha > 0),$$

$$\mathrm{BMO}_A(B), \quad \mathrm{VMO}_A(B), \quad C(L^1(S)), \quad C(M(S)).$$

The canonical solvability of the Gleason problem is also known in the spaces $\mathscr{H}_\alpha^p(B)$ $(\alpha, p > 0)$ and $A^m(B)$ $(m \leq +\infty)$.

We remark that

$$g_j(0, z) = \int_0^1 (D_j f)(tz)\, dt \quad (1 \leq j \leq n).$$

Using precisely this equality, Lejbenzon obtained the first solution to Gleason's problem for the space $A(B)$ and $a = 0$.

If the space $X \subset H(B)$ allows multiplication by coordinate functions (i.e. $z_j X \subset X$ for all $j \in \{1, 2, \dots, n\}$), then the canonical solvability in X of the Gleason problem for $a = 0$ is clearly equivalent to the following property of the space X:

$$\mathscr{R}_0 f \in X \Leftrightarrow D_j f \in X \qquad \text{for all } j \in \{1, 2, \dots, n\}, \tag{16}$$

for $f \in H(B)$.

Making use of (6) from Chap. 1, it is easy to verify that

$$z_j(T_{\bar{z}_j} f)(z) = \sum_{\substack{\alpha \in z_+^n \\ \alpha \neq 0}} \frac{\alpha_j}{|\alpha| + n - 1} \frac{(D^\alpha f)(0)}{\alpha!} z^\alpha.$$

This equality together with (14) shows that canonical solvability of the Gleason problem for a space X at the point $a = 0$ is "almost equivalent" to the following assertion:

$$T_{\bar{z}_j} X \subset X \qquad \text{for all } j \in \{1, 2, \dots, n\}. \tag{17}$$

In particular from (15) for $a = 0$, it is easy to see that (17) implies the canonical solvability of the Gleason problem for $a = 0$ if the space X sustains multiplication by the coordinate functions.

Chapter 4
Zeros of Functions Holomorphic in the Ball

§ 1. Characterization of Zeros of Functions in the Smirnov, Nevanlinna, and Nevanlinna-Dzhrbashyan Classes

1.1. One-Dimensional Results. We recall, first of all, the well known characterizations of zeros of functions in the Nevanlinna, Smirnov, and Hardy spaces. Let $\{a_n\}_{n\geq 1}$ be a sequence in the unit disc satisfying the *Blaschke condition*

$$\sum_{n\geq 1} (1 - |a_n|) < +\infty. \tag{1}$$

Then the Blaschke product

$$b(z) = \prod_{n\geq 1} h(a_n, z), \tag{2}$$

where

$$h(a, z) = \begin{cases} \dfrac{a - z}{1 - \bar{a}z} \cdot \dfrac{\bar{a}}{|a|}, & a \neq 0, \\ z, & a = 0, \end{cases}$$

is a bounded holomorphic function in the unit disc $\mathbb{D}$, having $\{a_n\}_{n\geq 1}$ as its precise set of zeros (counting multiplicities).

Conversely, the zero set (counted according to multiplicities) of a non-zero function of Nevanlinna class satisfies the Blaschke condition (1). Thus, the Blaschke condition gives a complete characterization of zero sets of functions in all classes X such that $H^\infty \subset X \subset N$.

We turn now to zero sets of functions in the Nevanlinna-Dzhrbashyan class. Let us denote by N_α the set of all functions f holomorphic in the disc $\mathbb{D}$ such that

$$\int_{\mathbb{D}} (1 - |z|)^{\alpha - 1} \log^+ |f(z)| \, dv_1(z) < +\infty \quad (\alpha > 0).$$

A sequence $\{a_n\}_{n\geq 1}$ of points in the unit disc is the set of zeros (counting multiplicities) of a function in the class N_α if and only if

$$\sum_{n\geq 1} (1 - |a_n|)^{1+\alpha} < +\infty \tag{3}$$

(see, for example, Dautov-Khenkin (1978)).

Korenblum (1975) characterized zero sets of functions in the class $\mathscr{D}'_A(\mathbb{D})$. However, this characterization is rather complicated. We remark that the Blaschke condition is no longer necessary here, however, it is still necessary if all of the zeros lie on a line. Moreover, from what we have said above, it follows that condition (3) is necessary for all $\alpha > 0$.

1.2. The Khenkin-Skoda Theorem. Throughtout the rest of this chapter, τ will denote $(2n - 2)$-dimensional Hausdorff measure h_{2n-2} in $\mathbb{C}^n$. Let M be an analytic subset of the ball B of dimension $(n - 1)$ at each of its points (see Shabat (1976)). In other words, M is the zero set of some non-zero holomorphic function in B. Let us denote by $\overset{\circ}{M}$ the set of all regular points of the set M, i.e. the set of all points of M in the neighborhood of which M is a complex manifold (of dimension $n - 1$). The set $M \backslash \overset{\circ}{M}$ of all critical points of M is an analytic set of dimension at most $n - 2$. Consequently, $\tau(M \backslash \overset{\circ}{M}) = 0$. By a *divisor* on M we shall mean a locally constant function $k : \overset{\circ}{M} \to \mathbb{N}$.

To each function f holomorphic in the ball B and each point $a \in B$, we may associate the *degree* $k_f(a)$ of the zero of f at the point a (if $f(a) \neq 0$, then $k_f(a) \overset{\text{def}}{=} 0$) defined as follows:

$$k_f(a) \overset{\text{def}}{=} \inf\{|\alpha| : (D^\alpha f)(a) \neq 0\}$$

Thus, the function k_f maps B into $Z_+ \cup \{\infty\}$.

We shall say that a divisor $k : \overset{\circ}{M} \to \mathbb{N}$ is the divisor of the function $f \in H(B)$ if $f^{-1}(0) = M$ and $k_f|_{\overset{\circ}{M}} = k$.

Of course, each divisor $k : \overset{\circ}{M} \to \mathbb{N}$ has a natural extension $\tilde{k}$ to all of M such that $k_f|_{\overset{\circ}{M}} = k$ implies $k_f|_M = \tilde{k}$. However for our purpose this is not essential since $\tau(M \backslash \overset{\circ}{M}) = 0$.

In the sequel, we shall say for brevity that M is an analytic set of dimension $n - 1$, omitting the words "at each of its points".

1.2.1. Theorem (G.M. Khenkin (1977a, b), Skoda (1976)). *Let M be an $(n - 1)$-dimensional analytic subset of the ball B and k a divisor on M. Then the following assertions are equivalent*:
1) *k is the divisor of some function $f \in N_*(B)$*;
2) *k is the divisor of some function $f \in N(B)$*;
3) *the divisor k satisfies the following Blaschke condition*

$$\int_M k(z)(1 - |z|) \, d\tau(z) < +\infty.$$

The necessity of the Blaschke condition is proved analogously to the one dimensional case. The essential diffulty in the multi-dimensional theorem is the proof of sufficiency, i.e. the construction of a holomorphic function having the given divisor.

1.3. Discussion of the Blaschke Condition. Let k be a divisor on an $(n - 1)$-dimensional analytic set M in B. To each point $\zeta \in S$ we associate the quantity

$$\mathscr{B}(k, \zeta) = \sum_{\substack{z \in D, \\ z\zeta \in M}} k(z\zeta)(1 - |z|).$$

We consider also the function $V_k : [0, 1) \to \mathbb{R}$,

$$V_k(r) = \int_{M \cap rB} k(z) \, d\tau(z).$$

1.3.1. Proposition. *The Blaschke condition is equivalent to each of the following two conditions*:
1) $\int_S \mathcal{B}(k, \zeta) \, d\sigma(\zeta) < +\infty$;
2) $\int_0^1 V_k(t) \, dt < +\infty$.

We mention also two consequences of the Khenkin-Skoda Theorem.

1.3.2. Theorem. *A divisor of a function $f \in H(B)$ is the divisor of some function $g \in N_*(B)$ if and only if*

$$\sup_{0<r<1} \int_S \log |f_r| \, d\sigma < +\infty.$$

1.3.3. Theorem. *Let k_1, and k_2 be divisors on $(n-1)$-dimensional analytic subsets M_1 and M_2 of the ball B. Suppose that $M_1 \subset M_2$ and $k_1 \le k_2|_{\dot{M}_1}$. If the divisor k_2 is the divisor of some function in $N_*(B)$, then the same is true of the divisor k_1.*

We remark that the analog of Theorem 1.3.3 for $H^\infty(B)$ $(n \ge 2)$ fails (see 7.3.6 in Rudin (1980) and Amar (1982)).

1.4. The Khenkin-Dautov Theorem. Let us denote by $N_\alpha(B)$ the space of all holomorphic funcions f in the ball such that

$$\int_B (\log^+ |f|)(1 - |z|)^{\alpha-1} \, dv(z) < +\infty,$$

where $\alpha > 0$.

1.4.1. Theorem (Dautov-Khenkin (1978)). *Let M be an $(n-1)$-dimensional analytic subset of the ball B and $\alpha > 0$. Then a divisor k on M is the divisor of some function $f \in N_\alpha(B)$ if and only if*

$$\int_M k(z)(1 - |z|)^{\alpha+1} \, d\tau(z) < +\infty.$$

We remark that if α is a positive integer, this theorem reduces to Theorem 1.2.1; it is sufficient to notice that if $f \in N(B^{n+\alpha})$, then

$$f(z_1, z_2, \ldots, z_n, 0, \ldots, 0) \in N_\alpha(B^n).$$

1.4.2. Proposition. *Let k be a divisor on an $(n-1)$-dimensional analytic subset M of B. The following assertions are equivalent*
1) $\int_M k(z)(1 - |z|)^{\alpha+1} \, d\tau(z) < +\infty$;
2) $\int_0^1 V_k(t)(1 - t)^\alpha \, dt < +\infty$;
3) $\int_G \mathcal{B}_\alpha(k, \zeta) \, d\sigma(\zeta) < +\infty$,
where

$$\mathcal{B}_\alpha(k, \zeta) = \sum_{\substack{z \in D \\ z\zeta \in M}} (1 - |z|)^{\alpha+1} k(z\zeta).$$

Moreover, there are analogs to Theorems 1.3.2 and 1.3.3. We present here only the analog of Theorem 1.3.2.

1.4.3. Proposition. *A divisor of a function $f \in H(B)$ is the divisor of some function $g \in N_\alpha(B)$ if and only if*

$$\int_0^1 (1 - r)^{\alpha-1} \left(\int_S \log |f_r| \, d\sigma \right) dr < +\infty.$$

We remark also that from the easy part (i.e. the necessity) of Theorem 1.4.1, we have the following

1.4.4. Proposition. *If $k : \mathring{M} \to \mathbb{N}$ is the divisor of a function $f \in \mathscr{D}'_A(B)$, then*

$$\int_M (1 - |z|)^{\alpha+1} k(z) \, d\tau(z) < +\infty,$$

for all $\alpha > 0$. The converse is false, even in the one-dimensional case (see 1.1).

§2. Zeros of Functions in the Hardy Spaces $H^p(B)$

The problem of characterizing the zero sets of functions in the Hardy classes $H^p(B)$ $(0 < p < +\infty)$ appears at the present to be very difficult. In any case, in view of Theorem 1.3.1 of Chap. 2, such a characterization would follow from a (at present unknown) characterization of zeros of functions in the one-dimensional classes $\mathscr{H}_\alpha^p$ for an integer α. From this remark it follows that the zero sets of functions in the $H^p(B)$ classes are different for different p.

Also unknown is a characterization of the zeros of functions in the class $\bigcup_{p>0} H^p(B)$. We present here a sufficient condition obtained by Varopoulos (1980).

2.1. Uniform Blaschke Condition. Let k be a divisor on an $(n - 1)$-dimensional analytic subset M of the ball B. We shall say that k satisfies the *uniform Blaschke condition* if

$$\sup_{\varphi \in \text{Aut}(B)} \int_{\varphi^{-1}(M)} (k \circ \varphi)(z)(1 - |z|) \, d\tau(z)) < +\infty.$$

In the one-dimensional case the uniform Blaschke condition is equivalent to the discrete measure[8] $(1 - |z|)k(z) \, d\tau(z)$ being a Carleson measure, which in turn is equivalent to the corresponding Blaschke product

$$\prod_{a \in M \setminus \{0\}} \left(\frac{a - z}{1 - \bar{a}z} \cdot \frac{\bar{a}}{|a|} \right)^{k(a)}$$

---

[8] Here and hereafter it is convenient to consider that k has been extended as the zero function on $B \setminus M$.

being a finite product of interpolating Blaschke products (see Garnett (1981)). In
the multidimensional case the uniform Blaschke condition is equivalent to

$$k(z)\left((1 - |z|) + \left(1 - \frac{|\langle z, n(z)\rangle|^2}{|z|^2}\right)\right) d\tau(z), \tag{4}$$

being a Carleson measure, where $n(z)$ is the unit vector orthogonal to the mani-
fold $\overset{\circ}{M}$ at the point $z \in \overset{\circ}{M}$. The measure (4) first appeared in the work of
Malliavin (1974). From the results of Malliavin (1974) and the Khenkin-Skoda
Theorem, it follows that the finiteness of the measure (4) is equivalent to the
Blaschke condition, i.e. the finiteness of the measure

$$k(z)(1 - |z|) \, d\tau(z). \tag{5}$$

However, the condition that the measure (4) be a Carleson measure is, in general, not equivalent to the condition that the measure (5) be a Carleson measure
(see Varopoulos (1980)). By the same token in the multidimensional situation,
the uniform Blaschke condition may fail even if the measure (5) is a Carleson
measure.

2.1.1. Theorem (Varopoulos (1980)). *If a divisor k satisfies the uniform
Blaschke condition, then it is the divisor of some function in the family*

$$\{f \in N_*(B) : \log |f| \in \mathrm{BMO}(S)\} \subset \bigcup_{p > 0} H^p(B).$$

It is worth mentioning that for each $p > 0$ there exists an analytic set of
dimension $n - 1$ which is a set of uniqueness for $H^p(B)$ (i.e. $f \in H^p(B)$ and $f|_M \equiv
0 \Rightarrow f \equiv 0$) and which nevertheless satisfies the uniform Blaschke condition.[9]

2.2. Piecewise-Linear Analytic Sets. We now direct our attention in more
detail to the situation when M is a *piecewise-linear analytic set* of dimension
$n - 1$, i.e.

$$M = \bigcup_{j \in J} (l_j \cap B),$$

where the l_j are complex hyperplanes in $\mathbb{C}^n$ and diam $(l_j \cap B) \to 0$. We shall
assume that $0 \notin M$. Then for each hyperplane l_j, there is a unique point $a_j \in B$
such that

$$l_j = \{z \in \mathbb{C}^n : \langle z, a_j \rangle = |a_j|^2\}.$$

Here, as in the one-dimensional case, instead of considering a divisor on M, we
shall consider that the sequence $\{l_j\}_{j \in J}$ (or what amounts to the same, $\{a_j\}_{j \in J}$)
allows repetitions.

The Blaschke condition for M becomes:

$$\sum_{j \in J} (1 - |a_j|)^n < +\infty.$$

---

[9] More precisely, the divisor which is identically one on M satisfies the uniform Blaschke condition.

The Dautov-Khenkin condition for M becomes

$$\sum_{j \in J} (1 - |a_j|)^{n+\alpha} < +\infty.$$

The uniform Blaschke condition for M becomes the requirement that the measure[10]

$$\sum_{j \in J} 1_{(l_j \cap B)} \cdot \tau$$

be a Carleson measure.

2.2.1. Theorem (Varopoulos (1980)). *Let $\Lambda \subset B$. If*

$$\sum_{a \in \Lambda} (1 - |a_j|)^n \delta_a$$

is a Carleson measure[11], then there exists a piecewise-linear $(n - 1)$-dimensional analytic set containing Λ and satisfying the uniform Blaschke condition.

We remark that under the conditions of Theorem 2.2.1, the set

$$\bigcup_{a \in \Lambda \setminus \{0\}} \{z \in \mathbb{C}^n : \langle z, a \rangle = |a|^2\}$$

may not satisfy the uniform Blaschke condition. For example, such is the case if the set

$$\{a \in \Lambda : \langle \zeta, a \rangle = |a|^2\}$$

is infinite for some $\zeta \in S$ (see Varopoulos (1980)).

From Theorems 2.1.1 and 2.2.1 we have the following.

2.2.2. Corollary (Varopoulos (1980)). *Under the conditions of Theorem 2.1.1 there exists a function*

$$f \in \bigcup_{p > 0} H^p(B)$$

not identically zero such that $f^{-1}(0) \supset \Lambda$.

One can find other examples of applications of Theorem 2.2.1 in the works of Varopoulos (1980) and Amar (1982).

2.3. Zeros of Bounded Holomorphic Functions. If k is a divisor of a bounded holomorphic function not identically zero in the ball B, then, in addition to the Blaschke condition, the divisor k satisfies the following additional condition (absent in general for divisors of H^p-functions ($p < +\infty$)).

If l is a complex affine subspace of $\mathbb{C}^n$, then either $l \cap B \subset M$ or the divisor $k|_{l \cap M}$ satisfies the Blaschke condition (in the ball $l \cap B$).

---

[10] Here 1_E signifies the signifies the characteristic function of the set E.
[11] The symbol δ_a denotes the δ-measure at the point a.

However, as the following theorem shows, this additional necessary condition for divisors of H^∞-functions does not superimpose (as compared to the Blaschke condition on the ball B) any additional restrictions on the "massiveness" of the set of zeros of bounded holomorphic functions.

2.3.1. Theorem (Hakim-Sibony (1982a)). *Let* $h: (0, 1] \to (0, +\infty)$ *be a function infinitely large at zero. Then there exists a function* $f \in H^\infty(B)$ *whose divisor has the following property*

$$\int k(z) \cdot (1 - |z|) h(1 - |z|) \, d\tau(z) = +\infty.$$

We remark that we may additionally require that the zero set of the function f be piecewise-linear and that $f \in A(B)$. Amar (1982) and Alexander (1982) showed independently that for $h(t) = \dfrac{1}{t}$ we may also require that $f \in A^n(B)$ $(n < +\infty)$. In the case of a piecewise-linear analytic set

$$M = \bigcup_{j \in J} \{z \in \mathbb{C}^n : \langle z, a_j \rangle = |a_j|^2\}$$

the inequality

$$\sum_{j \in J} (1 - |a_j|) < +\infty. \tag{6}$$

is clearly a sufficient condition to be the divisor of an H^∞-function. As the required function $f \in H^\infty(B)$, we may take the analog

$$f(z) = \prod_{j \in J} \frac{\langle a_j - z, a_j \rangle}{1 - \langle z, a_j \rangle}$$

of the Blaschke product.

It is useful to remark that it is not possible to weaken condition (6) on the terms $|a_j|$ since it is a necessary condition when all the points lie on a complex line.

The following beautiful result is unfortunately so far proven only for $n \le 3$.

2.3.2. Theorem (Berndtsson (1980)). *Let* k *be a divisor on an analytic subset* M *of the ball* B. *Set*

$$V(r) = \int_{M \cap rB} k(z) \, d\tau(z) \quad (0 < r < 1).$$

Suppose the $(n - 2)$-*nd derivative of the function* V *is bounded. Then if* $n = 2$ *or* $n = 3$, *the function* k *is the divisor of some function in* $H^\infty(B)$.

We remark that the boundedness of the $(n - 2)$-nd derivative of the function V for a piecewise-linear set

$$M = \bigcup_{j \in J} \{z \in \mathbb{C}^n : \langle z, a_j \rangle = |a_j|^2\}$$

is equivalent to condition (6).

The methods of Berndtsson (1980) yield some sufficient (but not so elegant) conditions for divisors of H^∞-functions also for $n \geq 4$.

Chapter 5
Interpolation, Peak Sets, A-Measures and P-Measures

§ 1. Representing Measures and A-Measures

1.1. A-Measures, Representing Measures, Totally Singular Measures and Their Properties

1.1.1. Definition. A measure $\mu \in M(S)$ is called an A-measure (*analytic measure, L-measure, or Khenkin measure*), if $\lim\limits_{n \to \infty} \int_S f_n \, d\mu = 0$ for any bounded sequence $\{f_n\}_{n \geq 1}$ in $A(B)$ which tends to zero everywhere in B. We denote the set of all A-measures by $HM(S)$.

In the one-dimensional case (as is well known, see for example Hoffman (1962)), any closed subset $F \subset \mathbb{T}$ of zero Lebesgue measure is a peak set for some function $f \in A(D)$. This means that $f|_F \equiv 1$ and $|f| < 1$ everywhere on $\overline{\mathbb{D}} \backslash F$. If $\mu \in HM(S)$, then $\mu(F) = \lim\limits_{n \to \infty} \int f^n \, d\mu = 0$. Consequently, in the one-dimensional case, $HM(\mathbb{T}) \subset L^1(\mathbb{T})$. In fact, $HM(\mathbb{T}) = L^1(\mathbb{T})$, for the reverse inclusion $L^1(S) \subset HM(S)$ holds for all $n \geq 1$ (see Rudin (1980)).

1.1.2. Definition. A positive measure $\mu \in M(S)$ is called a *representing measure* for the point $z \in B$ if $f(z) = \int_S f \, d\mu$ for all $f \in A(B)$. We denote by M_z the set of all representing measures for the point z.

For $n = 1$ each point $z \in \mathbb{D}$ has a unique representing measure, the Poisson kernel $P(z, \cdot) \cdot m$. In the multidimensional case each point $z \in B$ has many representing measures, for example, the Poisson kernel $P_h(z, \cdot) \cdot \sigma$, and the invariant Poisson kernel $P(z, \cdot)\sigma$. There are also singular representing measures: if $\zeta \in S$ then

$$f(0) = \int_{\mathbb{T}} f(\xi\zeta) \, dm(\xi),$$

for all $f \in A(B)$.

Clearly, any representing measure is an A-measure.

1.1.3. Definition. A measure $\mu \in M(S)$ us said to be *totally singular* if it is singular with respect to each representing measure (or, equivalently, if it is singular with respect to each measure in M_0). We shall denote by $TM(S)$ the set of all totally singular measures in $M(S)$. A Borel set $E \subset S$ is said to be a *null set* if $m(E) = 0$ for all $\mu \in M_0$.

In the one-dimensional case, total singularity is equivalent to singularity and a null set is just a set of Lebesgue measure zero on $\mathbb{T}$.

From the general Theorem of Glicksberg, König, and Seever (see Theorem 9.4.4 in Rudin (1980)) we deduce the following.

1.1.4. Theorem. *Any measure $\mu \in M(S)$ can be uniquely represented in the form $\mu = \mu_a + \mu_s$, where the measure μ_a is absolutely continuous with respect to some measure in M_0 and the measure μ_s is concentrated on a null set of type F_σ.*

We denote by $A(B)^\perp$ the set of all measures $\mu \in M(S)$ such that $\int f \, d\mu = 0$ for all $f \in A(B)$.

1.1.5. Theorem (Val'skij (1971)).

$$HM(S) = A(B)^\perp + L^1(S).$$

In other words each measure $\mu \in HM(S)$ determines a functional on $A(B)$ which extends to a functional on H^∞ which is continuous with respect to the weak topology $\sigma(L^\infty, L^1)$ on $H^\infty(B)$.

1.1.6. Corollary. *Let $\mu \in M(S)$. If m is an A-measure, then*

$$\lim_{y \to +\infty} y\sigma\{\zeta \in S : |(C\mu)(\zeta)| > y\} = 0.$$

The converse is false for $n \geq 2$. In order to show this, we start with a singular measure μ on the circle $\mathbb{T}$ such that

$$\int_\mathbb{T} \frac{d\mu(\zeta)}{(1 - \zeta z)^2} = o\left(\frac{1}{1 - |z|}\right) \quad (|z| \to 1-).$$

It is known that such a measure μ exists (see Piranian (1966)). Now in order to obtain our required counterexample, we have only to "transfer" the measure μ to the sphere S with the help of the imbedding $z \mapsto (z, 0, \ldots, 0) \, (z \in \mathbb{T})$.

1.1.7. Theorem. *In order that a measure $\mu \in M(S)$ be an A-measure, it is necessary and sufficient that it be absolutely continuous with respect to some representing measure (or measure in M_0).*

This theorem was proved by Khenkin and then reproved by several authors (see Chirka-Khenkin (1975)).

1.1.8. Corollary. *If $\mu \in A(B)^\perp$, μ is absolutely continuous with respect to some measure in M_0.*

For $n = 1$ this corollary is well known from the F. and M. Riesz Theorem (Hoffman (1962)).

Theorem 1.1.7 allows us to reformulate Theorem 1.1.4 in the following form.

1.1.9. Theorem. $M(S) = HM(S) \oplus TM(S)$.

1.2. A-Measures and Boundary Behavior of Bounded Holomorphic Functions

1.2.1. Theorem (see 11.3 in Rudin (1980)). *Let $\mu \in M(S)$. Then the following assertions are equivalent*:

1) *for any function $f \in H^\infty(B)$, the limit $\lim_{r \to 1-} f_r$ exists in $L^\infty(|\mu|)$ in the $\sigma(L^\infty, L^1)$-topology*;

2) *for any function $f \in H^\infty(B)$, the limit $\lim_{r \to 1-} \int f_r \, d\mu$ exists*;

3) $\mu \in HM(S)$.

It is no known whether $\lim_{r \to 1-} f_r$ exists μ-almost everywhere for all $f \in H^\infty(B)$ and all $\mu \in HM(S)$. In other words, is the set of all $\zeta \in S$ such that $\lim_{r \to 1-} f(r\zeta)$, fails to exist necessarily a null set for all functions $f \in H^\infty(B)$?

1.2.2. Theorem (Rudin (1980), Nagel-Wainger (1981)). *Let $\varphi : [0, 1] \to S$ be an absolutely continuous curve. Suppose that $\mathrm{Im}\langle \varphi', \varphi \rangle > 0$ almost everywhere on $[0, 1]$. Then any function $f \in H^\infty(B)$ has a B-limit at $\varphi(t)$ for almost all $t \in [0, 1]$.*

In Nagel-Wainger (1981) there is also an analogous assertion proved under weaker assumptions on the curve φ.

In view of Theorems 1.2.1 and 1.2.2 we can construct examples of A-measures.

1.3. A-Measures and Isomorphism Classification of Banach Spaces of Analytic Functions.

The notion of A-measure was introduced by Khenkin (1968) in order to show the non-isomorphism of the spaces $A(B)$ and $A(\mathbb{D}^k)$ $(k \geq 2)$. Here, $A(\mathbb{D}^k)$ denotes the space of all holomorphic functions in the polydisc $\mathbb{D}^k$ which are continuous up to the boundary. Subsequently, a stronger result was obtained by Pelcziński (1977).

We now present several more results on isomorphism or non-isomorphism of Banach spaces of holomorphic functions.

The space $H^p(B)$ is isomorphic to the space $L^p(S)$ for $p \in (1, +\infty)$ (see Boas (1955), Aleksandrov (1982), and Wojtaszczyk (1983)). Analogous assertions are also true for Hardy spaces $H^p(\mathbb{D}^k)$ on the polydisc $\mathbb{D}^k$ $(1 < p < +\infty)$.

Wojtaszczyk (1983) showed that $H^1(B_n)$ is isomorphic to H^1 for all n. It is interesting that for the polydisc, the situation is completely different: Bourgain (1982, 1983a) showed that if $H^1(\mathbb{D}^k)$ is isomorphic to $H^1(\mathbb{D}^j)$, then $k = j$. Mityagin and Pelcziński (see Pelcziński (1977)) showed that $A(B_n)$ is not isomorphic to $A(D)$ for $n \geq 2$. Bourgain (1983b) showed that if $A(\mathbb{D}^k)$ is isomorphic to $A(\mathbb{D}^j)$ then $k = j$. He also proves substantially more general assertions.

So far as the (non Banach) spaces $H(B)$ and $H(\mathbb{D}^k)$ of all holomorphic functions on the ball and polydisc are concerned, we have the following assertions: $H(B_n)$ is isomorphic to $H(\mathbb{D}^n)$ for all n; $H(B_n)$ is not isomorphic to $H(B_m)$ if $n \neq m$ (see Khenkin-Mityagin (1971)).

§2. Null Sets and Interpolation on the Sphere S by Functions in the Class $A(B)$

2.1. Z-Sets, P-Sets, I-Sets, and Null Sets. From Theorem 1.1.7 it follows that a Borel set $E \subset S$ is a null set if and only if $\mu(E) = 0$ (or $|\mu|(E) = 0$) for every A-measure μ. From this and from Theorem 1.5 there follows another necessary and sufficient condition in order that E be a null set: $|\mu|(E) = 0$ for all $\mu \in A(B)^{\perp}$.

2.1.1. Theorem. *Let K be a compact subset of the sphere S. Then the following assertions are equivalent*:

1) *K is the set of zeros (Z-set) of some function in $A(B)$, i.e. there exists a function $f \in A(B)$ such that $f|_K \equiv 0$ and $f(z) \neq 0$ for all $z \in \bar{B} \backslash K$*;

2) *K is a peak set (P-set) for $A(B)$, i.e. there is a function $f \in A(B)$ such that $f|_K \equiv 1$ and $|f| < 1$ for all $\bar{B} \backslash K$*;

3) *K is an interpolation set (I-set), i.e. for each function $h \in C(K)$, there is a function $f \in A(B)$ such that $f|_K \equiv h$*;

4) *K is a null set*.

This theorem has a rich history already in the one-dimensional case. Various parts of the one dimensional case where obtained in turn by Fatou, the brothers F. and M. Riesz, Rudin, and Carleson. Abtract versions of certain parts of Theorem 2.1.1 were obtained by Bishop and Varopoulos.

In Rudin (1980) one can find the proof of Theorem 2.1.1 as well as more detailed historical remarks.

2.2. Examples and Properties of I-Sets

1. The real sphere defined as $S_R = S \cap \mathbb{R}^n$ is the Z-set of the function

$$\sum_{j=1}^{n} z_j^2 - 1.$$

2. The torus $S_T = \left\{ z \in S : \prod_{j=1}^{n} z_j = n^{-n/2} \right\}$ is the Z-set of the function $\prod_{j=1}^{n} z_j - n^{-n/2}$.

We remark that S_R and S_T are smooth manifolds of dimension $n - 1$. It is convenient to bear in mind that for $n = 2$ there exists a unitary transformation of $\mathbb{C}^n$ which transforms S_R to S_T.

3. Davie and Öksendal (1972) (see also Rudin (1980)) showed that if $h_1(K, S) = 0$ (see Chap. 1, 4), where K is a compact subset of the sphere, then K is an I-set. In the one-dimensional case the converse is also true.

4. (See Rudin (1980).) The trace of a C^1-curve $\gamma : [0, 1] \rightarrow S$ is an I-set if and only if $\langle \gamma'(t), \gamma(t) \rangle = 0$ for all $t \in [0, 1]$. The necessity of this condition follows from Theorem 1.2.2, and the sufficiency from 3.

5. (Tumanov-Khenkin-Burns-Stout-Rudin (see Khenkin-Leiterer (1984) and Rudin (1980))) Let M be a C^1-smooth manifold. Then every compact subset K of M is an I-set if and only if M is integral. Notice that the necessity follows from 4.

6. A countable union of I-sets is an I-set if it is compact.

7. Stout (1982) showed that the topological dimension of any I-set is at most $n - 1$.

8. Henrikson (1982) constructed an example of an I-set K such that $\dim(K, \mathbb{C}^n) = 2n - 1$ and consequently $\dim(K, S) = n$ (see Chap. 1, 4). The first non-trivial result in this direction had been obtained by Tumanov (1977).

2.3. Boundary Uniqueness Sets. A subset K of the sphere S is called a *uniqueness set* (for $A(B)$) if any function $f \in A(B)$ which vanishes on K vanishes identically.

Here are some examples of sets of uniqueness and sets of non-uniqueness.

1. Any Z-set is not a set of uniqueness.

2. The sphere $\{z \in S : z_n = 0\}$ of dimension $2n - 3$ is neither a Z-set nor a set of uniqueness ($n \geq 2$).

3. Suppose $\gamma : [0, 1] \to S$ is a C^1 integral curve, i.e. $\langle \gamma', \gamma \rangle \equiv 0$. Then, because of Example 4 of 2.2, $\gamma([0, 1])$ is a set of non-uniqueness.

4. Consider the curve

$$\gamma(t) = n^{-1/2}(e^{i\alpha_1 t}, e^{i\alpha_2 t}, \dots, e^{i\alpha_n t}) \quad (\alpha_j > 0).$$

This curve is not integral. The set $\gamma([0, 1])$ is a set of uniqueness if and only if $\{\alpha_1, \alpha_2, \dots, \alpha_n\}$ is a $\mathbb{Q}$-linearly independent system of numbers (see 10.6.2 in Rudin (1980)).

5. If $\sigma(K) > 0$ then K is a set of uniqueness. In the one-dimensional case, for compact sets the converse is also true.

6. (Pinchuk (1974)). Let M be a C^2 submanifold of S. If M is generic at some point, then M is a set of uniqueness. In particular, if $\dim M = 2n - 2$, then M is a set of uniqueness. In the survey by Chirka and Khenkin (1975), it is shown that an analogous assertion is true also for C^1 submanifolds.

Sadullaev (1976a) showed that if M is generic at each of its points and $\sigma_M(K) > 0$ for some Borel set $K \subset M$, then K is a set of uniqueness. (Here σ_M signifies Lebesgue measure on M.)

Jöricke (1982) obtained a more precise quantitative version of Pinchuk's (1974) result in the spirit of the Two Constants Theorem.

All of the assertions of 6 can be proved by using a modification of the method of "gluing analytic discs" which originated with Bishop (1965).

2.4. Maximum Modulus Sets. There is also a definite interest in studying maximum modulus sets (M-sets) for $A(B)$. A subset K of the sphere S is called a M-set if there exists a non-constant function $f \in A(B)$ such that

$$K = \left\{ \zeta \in S : |f(\zeta)| = \max_S |f| \right\}.$$

We list some examples and properties of M-sets.

1. Any Z-set is an M-set.

2. The sets $\mathbb{T} \cdot S_R$ and $\mathbb{T} \cdot S_\mathbb{T}$ are M-sets respectively for the functions $\sum\limits_{j=1}^{n} z_j^2$ and $\prod\limits_{j=1}^{n} z_j$.

3. (Duchamp-Stout (1981)). If $f \in A(B)$, $|f| \le 1$, $\{\zeta \in S : |f(\zeta)| = 1\} \supset \mathbb{T} \cdot S_\mathbb{T}$ then $f = b\left(n^{n/2} \prod\limits_{j=1}^{n} z_j \right)$ for some finite Blaschke product b. An analogous assertion is true also for the set $\mathbb{T} \cdot S_R$; $\left(\text{in this case, } f = b\left(\sum\limits_{j=1}^{n} z_j^2 \right) \right)$.

From this it follows that the union of two (even disjoint) M-sets need not be an M-set.

4. (Duchamp-Stout (1981)). The topological dimension of an M-set is at most n. Example 2 shows that this estimate is sharp.

5. (Aleksandrov (1983, 1984)). There exists an M-set of positive Lebesgue measure on S.

6. If K is the M-set of a function $f \in \Lambda_A^\alpha(B)$ and $\alpha > 1/2$, then $\sigma(K) = 0$ (11.4, Rudin (1980)). With stronger smoothness assumptions, this was shown by Tumanov and Sibony (see Chap. 3, Khenkin-Leiterer (1984)). Therein, assertion 6 is also announced for $\alpha = 1/2$.

2.5. Interpolation Within the Ball by Functions in the Classes $A(B)$ and $H^p(B)$

2.5.1. Theorem (Khenkin-Leiterer (1984)). *Let M be a closed complex submanifold of the ball rB ($r > 1$). Let $f : M \cap B \to \mathbb{C}$ be a function holomorphic in $M \cap B$ and bounded (respectively uniformly continuous). In this case we write $f \in H^\infty(M \cap B)$ (respectively $f \in A(M \cap B)$). Then there exists a function $f \in H^\infty(B)$ (respectively $F \in A(B)$) such that $F|_{M \cap B} \equiv f$. There exists a continuous linear operator $R : H^\infty(M \cap B) \to H^\infty(B)$ which generates the function F from the function f.*

If, moreover, the manifold intersects the sphere transversally, then we may require additionally that $R(A(M \cap B)) \subset A(B)$ (Khenkin (1972)).

The case of holomorphic functions smooth up to the boundary is considered by Jaćobczak (1983).

Analogs of Theorem 2.5.1 for the spaces $H^p(B)$ and $\mathscr{H}_\alpha^p(B)$ were obtained by Cumenge (1983) (in the case of transversal intersection of M with S) and Amar (1983b) (in the general case). The results of Cumenge and Amar can be viewed as far-reaching generalizations of Theorem 1.3.1 of Chap. 2.

§3. P-Measures

3.1. Integral Representations of P-Measures

3.1.1. Definition. We shall call a measures $\mu \in M(S)$ a *P-measure* if its (invariant) Poisson integral is a plurisubharmonic function in the ball B.

We remark (see Chap. 2, 2.5) that a measure $\mu \in M(S)$ is a *P-measure* if its Poisson integral is identically equal to its invariant Poisson integral. Moreover, a measure $\mu \in M(S)$ is a *P-measure* if and only if $\int_S f \, d\mu = 0$ for all real functions $f \in C(S)$ such that $f \cdot \sigma \in A(B)^\perp$. We denote the set of all P-measures by $PM(S)$.

To every point $\zeta \in \mathbb{P}^{n-1}$ there corresponds a circle $\mathbb{T}_\zeta = \{\xi \in S : \pi(\xi) = \zeta\}$ and a disc $\mathbb{D}_\zeta = \{\xi \in B : \pi(\xi) = \zeta\} \cup \{0\}$. It is easy to see that if u is the Poisson integral of a measure $\mu \in PM(S)$, then for almost all $\zeta \in \mathbb{P}^{n-1}$ the function $u|_{\mathbb{D}_\zeta}$ is the Poisson integral of a measure $\mu_\zeta \in M(\mathbb{T}_\zeta)$ and we have

$$\mu = \int_{\mathbb{P}^{n-1}} \mu_\zeta \, d\hat{\sigma}(\zeta) \tag{1}$$

in the following weak sense: the function $\zeta \mapsto \int_{\mathbb{T}_\zeta} f \, d\mu_\zeta$ is summable in $\mathbb{P}^{n-1}$ and

$$\int_S f \, d\mu = \int_{\mathbb{P}^{n-1}} \left(\int_{\mathbb{T}_\zeta} f \, d\mu_\zeta \right) d\hat{\sigma}(\zeta)$$

for all $f \in C(S)$. We remark that in the same weak sense, we have the following identities:

$$|\mu| = \int |\mu_\zeta| \, d\hat{\sigma}(\zeta), \tag{2}$$

$$\mu^a = \int \mu_\zeta^a \, d\hat{\sigma}(\zeta), \tag{3}$$

$$\mu^s = \int \mu_\zeta^s \, d\hat{\sigma}(\zeta). \tag{4}$$

Here, μ^a is the absolutely continuous (with respect to σ) part of the measure μ, μ^s is the singular part, μ_ζ^a is the absolutely continuous (with respect to Lebesgue measure on $\mathbb{T}_\zeta$) part of the measure μ_ζ, and μ_ζ^s is the singular part.

We remark that if the measure $\mu \in PM(S)$ is positive, then the measure μ_ζ is defined for all $\zeta \in \mathbb{P}^{n-1}$ and the mapping $\zeta \mapsto \mu_\zeta$ is continuous from $\mathbb{P}^{n-1}$ to $M(S)$ (we identify $M(\mathbb{T}_\zeta)$ with the set of all measures $\mu \in M(S)$ for which supp $\mu \subset \mathbb{T}_\zeta$) endowed with the weak-* topology $\sigma(M(S), C(S))$.

We introduce several results which follow easily from the integral representations (1)–(4).

3.1.2. Theorem. *Let E be a Borel subset of the sphere S and $h_{2n-2}(E) = 0$. Then $|\mu|(\mathbb{T} \cdot E) = 0$ for each measure $\mu \in PM(S)$.*

A somewhat weaker result was obtained by Forelli (1974). One can find other results of the same kind in the works of Forelli (1974, 1975).

3.2. The Khrushchev-Vinogradov Asymptotic Formula. Let $\mu \in PM(S)$. Then

$$\lim_{y \to +\infty} y\sigma\{\zeta \in S : |(C\mu)(\zeta)| > y\} = \frac{1}{\pi} \|\mu^s\|. \tag{5}$$

In the one-dimensional case this result was obtained by Khrushchev and Vinogradov (1981). The multidimensional case reduces to the one-dimensional case with the help of formulas (1) to (4).

Since in the multidimensional case there exist singular A-measures, Corollary 1.1.6 shows that for $n \geq 2$ formula (5) does not always hold for measures $\mu \in M(S)$. We remark further (see Rudin (1980)) that if μ is a δ-measure on the sphere S, then

$$\lim_{y \to +\infty} y\sigma\{\zeta \in S : |(C\mu)(\zeta)| > y\} = \frac{n!}{4\left(\Gamma\left(\dfrac{n}{2} + 1\right)\right)^2} > \frac{1}{\pi},$$

for $n \geq 2$.

One can show that

$$\lim_{y \to +\infty} \sup y\sigma\{\zeta \in S : |(C\mu)(\zeta)| > y\} \leq \frac{n!}{4\left(\Gamma\left(\dfrac{n}{2} + 1\right)\right)^2} \|\mu^s\|$$

for each measure $\mu \in M(S)$.

From (5) and Corollary 1.1.6 we have the following.

3.2.1. Proposition. $HM(S) \cap PM(S) \subset L^1(S)$.

We can amplify somewhat on the results presented earlier for positive P-measures.

3.2.2. Boole's Formula (see Khrushchev-Vinogradov (1981)). *If μ is a positive singular measure in $PM(S)$, then*

$$\sigma\{\zeta \in S : |2(C\mu)(\zeta) - \mu(S)| > y\} = \frac{2}{\pi} \text{arctg} \frac{\|\mu\|}{y}.$$

Several of the "one-dimensional" proofs of this formula set forth in Khrushchev-Vinogradov (1981) go through in the multidimensional case.

3.3. "Smoothness" and "Regularity" Properties of P-Measures. Forelli (1974) has shown that if e is a Borel subset of projective space $\mathbb{P}^{n-1}$, then

$$\mu(\pi^{-1}(e)) = \hat{\sigma}(e) \cdot \mu(S) \tag{6}$$

for any measure $\mu \in PM(S)$. From this it follows that

$$\mu(B(\zeta, r)) \leq C_n \cdot r^{2n-2}\mu(S), \tag{7}$$

for each positive measure $\mu \in PM(S)$, each $r \in (0, 2)$, and each $\zeta \in S$. Thus, we also have the following:

$$\mu(Q(\zeta, r)) \leq C_n \cdot r^{n-1}\mu(S), \tag{8}$$

for each positive measure $\mu \in PM(S)$, each $r \in (0, 2)$, and each $\zeta \in S$.

Along with formula (6), we mention also (see Forelli (1974)) one more formula for P-measures:

$$\mu(P^{-1}(E) \cap S) = \int_{P^{-1}(E) \cap S} u(P\zeta) \, d\sigma(\zeta), \tag{9}$$

where u is the (invariant) Poisson integral of the measure μ and P is the orthogonal projection of $\mathbb{C}^n$ onto $\mathbb{C}^k$ ($k \leq n - 1$).

From (7) and (8) we may deduce the following.

3.3.1. Proposition. *If M is a C^1-submanifold of S with* dim $M \leq 2n - 2$, *then* $\mu(M) = 0$ *for each positive measure $\mu \in PM(S)$.*

If dim $M \leq 2n - 3$, then in view of Theorem 3.1.2 a stronger assertion holds: $|\mu|(M) = 0$ for each $\mu \in PM(S)$. It is unknown to the author whether this is true when dim $M = 2n - 2$.

The following lower estimates hold:

$$\mu(B(\zeta, r)) \geq C_n \cdot r^{2n+1} \mu(S), \tag{10}$$

$$\mu(Q(\zeta, r)) \geq C_n r^{n+1} \mu(S), \tag{11}$$

for each positive measure $\mu \in PM(S)$, each $r \in (0, 2)$, and each $\zeta \in S$ ($n \geq 2$).

The estimates (7), (8), (10), and (11) are best possible. Indeed, if the Poisson integral of the measure μ is equal to $\operatorname{Re} \dfrac{1 + z_1}{1 - z_1}$ $\left(\text{i.e. } \mu = \left(\operatorname{Re} \dfrac{1 + z_1}{1 - z_1} \right) \sigma \text{ for } n \geq 2 \right)$, then

$$\mu(B(e_1, r)) \asymp r^{2n-2} \quad (r \to 0),$$

$$\mu(Q(e_1, r)) \asymp r^{n-1} \quad (r \to 0),$$

$$\mu(B(-e_1, r)) \asymp r^{2n+1} \quad (r \to 0) \quad (n \geq 2),$$

$$\mu(Q(-e_1, r)) \asymp r^{n+1} \quad (r \to 0) \quad (n \geq 2).$$

All of the estimates of this section easily reduce to a few estimates for harmonic functions of two real variables with the help of (9) and (1) of Chapter 1.

We remark that for a signed measure $\mu \in PM(S)$, the functions $r \mapsto |\mu|(Q(\zeta, r))$ and $r \mapsto |\mu|(B(\zeta, r))$ can tend to zero rather quickly but not arbitrarily quickly ($n \geq 2$).

3.3.2. Theorem (Jöricke (1971)). *If $\mu \in PM(S)$ is not the zero measure and $n \geq 2$, then*

$$\int_0 \log(|\mu|(B(\zeta, r))) \, dr > -\infty,$$

$$\int_0 \log(|\mu|(Q(\zeta, r))) \, dr > -\infty,$$

for all $\zeta \in S$.

Jöricke (1971) shows also that this result is in some sense best possible.

3.3.3. Corollary (Forelli (1974)). *If $\mu \in P(S)$ is not the zero measure and $n \geq 2$, then* supp $\mu = S$.

We remark further that, for $n \geq 2$, any positive measure μ in $PM(S)$ has the following homogeniety property (with respect to the quasimetric d):

$$\mu(Q(\zeta, 2r)) \leq C_n \mu(Q(\zeta, r)) \quad (n \geq 2),$$

for all $\zeta \in S$ and all $r > 0$. In other words, for $n \geq 2$, the space (S, d, μ) is a space of homogeneous type (see Coifman-Weiss (1977)). For the usual Euclidean metric on the sphere S, the situation is in general different: if the Poisson integral of the measure μ is Re $\dfrac{1 + z_1}{1 - z_1}$, then

$$\sup_{\zeta \in S, r > 0} \frac{\mu(B(\zeta, 2r))}{\mu(B(\zeta, r))} = +\infty.$$

Let u be the Poisson integral of a positive measure $\mu \in PM(S)$. Consider the set

$$X_\mu \overset{\text{def}}{=} \left\{ \zeta \in S : \lim_{r \to 1-} u(r\zeta) = +\infty \right\}.$$

It is well known (see 5.2.7, 5.4.11, and 5.4.12 in Rudin (1980)) that the singular part μ^s of the measure μ is concentrated on the set X_μ. It is easy to see that $\rho(X_\mu) = 0$ for each $\rho \in M_0$. Consequently the measure μ^s is singular with respect to each measure in M_0. It is unknown to the author whether the analogous assertion is true for all measures in $PM(S)$.

§4. *P*-Measures and the Boundary Behavior of Holomorphic Functions

4.1. *P*-Measures and the Hardy-Lumer Class. Let us denote by $LH^p(B)$ the set of all functions f holomorphic in the ball B for which there exists a pluriharmonic function u such that $|f|^p \leq u$ everywhere in B. We set

$$\|f\|_p \overset{\text{def}}{=} \inf u(0)^{1/p},$$

where the infimum is taken over all positive pluriharmonic functions u. The classes $LH^p(B)$ were introduced by Lumer (1971). For $n = 1$, $LH^p(B) = H^p(B)$. The spaces $LH^p(B)$ are Banach spaces for $p \geq 1$ and Polish spaces for $p < 1$. It is not difficult to see (see 7.4.2 in Rudin (1980)) that the zeros of functions in the class $LH^p(B)$ have the same structure as the zeros of bounded holomorphic functions in the ball. However, linear-topological properties of the class $LH^p(B)$, for $n \geq 2$, are rather "bad": the spaces $LH^p(B)$ are not separable and the space $LH^2(B)$ is not isomorphic to a Hilbert space (see 7.4.6 in Rudin (1980)).

We remark further that if $\text{Re}f > 0$ ($f \in H(B)$), then $f \in LH^p(B)$ for all $p < 1$. Let us denote by CM_0 the space of all functions $f \in C(S)$ such that $f \cdot \sigma \in M_0$.

4.1.1. Theorem. *Let f be a function holomorphic in the ball B and let $0 < p < +\infty$. Then $f \in LH^p(B)$ if and only if $f \in N_*(B)$ and*

$$\mathscr{M}(|f|^p) \overset{\text{def}}{=} \sup d\left\{\int_S |f|^p \varphi \, d\sigma : \varphi \in CM_0\right\} < +\infty;$$

moreover $\mathscr{M}(|f|^p) = \|f\|_p^p$.

A similar characterization of functions in the class $LH^p(B)$ was obtained by Lumer (1971). However, in Theorem 4.1.1 functions in the class $LH^p(B)$ are characterized by their boundary values, whereas Lumer's characterization is in terms of their values inside the ball B.

Theorem 4.1.1 is implicitly proved in 9.7 of Rudin (1980). It follows from the following assertion which is also implicitly proved in 9.7 of Rudin (1980).

4.1.2. Theorem. *Let v be a positive measure in $M(S)$. Then*

$$\sup\left\{\int \varphi \, dv : \varphi \in CM_0\right\} = \inf\{\mu(S) : \mu \in PM(S), \mu \geq v\}. \qquad (12)$$

In particular, the existence of a measure $\mu \in PM(S)$ such that $\mu \geq v$ is equivalent to the inequality

$$\sup\left\{\int \varphi \, dv : \varphi \in CM_0\right\} < +\infty.$$

It is easily seen that the infimum on the right side of (12) is attained if both sides of (12) are finite. For $n \geq 2$, the measure for which this infimum is attained may not be unique. For example, if

$$v = \left(\min\left(\text{Re}\,\frac{1 + z_1}{1 - z_1}, \text{Re}\,\frac{1 - z_1}{1 + z_1}\right)\right)\sigma,$$

then

$$\sup\left\{\int \varphi \, dv : \varphi \in CM_0\right\} = 1,$$

and for all $\alpha \in [0, 1]$ the measure $\mu_\alpha = \left(\alpha \cdot \text{Re}\,\frac{1 + z_1}{1 - z_1} + (1 - \alpha)\,\text{Re}\,\frac{1 - z_1}{1 + z_1}\right)\cdot\sigma$ is a P-measure, $\mu_\alpha \geq v$, and $\mu_\alpha(S) = 1$ ($n \geq 2$).

Theorem 4.1.2 is an easy corollary of (2) in 9.7.4 of Rudin (1980).

We remark also that from Theorem 4.1.2 and from the Minimax Theorem (see, e.g., 9.4.2 in Rudin (1980)), we deduce the following.

4.1.3. Theorem. *Let K be a compact subset of the sphere S. Let W denote the set of all probability measures in $PM(S)$. Then*

$$\sup\{\mu(K) : \mu \in W\} = \inf\left\{\frac{1}{\min_K \varphi} : \varphi \in CM_0\right\}.$$

The extreme points of the compact set W are studied in the works of Forelli (1977, 1979) (see also 19.2 in Rudin (1980)).

4.2. P-Measures and Boundary Values of Holomorphic Functions.

Throughout the present section φ will denote a positive lower semicontinuous function defined on the ball $\bar{B}$ ($\bar{B} = B \cup S$).

4.2.1. Theorem (Aleksandrov (1982, 1984)). *If $\varphi|_S \in L^1(S)$ then there is a positive singular measure $\mu \in M(S)$ such that*

$$\mu(S) = \int_S \varphi \, d\sigma, \quad \varphi\sigma - \mu \in PM(S)$$

and the Poisson integral[12] *of the measure $\varphi\sigma - \mu$ is dominated by φ everywhere in the ball B.*

4.2.2. Corollary. *There exists a positive singular P-measure on the sphere S.*

4.2.3. Corollary (Aleksandrov (1982, 1984), LÖW (1982, 1984)). *If $\log \sigma|_S \in L^1(S)$ then there exists a function $f \in N_*(B)$ such that $|f| \leq \varphi$ everywhere in B and $|f| = \varphi$ almost everywhere on S.*

4.2.4. Corollary (Aleksandrov (1982), Löw (1982)). *There exists a nonconstant inner function $f : B \to \mathbb{C}$.*

We recall that a function f is an *inner function* if

$$f \in H^\infty(B) \text{ and } |f| = 1 \text{ almost everywhere on } S.$$

The analog of Theorem 4.2.1 for the polydisc was proved by Rudin (1969).

4.2.5. Theorem (Aleksandrov (1982, 1984)). *If $\log \varphi|_S \in L^1(S)$, then the closure of the set $\{f \in N_*(B) : |f| \leq \varphi$ in B and $|f| = \varphi$ almost everyhwere on $S\}$ in the topology of uniform convergence on compact subsets of the ball B contains the set $\{f \in A(B) : |f| \leq \varphi$ everywhere on $\bar{B}\}$.*

This theorem can be seen as a multidimensional variant of Schur's Theorem on approximation by inner functions.

Let us denote by $\tilde{A}(B)$ the set of all functions $f \in H^\infty(B)$ such that the limit $\lim_{\substack{\xi \to \zeta \\ \xi \in B}} f(\xi)$ exists for almost all $\zeta \in S$.

4.2.6. Theorem (Aleksandrov (1982, 1984)). *Let $f, g \in \tilde{A}(B)$. Suppose there exists a function $h \in \tilde{A}(B)$ different from zero such that $|f| + |h| \leq \varphi$ everywhere in B and almost everywhere on S. Then if $\log \varphi|_S \in L^1(S)$ there exists a function $F \in N_*(B)$ such that $|F| < \varphi$ everywhere in B, $|F| = \varphi$ almost everywhere on S, and $(F - f)g^{-1} \in N_*(B)$.*

---

[12] Since $\varphi\sigma - \mu \in PM(S)$, it makes no difference which Poisson kernel we consider (invariant or "harmonic").

This theorem can be viewed as a multidimensional variant of the Pick-Nevanlinna Interpolation Theorem. In the one-dimensional case stronger assertions hold (see Garnett (1981)). For example, in Theorem 4.2.6, instead of the space $\tilde{A}(B)$, one can consider the space $H^\infty(B)$. In the multidimensional situation, such is not the case, even for $\varphi \equiv 1$ and $f \equiv 0$. We may construct the appropriate counterexample using Theorem 4.3.1 below; see also Rudin (1983).

4.2.7. Theorem (Aleksandrov (1983, 1984)). *Let $\varepsilon > 0$. Then there is a function $f \in A(B)$ such that $|f| \le \varphi$ everywhere in B and $\sigma\{\zeta \in S : |f(\zeta)| \ne \varphi(\zeta)\} < \varepsilon$.*

4.2.8. Corollary (Aleksandrov (1983)). *There exists a non-constant function $f \in A(B)$ such that $|f| \le 1$ everywhere in B and $\sigma\{|f| = 1\} > 0$.*

4.2.9. Corollary (Aleksandrov (1983)). *There exists a function $f \in A(B)$ such that $f(z) \ne 0$ for all $z \in B$ and $\dfrac{1}{f} \notin N_*(B)$.*

It is unknown to the author whether this can occur if $\dfrac{1}{f} \in L^1(S)$ $(n \ge 2)$.

4.2.10. Theorem (Hakim-Sibony (1982b) and Hakim (1982/83)). *Let $\varepsilon > 0$ and $\varphi \in C(\overline{B})$. Then there exists a compact set $K \subset S$ and a function f holomorphic in some neighborhood of the set $\overline{B}\backslash K$ such that $\sigma(K) = 0$, $|f| \le \varphi$ everywhere in B, and $|f| \ge \varphi - \varepsilon$ everywhere on $S\backslash K$.*

Further results[13] in this direction (in particular, assertions 4.2.1 through 4.2.10 for strictly pseudoconvex classical domains and several others) can be found in Aleksandrov (1982, 1984), Löw (1982, 1984), Hakim-Sibony (1982b, 1983), Rudin (1983), Tomaszewski (1984), and Hakim (1982/83).

4.3. LSC-Property[14]. Let Φ be a real continuous (not necessarily linear) functional on the space of all functions holomorphic in some neighborhood of the disc $\mathbb{D}$. To each function f holomorphic in the ball B, we associate the function

$$\Phi_f : S \to \mathbb{R} \cup \{+\infty\}, \quad \Phi_f(\zeta) \overset{\text{def}}{=} \sup_{0 < r < 1} \Phi(f_r^\zeta),$$

where $f_r^\zeta(z) = f(r\zeta z)$ $(z \in \mathbb{D})$. It is easy to see that the function Φ_f is lower semicontinuous.

This simple assertion has a whole series of interesting consequences (see Rudin (1983)). As Φ, it is particularly useful to take functionals which enjoy the following monotonicity property

$$\Phi(f_r) \le \Phi(f) \quad (0 < r < 1),$$

---

[13] See also the references to applications in the footnotes to Sect. 3.6.
[14] This notion was introduced by Rudin (1983): LSC = Lower Semicontinuous.

for example,

$$\int_T \log(1 + |f|) \, dm, \quad \|f\|_{H^p} \quad (0 < p \le +\infty), \quad \int_T |\mathrm{Re}\, f|^p \, dm \quad (p \ge 1).$$

Thus, if we take as Φ the first of these functionals and invoke Corollary 4.2.3, then we obtain the following assertion.

4.3.1. Theorem. *Let ψ be a measurable T-invariant function on the sphere S. Then the following assertions are equivalent:*

1) there exists a function $f \in N_*(B)$ distinct from zero such that $|f| = \psi$ almost everywhere on S;

2) there exists a homogeneous polynomial $F \in C[z_1, z_2, \ldots, z_n]$ distinct from zero such that $|F| = \psi$ almost everywhere on S, $\log \psi \in L^1(S)$, and the function ψ agrees almost everywhere on S with a lower semicontinuous function.

This theorem shows that it is not possible to drop the condition of lower semicontiniuity in Theorems 4.2.1, 4.2.5, and 4.2.6. For other results on the necessity of lower semicontinuity in the multidimensional case see Hakim-Sibony (1983).

Various modifications of the LSC-property are possible (for example, by considering analytic discs parallel to the vector e_n, we may analogously consider a functional $\Phi_f : B_{n-1} \to (-\infty, +\infty]$; moreover, defining Φ_f somewhat differently, one can guarantee that this function belongs to the first Baire class (see Jewell (1980)).

Using the LSC-property, Rudin (1983) showed several results on the impossibility of approximation in the L^∞-norm. See also Jewell (1980) where one of the above described modifactions of the LSC-property is exploited.

4.4. Outer Functions. Let f be a non-zero function in $H^\infty(B)$. Consider the following for the function f:

a) for each function $g \in H^\infty(B)$, from the inequality $|g| \le |f|$ almost everywhere on S, it follows that $|g| \le |f|$ everywhere in B;

b) $\log|f(0)| = \displaystyle\int_S \log|f| \, d\sigma$;

c) $\dfrac{1}{f} \in N_*(B)$;

d) $f \cdot A(B)$ is dense in $N_*(B)$;

e) $f \cdot A(B)$ is dense in $H^2(B)$.

In the one-dimensional case it is not hard to verify that e) $\Rightarrow$ d) $\Longleftrightarrow$ c) $\Longleftrightarrow$ b) $\Rightarrow$ a). The example $f(z) = \exp \dfrac{z_1 + 1}{z_1 - 1}$ (see 4.4.8 in Rudin (1969)) shows that property d) does not imply property e) $(n \ge 2)$. The function $f(z) = z_1$ shows that b) does not follow from a).

It is not difficult to prove that any function in class $A(B)$ satisfies property a). We remark also that if $\|f\|_{H^1(B)} = 1$ and f satisfies property a), then f is an extreme point of the unit ball in the space $H^1(B)$.

Let us denote by $\sigma(f)$ the spectrum of a function $f \in L^{\infty}(S)$, i.e.

$$\sigma(f) = \{\lambda \in \mathbb{C} : (\lambda - f)^{-1} \notin L^{\infty}(S)\}.$$

From the Hartogs Theorem on the removability of compact singularities, it follows that

$$f(\bar{B}) = \sigma(f), \tag{13}$$

for all $f \in A(B)$ ($n \geq 2$). From the existence of non-constant inner functions in the ball, it follow that (13) does not in general hold for functions $f \in H^{\infty}(B)$. If property a) holds for all functions of the form $f + c$ (where $c \in \mathbb{C}$), then (13) clearly holds. Tamm (1982) proved (13) for all functions $f \in H^{\infty}(B)$ such that

$$\|f - f_r\|_{H^p(B)} = o((1 - r)^{1/2p})(r \to 1-) \quad (p > 0).$$

In Tamm (1982) the sharpness of this result is asserted (at least for $p = 2$).

Improving on a result of Sadullaev (1976b), Rudin (1983) showed that non-constant inner functions have extremely pathological boundary behavior.

§ 5. Peak Sets for Smooth Functions

5.1. Peak Sets and Local Peak Sets. A set $K \subset S$ is called a *peak set* for $A^m(B)$ if there exists a function $f \in A^m(B)$ such that $f|_K \equiv 1$ and $|f| < 1$ everywhere on $\bar{B} \backslash K$. A set $K \subset S$ is called a *local peak set* for $A^m(B)$ if each point of K has a closed neighborhood V such that $V \cap K$ is a peak set for $A^m(B)$.

It is easy to see that the real sphere S_R and S_T (see Sect. 2) are peak sets for $A^{\infty}(B)$; as functions $f \in A^{\infty}(B)$ we may take

$$\frac{1}{2}\left(\sum_{j=1}^{n} z_j^2 + 1\right) \text{ and } \frac{1}{2}\left(n^{n/2} \prod_{j=1}^{n} z_j + 1\right).$$

5.1.1. Theorem. *Let K be a compact subset of the sphere S. The following assertions are equivalent:*

1) *K is a peak set for $A^{\infty}(B)$;*
2) *K is a local peak set for $A^{\infty}(B)$;*
3) *each point of K has a neighborhood V such that $K \cap V$ is contained in some totally-real C^{∞}-manifold of dimension $n - 1$ and having the following property:*

$$T_p(M) \subset T_p^C(S)$$

for all $p \in K \cap V$;

4) *each point of K has a neighborhood V such that $K \cap V$ is contained in an integral C^{∞}-manifold.*

The implicaton 1) $\Rightarrow$ 2) is trivial. Hakim and Sibony (1978) proved the implication 4) $\Rightarrow$ 2) $\Rightarrow$ 3). Chaumat and Chollet (1979) proved the implication 3) $\Rightarrow$ 4). Fornaess and Henriksen (1982) proved the implication 2) $\Rightarrow$ 1).

It is easy to see that, for $n = 2$, the existence of an integral C^∞-manifold M such that $K \subset M$ follows from 4). Fornaess and Henriksen (1982) showed the analogous assertion for $n = 3$. Chaumat and Chollet (1982) showed that for $n \geq 4$ this is generally not the case, if instead of the ball one considers a strictly pseudoconvex domain with smooth boundary.

5.2. Peak Sets and Interpolation. A compact subset K of the sphere S is called an *interpolation set* (for $A^\infty(B)$) of order r ($r \in \mathbb{N} \cup \{\infty\}$), if for each function $f \in C^\infty(\mathbb{C}^n)$ such that $D^\alpha \bar{\partial} f = 0$ everywhere on K for each multi-index $\alpha \in \mathbb{Z}_+^{2n}$, with $|\alpha| < r$, there exists a function $F \in A^\infty(B)$ for which $D^\alpha F = D^\alpha f$ everywhere on K, for all multi-indeces $\alpha \in \mathbb{Z}_+^{2n}$ with $|\alpha| \leq r$. For $n \geq 2$, the sphere S is an interpolation set of order r for all $r \in \mathbb{N} \cup \{\infty\}$.

5.2.1. Theorem (Chaumat-Chollet (1980)). *Let K be a peak set for $A^\infty(B)$. Then K is an interpolation set of order r for all $r \in \mathbb{N} \cup \{\infty\}$.*

Let Γ be a C^∞-smooth curve on the sphere S. Suppose that $T_p(\Gamma) \not\subset T_p^C(S)$, for all $p \in \Gamma$. In the paper of Chaumat-Chollet (1983) the following two problems are solved:

1. Which compact subsets of the curve Γ coincide with the set $f^{-1}(0)$ for some function $f \in A^\infty(B)$?

2. Which compact subsets of the curve Γ are interpolation sets of infinite order for $A^\infty(B)$?

The answers to both of these questions (see Chaumat-Chollet (1983)) are respectively analogous to the one-dimensional results (see Garnett (1981)).

5.3. Finitely Generated Ideals in the Algebra $A^\infty(B)$. Let M be a closed C^∞-submanifold of some neighborhood of the ball $\bar{B}$;

$$\bar{B} \cap M = \{z \in \bar{B} : g_1(z) = g_2(z) = \cdots = g_k(z) = 0\},$$

where $g_j \in A^\infty(B)$. When does the ideal in $A^\infty(B)$ generated (in the algebraic sense) by the functions $g_1, g_2 \ldots g_k$, coincide with the ideal $\{f \in A^\infty(B) : f|_{\bar{B} \cap M} \equiv 0\}$? By way of an answer to this question we refer the reader to Bartolomeis-Tomassini (1981) and Amar (1983a).

Update on Problems from Rudin's Book (1980) Solved up to the Present Time

1°. 7.1.7 and 19.1.9.[15]. Let $n \geq 2$. Does there exist a constant $C = C(n)$ such that

$$\int_S |g| \, d\sigma \leq C \int_S |\operatorname{Re} g| \, d\sigma$$

for all functions $g \in A(B)$ whose value at the origin is real?

---

[15] Here and in the sequel a.b.c. denotes the corresponding section in Rudin (1980).

A negative answer to this question follows easily from the existence of a nonconstant inner function on the ball (see Chap. 5, 4). Indeed, if I is an inner function and $I(0) = 0$, then

$$\int_S \left| \operatorname{Re} \frac{1 + I_r}{1 - I_r} \right| d\sigma = \int_S \operatorname{Re} \frac{1 + I_r}{1 - I_r} d\sigma = 1$$

for all $r \in (0, 1)$. However, $i\dfrac{1 + I}{1 - I} \notin H^1(B)$ since any function in $H^1(B)$ which is real almost everywhere on S is constant. Consequently,

$$\lim_{r \to 1-} \int_S \left| \frac{1 + I_r}{1 - I_r} \right| d\sigma = +\infty.$$

We remark that a negative answer to this question can be obtained without using inner functions (see A.B. Aleksandrov, H^p Hardy classes for $p < 1$ and semi-inner functions in the ball. Dokl. Akad. Nauk SSSR *262*, No 5, 1033–1036 (1982)).

2°. 7.3.6 and 19.3.2. Does there exist, for $n \geq 3$, a function $f \in A(B)$ not identically zero and such that the set $f^{-1}(0)$ has infinite $(2n - 2)$-dimensional Hausdorff measure?

A corresponding example for $n = 2$ is constructed in 7.3.6 of Rudin (1980) (the one-dimensional case is trivial). A positive answer to Problem 2 was obtained independently by Alexander and Amar (see Chap. 4, 2.3). Further results in this direction were obtained by Hakim and Sibony (Chap. 4, Theorem 2.3.1) and also by the author (A.B. Aleksandrov, The Blaschke condition and zeros of bounded holomorphic functions, in "Multidimensional Complex Analysis", Inst. Fiz. SO Akad. Nauk SSSR, Krasnoyarsk, 1985, 23–26). Amar used the theorem of Varopoulos (Chap. 4, Theorem 4.1.1). The construction of Amar gives a function having a piecewise-linear zero set. Alexander used the Ryll-Wojtaszczyk polynomials (Chap. 1, 3). The Ryll-Wojtaszczyk polynomials also allow one (see Ryll-Wojtaszczyk (1983)) to strengthen and extend, to arbitrary dimensions n, practically all results (see Chap. 7 in Rudin (1980)) based on 7.2.8 in Rudin (1980). We now enumerate several of these:

7.2.9. There exists a function $f \in H^2(B)$ (for $n = 2, 3$) for which almost no slice function has a Taylor expansion absolutely convergent in the closed unit disc. This result was essentially duplicated by Wojtaszczyk (1982) (see also Chap. 1, 3.6). He showed that with no restrictions on the dimension n, one can also require the following properties:

1) $f \in A(B)$;
2) $\sum_{k \geq 0} |\hat{f}_\zeta(k)|^p = +\infty$ for each $p < 2$ and for almost every $\zeta \in S$; where $\hat{f}_\zeta(k) \stackrel{\text{def}}{=} \dfrac{f_\zeta^{(k)}(0)}{k!}.$

7.2.10. There exists a function $f \in \mathscr{H}^2_{1/2}(B)$ (for $n = 2, 3$) such that

$$\sup_{0 < r < 1} |f(r\zeta)| = +\infty \tag{1}$$

for almost all $\zeta \in S$.

This assertion also can be strengthened for all n. In order to convince oneself of this, we consider the function

$$f = \sum_{k \geq 1} \frac{1}{\sqrt{k}} f_{2^k},$$

where f_p $(p > 0)$ are Ryll-Wojtaszsczyk polynomials (see Chap. 1, Theorem 3.6.1). It is not difficult to verify that the functon f satisfies (1) and $f \in \mathscr{H}^p_\alpha(B)$ for all α, $p \in (0, +\infty)$. This same function allows one also to strengthen the results of 7.2.11 in Rudin (1980).

$3°$. 10.5.7. In this section the conjecture was made that the topological dimension of any I-set is no greater than $n - 1$. This conjecture was proved by Stout (see Chap. 5, 2).

$4°$. 10.7.4. For $A^\infty(B)$, is every local Peak set global? A positive answer to this question was obtained by Fornaess and Henriksen (see Chap. 5, 5).

$5°$. 11.4.1 and 19.3.15. Let f be a non-constant function in $A(B)$, $n \geq 2$. Is it true that

$$\sigma \left\{ \zeta \in S : |f(\zeta)| = \max_{z \in S} |f(z)| \right\} = 0?$$

A negative answer to this question was obtained by the author (see Chap. 5, 4).

$6°$. 13.4.6 and 19.3.7. Let X be a closed subalgebra of the algebra $C(B)$ of all continuous functions on the ball B endowed with the topology of uniform convergence on compact subsets of the ball B. Suppose that X is $\mathscr{M}$-invariant, i.e. $f \circ \varphi \in X$ for all $f \in X$ and all $\varphi \in \text{Aut}(B)$. Is it then true that X coincides with one of the following five algebras

$$\{0\}, \; \mathbb{C}. \; H(B), \; \overline{H(B)}, \; C(B)?$$

This conjecture was proved by Rudin (W. Rudin, Moebius-invariant algebras in balls. Ann. Inst. Fourier, 33, No. 2, 19–41 (1983)). Rudin's result is new even in the one-dimensional case.

$7°$. 19.1.1. Do there exist non-constant inner functions in the ball B $(n \geq 2)$? A positive answer to this question was obtained independently by the author and by Löw (see Chap. 5, 4). The example of a non-constant inner function refutes essentially all conjectures in 19.1 of Rudin (1980) and also gives an example of a non-surjective isometric operator on the space $H^p(B)$ (19.3.8 in Rudin (1980)) and an example of a nontrivial inner mapping (19.3.9 in Rudin (1980)).

8°. 19.3.1. Let $n \geq 2$. Does there exist a function in $f \in H^1(B)$ which cannot be written as the product of two functions in $H^2(B)$? Gowda obtained a positive answer to this question. He showed that for $n \geq 2$ the set

$$\{gh : g, h \in H^2(B)\}$$

is a set of first category in $H^1(B)$ (see Chap. 3, 2).

References*

Ahern, P.R., Schneider, R. (1980): A smoothing property of the Henkin and Szegö projections. Duke Math. J. ● ● , No. 1, 135–143, Zbl.453.32004

Aizenberg, L.A., Yuzhakov, A.P. (1979): Integral representations and residues in multidimensional complex analysis. Novosibirsk: Nauka. 335 pp. Engl. transl.: Transl. Math. Monogr., Vol. 58, Providence, 283 pp. (1983), Zbl.445.32002

Aleksandrov, A.B. (1981): Essays on non locally convex Hardy classes. Lect. Notes Math. 864, 1–89. Berlin, Heidelberg, New York: Springer-Verlag, Zbl.482.46035

Aleksandrov, A.B. (1982): Existence of inner functions in the ball. Mat. Sb., Nov. Ser. 118, No. 2, 147–163. Engl. transl.: Math. USSR, Sb. 46, 143–159 (1983), Zbl.503.32001

Aleksandrov, A.B. (1983): On the boundary values of functions holomorphic in the ball. Dokl. Akad. Nauk SSSR 274, No. 4, 777–779. Engl. transl.: Sov. Math., Dokl. 28, 134–137 (1983), Zbl.543.32002

Aleksandrov, A.B. (1984): Inner functions on compact spaces. Funkts. Anal. Prilozh. 18, No. 2, 1–13. Engl. transl.: Funct. Anal. Appl. 18, 87–98 (1984), Zbl.574.32006

Alexander, H. (1982): On zero sets for the ball algebra. Proc. Am. Math. Soc. 86, No. 1, 71–74, Zbl.504.32005

Amar, E. (1982): Sur le volume des zéros des fonctions holomorphes et bornées dans la boule de $\mathbb{C}^n$. Proc. Am. Math. Soc. 85, No. 1, 47–52, Zbl.507.32001

Amar, E. (1983a): Non division dans $A^\infty(\Omega)$. C. R. Acad. Sci., Sér. 1, 296, No. 13, 541–544. See also: Math. Z. 188, 493–511 (1985), Zbl.547.32010

Amar, E. (1983b): Extension de fonctions holomorphes et courants. Bull. Sci. Math., II. Ser. 107, 25–48, Zbl.543.32007

Axler, Sh., Shapiro, J.H. (1983): Putnam's theorem, Alexander's spectral area estimate, and VMO. Math. Ann. 271, 161–183 (1985), Zbl.541.30021

Bartolomeis, P. de, Tomassini, G. (1981): Idéaux de type fini dans $A^\infty(D)$. C. R. Acad. Sci., Paris, Sér. I, 293, No 2, 133–134, Zbl.477.32016

Berndtsson, B. (1980): Integral formulas for the $\partial\bar{\partial}$-equation and zeros of bounded holomorphic functions in the unit ball. Math. Ann. 249, No. 2, 163–176, Zbl.414.31007

Bishop, E. (1965): Differentiable manifolds in complex Euclidean space. Duke Math. J. 32, No. 1, 1–21, Zbl.154,85

Boas, R. (1955): Isomorphism between H^p and L^p. Am. J. Math. 77, No. 4, 655–656, Zbl.65,345

Bourgain, J. (1982): The non-isomorphism of H^1-spaces in one and several variables. J. Funct. Anal. 46, No. 1, 45–57, Zbl.492.46043

---

*For the convenience of the reader, references to reviews in Zentralblatt für Mathematik (Zbl.), compiled using the MATH database, have, as far as possible, been included in this bibliography.

Bourgain, J. (1983a): The non-isomorphism of H^1-spaces in a different number of variables. Bull. Soc. Math. Belg., Ser. B *35*, No. 2, 127–136, Zbl.533.46036

Bourgain, J. (1983b): The dimension conjecture for polydisc algebras. Isr. J. Math. *48*, 289–304 (1984), Zbl.572.46047

Chaumat, J., Chollet, A.-M. (1979): Ensembles pics pour $A^\infty(D)$. Ann. Inst. Fourier *29*, No. 3, 171–200, Zbl.398.32004

Chaumat, J., Chollet, A.-M. (1980): Caractérisation et propriétés des ensembles localement pics de $A^\infty(D)$. Duke Math. J. *47*, No. 4, 763–787, Zbl.454.32013

Chaumat, J., Chollet, A.-M. (1982): Ensembles pics pour $A^\infty(D)$ non globalement inclus dans une variété intégrale. Math. Ann. *258*, No. 3, 243–252, Zbl.574.32023

Chaumat, J., Chollet, A.-M. (1983): Ensembles des zéros et d'interpolation le long de courbes. C. R. Acad. Sci., Paris, Sér. I *296*, No. 19, 789–792, Zbl.566.32010

Chirka, E.M. (1973): The theorems of Lindelöf and Fatou in $\mathbb{C}^n$. Mat. Sb., Nov. Ser. *92*, No. 4 622–644. Engl. transl.: Math. USSR, Sb. *21*, 619–639 (1975), Zbl.297.32001

Chirka, E.M., Khenkin, G.M. (= Henkin, G.M.) (1975): Boundary properties of holomorphic functions of several complex variables. In Itogi Nauki Tekh., Ser. Sovrem. Probl. Mat. *4*, 13–142. Engl. transl.: J. Sov. Math. *5*, 612–687 (1976), Zbl.375.32005

Coifman, R.R. (1974): A real variable characterization of H^p. Stud. Math. *51*, No. 3, 269–274, Zbl.289.46037

Coifman, R.R., Rochberg, R. (1980): Representation theorems for holomorphic and harmonic functons. Astérisque *77*, 12–66, Zbl.472.46040

Coifman, R.R., Rochberg, R., Weiss, G. (1976): Factorization theorems for Hardy spaces in several variables. Ann. Math., II, Ser. *103*, No. 3, 611–635, Zbl.326.32011

Coifman, R.R., Weiss, G. (1977): Extensions of Hardy spaces and their use in analysis. Bull. Am. Math. Soc. *83*, No. 4, 569–645, Zbl.358.30023

Cumenge, A. (1983): Extension dans des classes de Hardy de fonctions holomorphes et estimations de type "mesure de Carleson" pour l'équation $\bar{\partial}$. Ann. Inst. Fourier *33*, No. 3, 59–97, Zbl.487.32011

Dautov, Sh.A, Khenkin, G.M. (= Henkin, G.M.) (1978): Zeros of holomorphic functions of finite order and weighted estimates for solutions of the $\bar{\partial}$-equation. Mat. Sb. *107*, No. 2, 163–174. Engl. transl.: Math. USSR, Sb. *35*, 449–459 (1979), Zbl.392.32001

Davie, A.M., Jewell, N.P. (1977): Toeplitz operators in several complex variables. J. Funct. Anal. *26*, No. 4, 356–368, Zbl.374.47011

Davie, A., Öksendal, B. (1972): Peak interpolation sets for some algebras of analytic functions. Pac. J. Math. *41*, No. 1, 81–87, Zbl.218,292

Duchamp, Th., Stout, E.L. (1981): Maximum modulus sets. Ann. Inst. Fourier *31*, No. 3, 37–69, Zbl.439.32007

Duren, P.L. (1970): Theory of H^p Spaces. 258 pp. New York: Academic Press, Zbl.215,202

Duren, P.L., Romberg, B.W., Shields, A.L. (1969): Linear functionals on H^p spaces with $0 < p < 1$. J. Reine Angew. Math. *238*, No. 1, 32–60, Zbl.176,431

Fefferman, C., Stein, E.M. (1972): H^p spaces of several variables. Acta Math. *129*, No. 3–4, 137–193, Zbl.257.46078

Forelli, F. (1974): Measures whose Poisson integrals are pluriharmonic. Ill. J. Math. *18*, No. 3, 373–388, Zbl.296.31014

Forelli, F. (1975): Measures whose Poisson integrals are pluriharmonic II. III. J. Math. *19*, No. 4, 584–592, Zbl.329.31002

Forelli, F. (1977): A necessary condition on the extreme points of a class of holomorphic functions. Pac. J. Math. *73*, No. 1, 81–86, Zbl.346.32002

Forelli, F. (1979): Some extreme rays of the positive pluriharmonic functions. Can. J. Math. *31*, No. 1, 9–16, Zbl.373.31005

Forelli, F., Rudin, W. (1974): Projections on spaces of holomorphic functions in balls. Indiana Univ. Math. J. *24*, No. 6, 593–602, Zbl.297.47041

Fornaess, J.E., Henriksen, B.S. (1982): Characterization of global peak sets for $A^\infty(D)$. Math. Ann. *259*, No. 1, 125–130, Zbl.489.32010

Garnett, J.B. (1981): Bounded Analytic Functions. 467 pp. New York, London, Toronto, Sydney, San Francisco: Academic Press, Zbl.469.30024

Garnett, J.B., Latter, R.H. (1978): The atomic decomposition for Hardy spaces in several complex variables. Duke Math. J. 45, No. 4, 815–845, Zbl.403.32006

Gokhberg, I.Ts., Krupnik, N.Ya. (1973): Introduction to the Theory of One-Dimensional Singular Integral Operators. 426 pp. Kishinev: Shtiintsa. German transl.: Basel, Boston, Stuttgart: Birkhäuser 1979, Zbl.271.47017

Gowda, M.S. (1983): Nonfactorization theorems in weighted Bergman and Hardy spaces on the unit ball of $\mathbb{C}^n$ ($n > 1$). Trans. Am. Math. Soc. 277, No. 1, 203–212, Zbl.526.32005

Greiner, P.C., Stein, E.M. (1977): Estimates for the $\bar{\partial}$-Neumann problem. Math. Notes 19. Princeton, N.J.: Princeton Univ. Press, 194 pp., Zbl.354.35002

Hakim, M. (1983): Valeurs au bord de fonctions holomorphes bornées en plusieurs variables complexes. Sémin. Bourbaki, 35e année, Vol. 1982/83, Exp. No. 613, Astérisque 105–106, 293–305, Zbl.519.32007

Hakim, M., Sibony, N. (1978): Ensembles pics dans des domaines strictement pseudoconvexes. Duke Math. J. 45, No. 3, 601–617, Zbl.402.32008

Hakim, M., Sibony, N. (1982a): Ensemble des zéros d'une fonction holomorphe bornée dans la boule unité. Math. Ann. 260, No. 4, 469–474, Zbl.499.32006

Hakim, M., Sibony, N. (1982b): Fonctions holomorphes bornées sur la boule unité de $\mathbb{C}^n$. Invent. Math. 67, No. 2, 213–222, Zbl.475.32007

Hakim, M., Sibony, N. (1983): Valeurs au bord des modules de fonctions holomorphes. Math. Ann. 264, No. 2, 197–210, Zbl.514.32008

Henriksen, B.S. (1982): A peak set of Hausdorff dimension $2n - 1$ for the algebra $A(D)$ in the boundary of a domain D with C^∞-boundary in $\mathbb{C}^n$. Math. Ann. 259, No. 2, 271–277, Zbl.483.32011

Hoffman, K. (1962): Banach Spaces of Analytic Functions. Englewood Cliffs, N.J.: Prentice Hall, Inc., 217 pp., Zbl.117,340

Hörmander, L. (1967): L^p estimates for (pluri-) subharmonic functions. Math. Scand. 20, No. 1, 65–78, Zbl.156,122

Horowitz, C. (1977): Factorization theorems for functions in the Bergman spaces. Duke Math. J. 44, No. 1, 201–213, Zbl.362.30031

Hurewicz, W., Wallman, H. (1941): Dimension Theory. Princeton: Princeton University Press, 165 pp., Zbl.60,398

Jakóbczak, P. (1983): On the regularity of extension to strictly pseudoconvex domains of functions holomorphic in a submanifold in general position. Ann. Pol. Math. 42, 115–124, Zbl.552.32009

Janson, S. (1976): On functions with conditions on the mean oscillation. Ark. Mat. 14, No. 2, 189–196, Zbl.341.43005

Jewell, N.P. (1980): Toeplitz operators on the Bergman spaces and in several complex variables. Proc. Lond. Math. Soc., III. Ser. 41, No. 2, 193–216, Zbl.412.47014

Jöricke, B. (1981): On pseudoanalytic continuation and the behavior of the boundary values of analytic functions on small sets. Preprint, Akad. Wiss. DDR, Inst. Math. P-Math-22/81, 24 pp., Zbl.466.30005

Jöricke, B. (1982): The two-constants Theorem for functions of several complex variables. Math. Nachr. 107, 17–52 (Russian), Zbl.526.32003

Khenkin, G.M. (= Henkin, G.M.) (1968): The Banach space of analytic functions in the sphere and in the bicylinder are not isomorphic. Funkts. Anal. Prilozh. 2, No. 4, 82–91. Engl. transl.: Funct. Anal. Appl. 2, 334–341 (1968), Zbl.181,134

Khenkin, G.M. (= Henkin, G.M.) (1971): Approximation of functions in pseudoconvex domains and a theorem of Lejbenzon. Bull. Acad. Pol. Sci., Ser. Sci. Math. Astron. Phys. 19, No. 1, 37–42 (Russian), Zbl.214,337

Khenkin, G.M. (= Henkin, G.M.) (1972): Continuation of bounded holomorphic functions from submanifolds in general position to strictly pseudoconvex domains. Izv. Akad. Nauk SSSR, Ser. Mat. 36, No. 3, 540–567. Engl. transl.: Math. USSR, Izv. 6, 536–563 (1973), Zbl.249.32009

Khenkin, G.M. (= Henkin, G.M.) (1977a): The equation of H. Lewy and analysis on pseudoconvex manifolds I. Usp. Mat. Nauk *32*, No. 3, 57–118. English transl.: Russ. Math. Surv. *32*, No. 3, 59–130 (1977), Zbl.358.35057

Khenkin, G.M. (= Henkin, G.M.) (1977b): The equation of H. Lewy and analysis on pseudoconvex manifolds II. Mat. Sb., Nov. Ser. *102*, No. 1, 71–108. Engl. transl.: Math. USSR, Sb. *31*, 63–94 (1977), Zbl.388.35052

Khenkin, G.M. (= Henkin, G.M.), Leiterer, J. (1984): Theory of Functions on Complex Manifolds. 226 pp. Berlin: Akademie-Verlag, Zbl.573.32001

Khenkin, G.M. (= Henkin, G.M.), Mityagin, B.S. (1971): Linear problems of complex analysis. Usp. Mat. Nauk *26*, No. 4, 93–152. Engl. transl.: Russ. Math. Surv. *26* (1971), No. 4, 99–164 (1972), Zbl.245.46027

Khenkin, G.M. (= Henkin, G.M.), Tumanov, A.E. (1976): Interpolation submanifolds of pseudoconvex manifolds. Math. program. rel. Probl., Cent. Ehkon. Mat. Inst. Akad. Nauk SSSR, Mosk. 1974, 74–86. Engl. transl.: Transl., II. Ser., Am. Math. Soc. *115*, 59–69 (1980), Zbl.455.32009

Khrushchev, S.V., Peller, V.V. (1982): Hankel operators, best approximation and stationary Gaussian processes. Usp. Mat. Nauk *37*, No. 1, 53–124. Engl. transl.: Russ. Math. Surv. *37*, No. 1, 61–144 (1982), Zbl.497.60033

Khrushchev, S.V., Vinogradov, S.A. (1981): Free interpolation in the space of uniformly convergent Taylor series. Lect. Notes Math. *864*, 171–213, Zbl.463.30001

Koosis, P. (1980): Introduction to H^p-spaces. Lond. Math. Soc. Lect. Notes Ser. No. *40*, 376 pp., London: Cambridge Univ. Press. Zbl.435.30001

Koranyi, A., Vagi, S. (1971): Singular integrals in homogeneous spaces and some problems classical analysis. Ann. Sc. Norm. Super. Pisa, Sci. Fij. Mat., III, Ser. *25*, No. 4, 575–648, Zbl.291.43014

Korenblum, B. (1975): An extension of the Nevanlinna theory. Acta Math. *135*, No. 3–4, 187–219, Zbl.323.30030

Krantz, S.G. (1980): Holomorphic functions of bounded mean oscillation and mapping properties of the Szegö projection. Duke Math. J. *47*, No. 4, 743–761, Zbl.456.32004

Lindenstrauss, J., Pełczyński, A. (1971): Contributions to the theory of the classical Banach spaces. J. Funct. Anal. *8*, No. 2, 225–249, Zbl.224.46041

Löw, E. (1982): A construction of inner functions on the unit ball in $\mathbb{C}^n$. Invent. Math. *67*, No. 2, 223–229, Zbl.528.32006

Löw, E. (1984): Inner functions and boundary values in $H^\infty(\Omega)$ and $A(\Omega)$ is smoothly bounded pseudoconvex domains. Math. Z. *185*, No. 2, 191–210, Zbl.508.32005

Lumer, G. (1971): Espaces de Hardy en plusieurs variables complexes. C. R. Acad. Sci., Paris, Sér. A *273*, No. 3, 151–154, Zbl.216,161

Malliavin, P. (1974): Fonctions de Green d'un ouvert strictement pseudoconvex et classe de Nevanlinna. C. R. Acad. Sci., Paris, Sér. A *278*, No. 3, 141–144, Zbl.279.32005

McDonald, G. (1979): The maximal ideal space of $H^\infty + C$ on the ball in $\mathbb{C}^n$. Can. J. Math. *31*, No. 1, 79–86, Zbl.412.46045

Nagel, A., Wainger, S. (1981): Limits of bounded holomorphic functions along curves. Ann. Math. Stud. *100*, 327–344, Zbl.517.32003

Nikol'skij, N. (1986): Treatise on the Shift Operator. Spectral Function Theory. New York, Berlin, Heidelberg: Springer-Verlag, 505 pp. Enlarged English edition of the Russian original, Moscow: Nauka 1980, 304 pp., Zbl.587.47036

Pelczyński, A. (1977): Banach spaces of analytic functions and absolutely summing operators. Amer. Math. Soc., Providence, R. I, 92 pp., Zbl.475.46022

Pinchuk, S.I. (1974): A boundary uniqueness theorem for holomorphic functions of several complex variables. Mat. Zametki *15*, No. 2, 205–212. Engl. transl.: Math. Notes. *15*, 116–120 (1974), Zbl.285.32002

Pinchuk, S.I. (1986): Holomorphic mappings in $\mathbb{C}^n$ and the problem of holomorphic equivalence. Itogi Nauki Tekh., Ser. Sovrem. Probl. Mat., Fundam. Napravleniya 9, 195–223. English transl.: Several Complex Variables III, Encycl. Math. Sci. *9*, 173–199, Berlin, Heidelberg, New York: Springer-Verlag 1989, Zbl.658.32011

Piranian, G. (1966): Two monotonic, singular, uniformly almost smooth functions. Duke Math. J. *33*, No. 2, 255–262, Zbl.143,74

Power, S.C. (1985): Hörmander's Carleson theorem for the ball. Glasg. Math. J. *26*, No. 1, 13–17, Zbl.576.32007

Rham, G. de (1955): Variétés Différentiables. Paris: Hermann, 196 pp., Zbl.65,324

Rothschild, L.P., Stein, E.M. (1976): Hypoelliptic differential operators and nilpotent groups. Acta Math. *137*, No. 3–4, 247–320, Zbl.346.35030

Rudin, W. (1969): Function Theory in Polydiscs. New York: Benjamin, 188 pp., Zbl.177,341

Rudin, W. (1980): Function Theory in the Unit Ball of $\mathbb{C}^n$. New York, Berlin, Heidelberg: Springer-Verlag, 436 pp., Zbl.495.32001

Rudin, W. (1983): Inner functions in the unit ball of $\mathbb{C}^n$. J. Funct. Anal. *50*, No. 1, 100–126, Zbl.554.32002

Ryll, J., Wojtaszczyk, P. (1983): On homogeneous polynomials on a complex ball. Trans. Am. Math. Soc. *276*, No. 1, 107–116, Zbl.522.32004

Sadullaev, A. (1976a): A boundary uniqueness theorem in $\mathbb{C}^n$. Mat. Sb., Nov. Ser. *101*, No. 4, 568–583. Engl. transl.: Math. USSR, Sb. *30*, 501–514 (1978), Zbl.346.32024

Sadullaev, A. (1976b): On inner functions in $\mathbb{C}^n$. Mat. Zametki *19*. No. 1, 63–66. Engl. transl.: Math. Notes *19*, 37–38 (1976), Zbl.335.32002

Shabat, B.V. (1976): Introduction to Complex Analysis, Part II. Moscow: Nauka. 400 pp. French transl.: Moscow: MIR 1990, Zbl.578.32001; Zbl.188,379

Skoda, H. (1976): Valeurs au bord pour les solutions de l'opérateur d'' et caractérisation des zéros des fonctions de la classe de Nevanlinna. Bull. Soc. Math. Fr. *104*, No. 3, 225–299, Zbl.351.31007

Stein, E.M. (1970): Singular Integrals and Differentiability Properties of Functions. Princeton, N.J.: Princeton Univ. Press, 290 pp., Zbl.207,135

Stein, E.M. (1973): Singular integrals and estimates for the Cauchy-Riemann equations. Bull. Am. Math. Soc. *79*, No. 2, 440–445, Zbl.257.35040

Stout, E.L. (1982): The dimension of peak-interpolations sets. Proc. Am. Math. Soc. *86*, No. 3, 413–416, Zbl.502.32012

Szegö, G. (1959): Orthogonal Polynomials. Am. Math. Soc., Colloq. Publ. *23*, 421 pp., Zbl.89,275

Tamm, M. (1982): Sur l'image par une fonction holomorphe bornée du bord d'un domaine pseudoconvexe. C. R. Acad. Sci., Paris, Sér. I *294*, No. 16, 537–540, Zbl.497.32003

Tomaszewski, B. (1984): Interpolation and inner maps that preserve measure. J. Funct. Anal. *55*, No. 1, 63–67, Zbl.531.32002

Tumanov, A.E. (1977): A peak set of metric dimension 2.5 for the algebra of holomorphic functions on the three-dimensional sphere in $\mathbb{C}^n$. Izv. Akad. Nauk SSSR, Ser. Mat. *41*, No. 2, 370–377. Engl. transl.: Math. USSR, Izv. *11*, 353–359 (1977), Zbl.368.46048

Val'skij, R.Eh. (1971): On measures orthogonal to analytic functions in $\mathbb{C}^n$. Dokl. Akad. Nauk SSSR *198*, No. 3, 502–505. Engl. transl.: Sov. Math., Dokl. *12*, 808–812 (1971), Zbl.234.46056

Varopoulos, N.Th. (1980): Zeros of H^p functions in several complex variables, Pac. J. Math. *88*, No. 1, 189–246, Zbl.454.32006

Wells, R.O. jun. (1973): Differential Analysis on Complex Manifolds. Englewood Cliffs, N.J.: Prentice-Hall, Inc., 252 pp. 2nd ed.: New York, Berlin, Heidelberg: Springer-Verlag 1980, Zbl.262.32005

Widom, H. (1964): On the spectrum of Toeplitz operators. Pac. J. Math. *14*, No. 1, 365–375, Zbl.197,109

Wojtaszczyk, P. (1982): On functions in the ball algebra. Proc. Am. Math. Soc. *85*, No. 2, 184–186, Zbl.503.32005

Wojtaszczyk, P. (1983): Hardy spaces on the complex ball are isomorphic to Hardy spaces on the disc, $1 \leq p < +\infty$. Ann. Math., II. Ser. *118*, No. 1, 21–34, Zbl.546.32003

IV. Complex Analysis in the Future Tube

A.G. Sergeev, V.S. Vladimirov

Translated from the Russian
by P.M. Gauthier

Contents

Introduction

The future tube in $\mathbb{C}^{n+1}$ is the unbounded domain

$$\tau^+ = \{z = (z_0, \ldots, z_n) \in \mathbb{C}^{n+1} : (\operatorname{Im} z_0)^2 > (\operatorname{Im} z_1)^2 + \cdots + (\operatorname{Im} z_n)^2, \operatorname{Im} z_0 > 0\}.$$

In other words, τ^+ is a tube domain over the future cone $V^+ = \{y \in \mathbb{R}^{n+1} : y_0^2 > y_1^2 + \cdots + y_n^2, y_0 > 0\}$. The domain τ^+ is biholomorphically equivalent to a classical Cartan domain of the IVth type, hence to a bounded symmetric domain in $\mathbb{C}^{n+1}$. The future tube τ^+ in $\mathbb{C}^4$ ($n = 3$) is important in mathematical physics, especially in axiomatic quantum field theory, being the natural domain of definition of holomorphic relativistic fields. These specific features of the future tube motivated its investigation by mathematicians and physicists. Beginning with Elie Cartan's classification of bounded symmetric domains, these domains were examined in many papers where the complex structure of their boundaries, integral representations, boundary values of holomorphic functions and so on were considered. The proof of the "edge-of-the-wedge" theorem by N.N. Bogolubov generated the rapid development of applications of the theory of several complex variables to axiomatic quantum field theory. During this period the "C-convex hull" and "finite covariance" theorems were proved, the Jost-Lehmann-Dyson representation was found *et cetera*. Recently R. Penrose has proposed a transformation connecting holomorphic solutions of the basic equations of field theory with analytic sheaf cohomologies of domains in $\mathbb{CP}^3$. These two directions developed, to a large extent, independently from each other, and some important results obtained in axiomatic quantum field theory still remain unknown to specialists in several complex variables and differential geometry. One of the goals of this paper is to give a unified presentation of advances in complex analysis in the future tube and related domains achieved in both of these directions.

Among the main classes of bounded domains of holomorphy, the following classes are usually considered: strictly pseudoconvex domains, smooth weakly pseudoconvex domains, and analytic polyhedra. As basic examples of these classes we can consider respectively the unit ball $\{z \in \mathbb{C}^n : \sum |z_j|^2 < 1\}$, a domain $\{z \in \mathbb{C}^n : \sum |z_j|^{2p_j} < 1, p_j \geq 1$ and not all p_j equal to $1\}$, and the unit polydisc $\{z \in \mathbb{C}^n : |z_j| < 1\}$. These domains are distinguished by the complex geometry of their boundaries: the Levi form of the ball is non-degenerate (the complex tangent hyperplane has 2nd order contact with the boundary at each of its points), the boundary of the second domain has only finite type points (the complex tangent hyperplane has finite order of contact with the boundary at each of its points), the Levi form of the polydisc is identically zero on the smooth part of the boundary (the complex tangent hyperplane "sticks" to the boundary). All these domains belong to the general class of pseudo-convex polyhedra. The Levi form of the future tube τ^+ degenerates at all points of the smooth part of the boundary because any point of $\partial\tau^+$ is contained in a complex halfplane (complex light ray) lying entirely on the boundary. So from this point of view, the future tube is similar to analytic polyhedra. However for $n \geq 2$ there is a principal difference – the boundary of τ^+ cannot be "straightened" along complex light rays by a biholomorphic transformation. In other words, for $n \geq 2$, τ^+ is not (even locally) a pseudoconvex polyhedron.

There are two ways of generalizing the future tube: one way is to consider bounded symmetric domains in $\mathbb{C}^n$, the other is to consider arbitrary tube cones and tuboids. Some results of this survey are true for these generalizations and are formulated in their maximal generality though we are mainly interested in the future tube. We have included in our list of references related papers published after 1970 and several earlier papers (the reader can find references before 1970 in the books Vladimirov (1964, 1979), Rudin (1969), Stein-Weiss (1971), and the articles Vladimirov (1969c), 1983a, 1971, 1982), Zharinov (1983), Korányi (1972), Morimoto (1973), Stein (1971). This list is, of course, not complete and, to some extent, reflects the authors' interests. We provide all chapters with bibliographical notes where we have collected some easy-to-find books and papers, review articles and also some additional references.

Some notations: the Euclidean (complex linear) inner product of vectors z, $\zeta \in \mathbb{C}^n$ and Hermitian norm of z are denoted respectively by $(z, \zeta) = z_1\bar\zeta_1 + \cdots + z_n\bar\zeta_n$ and $|z|^2 = |z_1|^2 + \cdots + |z_n|^2$, the Lorentzian inner product of z, $\zeta \in \mathbb{C}^{n+1}$ and Lorentz norm of z are denoted by $z \cdot \zeta = z_0 \cdot \zeta_0 - z_1\zeta_1 - \cdots - z_n\zeta_n$ and $z^2 = z_0^2 - z_1^2 - \cdots - z_n^2$. A proper convex cone in $\mathbb{R}^n$ is denoted by C, V^+ is the future cone in $\mathbb{R}^{n+1}$, S^{n-1} is the unit sphere in $\mathbb{R}^n$. For an open Ω in $\mathbb{R}^n$, we call a *complex neighborhood of Ω* any domain of holomorphy $\tilde\Omega$ in $\mathbb{C}^n$ such that $\tilde\Omega \cap \mathbb{R}^n = \Omega$. The space of distributions in Ω is denoted by $\mathscr{D}'(\Omega)$, the space of tempered distributions in $\mathbb{R}^n$ by $\mathscr{S}'$, the space of bounded functions with compact supports in $\mathbb{R}^n$ by $L_c^\infty(\mathbb{R}^n)$. The space of holomorphic functions in a domain D in $\mathbb{C}^n$ is denoted by $\mathcal{O}(D)$, the sheaf of holomorphic functions in $\mathbb{C}^n$ by $\mathcal{O}$.

The authors are grateful to E.M. Chirka, Ju. N. Drozhzhinov, G.M. Khenkin, B.I. Zavialov and V.V. Zharinov for their remarks which helped the authors to improve the original text of the paper.

Chapter 1
Geometry of the Future Tube

§ 1. The Future Tube

1.1. Definition, Description of the Boundary. The future tube τ^+ is a domain in $\mathbb{C}^4$ of the form

$$\tau^+ = \{z \in \mathbb{C}^4 : y^2 = y_0^2 - y_1^2 - y_2^2 - y_3^2 > 0, y_0 > 0\},$$

where $z = (z_0, z_1, z_2, z_3)$, $z_j = x_j + iy_j$. Using the Lorentz inner product this definition can be rewritten in the form: $\tau^+ = \{z : y^2 > 0, y_0 > 0\}$. So, τ^+ is a tube domain over the *future cone*

$$V^+ = \{y \in \mathbb{R}^4 : y^2 > 0, y_0 > 0\}.$$

The section $\tau^+ \cap \{z : \operatorname{Re} z = x\}$ of τ^+ for arbitrary fixed $x \in \mathbb{R}^4$ coincides with the cone V^+, the section $\tau^+ \cap \{z : \operatorname{Im} z = y\}$ of τ^+ for arbitrary fixed $y \in V^+$ coincides with the whole of $\mathbb{R}^4$.

The boundary $\partial \tau^+$ of τ^+ consists of the smooth part

$$S = \{\zeta = \xi + i\eta \in \mathbb{C}^4 : \eta^2 = 0, \eta_0 > 0\}$$

and the *distinguished boundary*

$$M = \{\zeta \in \mathbb{C}^4 : \eta = 0\} = \mathbb{R}^4.$$

M is the set where the boundary $\partial \tau^+$ degenerates.

Through any point $\zeta \in S$ there passes a generator l_ζ of the cone

$$\Gamma^+ = \partial V^+ = \{\eta \in \mathbb{R}^4 : \eta^2 = 0, \eta_0 \geq 0\},$$

called a *(real) light ray*. The complexification λ_ζ of the ray l_ζ which coincides with a complex halfplane $\lambda_\zeta = \{\xi + \alpha \eta : \alpha \in \mathbb{C}, \operatorname{Im} \alpha > 0\}$ is called a *complex light ray* (Fig. 1). The complex light ray λ_ζ goes through the point ζ and lies entirely on S; the section $\lambda_\zeta \cap \{z : \operatorname{Re} z = x\}$ for any fixed $x \in \bar{\lambda}_\zeta \cap \mathbb{R}^4$ coincides

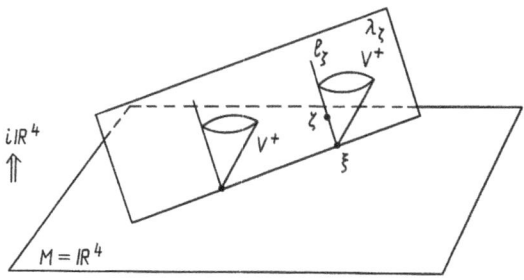

Fig. 1

with the generator l_ζ moved to the point x. At any point ξ of the distinguished boundary we have a collection of real and complex light rays parametrized by points of the 2-dimensional sphere

$$\{\xi + i\eta : \eta_0^2 = \eta_1^2 + \eta_2^2 + \eta_3^2 = 1\}.$$

The *future tube* $\tau^+ = \tau^+(n)$ in $\mathbb{C}^{n+1}$, $n \geq 0$, is defined by analogy with the future tube in $\mathbb{C}^4$ as a tube domain over the *future cone*

$$\tau^+(n) = \mathbb{R}^{n+1} + iV^+ = \{z \in \mathbb{C}^{n+1} : y^2 = y_0^2 - y_1^2 - \cdots - y_n^2 > 0, \, y_0 > 0\},$$

$$V^+ = V^+(n) = \{y \in \mathbb{R}^{n+1} : y^2 > 0, \, y_0 > 0\}.$$

Its boundary has the same structure as for $n = 3$, namely, through any point of the smooth part of the boundary $\partial\tau^+$ there passes a complex light ray lying entirely on the boundary and in any point of the distinguished boundary we have a collection of complex light rays parametrized by points of the $(n-1)$-dimensional sphere. We keep for $\tau^+(n)$ the notations introduced above for $n = 3$ (henceforth we omit usually the index "n" in the notation $\tau^+(n)$).

For $n = 0$ the domain $\tau^+(0)$ coincides with the upper halfplane $(z \in \mathbb{C}$: $\operatorname{Im} z > 0\}$; for $n = 1$ the domain $\tau^+(1)$ can be transformed by a linear transformation onto the domain $\{z \in \mathbb{C}^2 : y_0 > 0, \, y_1 > 0\}$ which is the unbounded realization of the bidisc. These cases are degenerate in the sense that for $n \geq 2$ the future tube $\tau^+(n)$ is not equivalent biholomorphically to the polydisc. We suppose in the sequel that $n \geq 2$.

1.2. Tangent Bundle, Levi Form. Denote by r the real function

$$r(z) = -y^2, \quad z \in \mathbb{C}^{n+1}.$$

This is a local defining function of τ^+ in the sense that

$$\tau^+ = \{z : r(z) < 0\} \cap \{y_0 > 0\}.$$

Differentials of this function having the form

$$dr(z) = \sum_{j=0}^{n} \frac{\partial r(z)}{\partial z_j} dz_j + \sum_{j=0}^{n} \frac{\partial r(z)}{\partial \bar{z}_j} d\bar{z}_j = -2y \cdot dy$$

$$= -2(y_0 \, dy_0 - y_1 \, dy_1 - \cdots - y_n \, dy_n),$$

$$\partial r(z) = \sum_{j=0}^{n} \frac{\partial r(z)}{\partial z_j} dz_j = iy \cdot dz = i(y_0 \, dz_0 - y_1 \, dz_1 - \cdots - y_n \, dz_n),$$

are non-degenerate at all points $z \in \overline{\tau^+} \backslash M$.

Let us consider in more detail the structure of the boundary $\partial\tau^+$ at points $\zeta \in S$. Denote by $T_\zeta S$ the tangent space of S at ζ and by $T_\zeta^c S$ the *complex tangent space* of S at ζ. The latter space is defined as the linear space of tangent vectors $Z = \sum Z^j \, \partial z_j \in T_\zeta \mathbb{C}^{n+1}$ satisfying the condition

$$(\partial r(\zeta), Z) = i(\eta_0 Z^0 - \eta_1 Z^1 - \cdots - \eta_n Z^n) = 0, \quad \zeta = \xi + i\eta \in S.$$

Consider the following vectors at a point $\zeta \in S$

$$Z_0 = \frac{\eta_0}{|\eta|}\frac{\partial}{\partial z_0} + \sum_{j=1}^{n}\frac{\eta_j}{|\eta|}\frac{\partial}{\partial z_j}, \quad Z_n = \frac{\eta_0}{|\eta|}\frac{\partial}{\partial z_0} - \sum_{j=1}^{n}\frac{\eta_j}{|\eta|}\frac{\partial}{\partial z_j},$$

$$Z_k = \sum_{j=1}^{n} X_k^j \frac{\partial}{\partial z_j}, \quad k = 1, \ldots, n-1,$$

where the vectors $X_k = (X_k^1, \ldots, X_k^n)$, $k = 1, \ldots, n-1$, form an orthonormal base in the hyperplane $\sum_{j=1}^{n} \eta_j X^j = 0$. The vectors $Z_0, Z_1, \ldots, Z_{n-1}$ generate the space $T_\zeta^c S$ and the vectors

$$X_0 = 2\,\text{Re}\,Z_0, \ldots, X_{n-1} = 2\,\text{Re}\,Z_{n-1}, \quad T = 2\,\text{Re}\,Z_n,$$

$$Y_0 = -2\,\text{Im}\,Z_0, \ldots, Y_{n-1} = -2\,\text{Im}\,Z_{n-1}$$

form an orthonormal base of the space $T_\zeta S$. The vector Z_0 points in the direction of the complex light ray λ_ζ.

The *Levi form* $\mathscr{L}_r$ at a point ζ is defined by

$$\mathscr{L}_r(Z, \overline{W}) = \partial\bar{\partial} r(\zeta)(Z, \overline{W}) = \frac{1}{2}(-Z^0\overline{W}^0 + Z^1\overline{W}^1 + \cdots + Z^n\overline{W}^n)$$

on vectors

$$Z = \sum_{j=0}^{n} Z^j \partial/\partial z_j, \quad W = \sum_{j=0}^{n} W^j \partial/\partial z_j.$$

The matrix of the restriction of the Levi form $\mathscr{L}_r(Z, \overline{Z})$ to the complex tangent space $T_\zeta^c S$ in the base $\{Z_0, Z_1, \ldots, Z_{n-1}\}$ is diagonal and has the form diag $(0, 1/2, \ldots, 1/2)$, i.e. the restriction of $\mathscr{L}_r$ to $T_\zeta^c S$ has one zero eigenvalue and a positive eigenvalue of multiplicity $n-1$.

1.3. Group Structure, Automorphisms.

The *Lorentz group* L consists of all linear transformations of $\mathbb{R}^{n+1}$ preserving the quadratic form $y^2 = y_0^2 - y_1^2 - \cdots - y_n^2$, $y \in \mathbb{R}^{n+1}$, and fixing the origin. Denote by $L^\uparrow$ the subgroup of L consisting of transformations preserving the cone V^+ (i.e. preserving the orientation of "time" y_0). Linear automorphisms of the future tube τ^+ are given by transformations of the form $z \to Az + b$ where A is a linear transformation of $\mathbb{R}^{n+1}$ preserving the cone V^+ and fixing the origin (in other words, A is a composition of transformation of $L^\uparrow$ and dilatations), and b is an arbitrary vector of $\mathbb{R}^{n+1}$. Transformations of this type exhaust all analytic automorphisms of τ^+ continuous in the closure $\bar{\tau}^+$. *Conformal transformations* of the space M with the metric y^2 are generated by Poincaré transformations $x \to Ax + b$ where $A \in L^\uparrow$, $b \in \mathbb{R}^{n+1}$, dilatations, and inversions (inversion with respect to the origin is given by $x \to x/x^2$). An arbitrary analytic automorphism of the future tube τ^+ is a composition of transformations of this type (cf. Vladimirov (1964) and also Sect. 2.2, 2.3 and Chap. 2, Sect. 4.3).

§2. The Future Tube as a Classical Domain

2.1. A Realization of the Future Tube as the Generalized Unit Disc. We construct here a biholomorphic map of the future tube $\tau^+ = \tau^+$ (3) onto a bounded homogeneous domain – the generalized unit disc. This map is a composition of two mappings. The first of them is a realization of τ^+ as the generalized upper halfplane. It is given by the formula

$$z \to \tilde{z} = \begin{pmatrix} z_0 + z_3 & z_1 - iz_2 \\ z_1 + iz_2 & z_0 - z_3 \end{pmatrix} = \sum_{k=0}^{3} z_k \sigma_k, \tag{1}$$

where σ_0 is the unit 2×2-matrix, σ_i for $i = 1, 2, 3$ are the Pauli matrices

$$\sigma_1 = \begin{pmatrix} 0 & 1 \\ 1 & 0 \end{pmatrix}, \quad \sigma_2 = \begin{pmatrix} 0 & -i \\ i & 0 \end{pmatrix}, \quad \sigma_3 = \begin{pmatrix} 1 & 0 \\ 0 & -1 \end{pmatrix}.$$

The mapping (1) biholomorphically maps the future tube τ^+ onto the *generalized upper halfplane* H consisting of complex 2×2-matrices $\tilde{z}$ with positive definite imaginary part $\text{Im } \tilde{z} = \dfrac{1}{2i}(\tilde{z} - \tilde{z}^*)$. The mapping inverse to (1) is given by the formula

$$\tilde{z} \to z = \left(\frac{1}{2} \text{Tr } \tilde{z}, \frac{1}{2} \text{Tr}(\tilde{z}\sigma_1), \frac{1}{2} \text{Tr}(\tilde{z}\sigma_2), \frac{1}{2} \text{Tr}(\tilde{z}\sigma_3) \right).$$

The mapping (1) has the following properties

$$\det \tilde{z} = z^2, \quad \det(\text{Im } \tilde{z}) = y^2.$$

Its extension to the distinguished boundary M maps M bijectively onto the space of Hermitian 2×2-matrices.

The second mapping is a realization of the generalized upper halfplane as the generalized unit disc. It is given by the *Cayley transform*

$$\tilde{z} \to Z = (I - i\tilde{z})^{-1}(I + i\tilde{z}), \tag{2}$$

mapping the generalized upper halfplane H biholomorphically onto the *generalized unit disc*

$$B = \{Z \in \mathbb{C}[2 \times 2] : ZZ^* < I\}.$$

In other words, B consists of complex 2×2-matrices Z such that the matrix $I - Z^*Z$ is positive definite. The inverse Cayley transform has the form

$$Z \to \tilde{z} = i(I - Z)(I + Z)^{-1}.$$

The composed mapping $z \to \tilde{z} \to Z$ maps the future tube τ^+ biholomorphically onto the generalized unit disc B and is given by the formula

$$z \to Z = \frac{1}{\Delta(z)} \begin{pmatrix} 1 + z^2 + 2iz_3 & 2(iz_1 + z_2) \\ 2(iz_1 - z_2) & 1 + z^2 - 2iz_3 \end{pmatrix}, \tag{3}$$

where $\Delta(z) = \det(I - i\bar{z}) = 1 - z^2 - 2iz_0 = -(z + \mathbf{i})^2$, $\mathbf{i} = (i, 0, 0, 0)$. We have

$$\det(I - ZZ^*) = \frac{16y^2}{|\Delta(z)|^2} = \frac{16y^2}{|(z + \mathbf{i})^2|^2}.$$

The extension of the mapping (3) to the distinguished boundary M maps M injectively into the distinguished boundary $U = \{Z : ZZ^* = I\}$ of the generalized unit disc B which coincides with the group $U(2)$ of unitary 2×2-matrices. The image of the mapping (3) coincides with the set $U \setminus U_0$ where

$$U_0 = \{X \in U : \det(I + X) = 0\}.$$

2.2. Geometry of the Generalized Unit Disc. The generalized unit disc B is a convex domain with the boundary given by

$$\partial B = \{Z \in \mathbb{C}[2 \times 2] : \det(I - Z^*Z) = 0, ZZ^* \leq I\}.$$

Note that the set $\{\det(I - Z^*Z) = 0\}$ has two parts – the bounded part consists of Z subject to the condition $ZZ^* \leq I$, and the unbounded one given by the inequality $Z^*Z \geq I$. These parts intersect in the distinguished boundary U.

In terms of the polar representation of matrices $Z \in \bar{B}$

$$Z = X\Lambda,$$

where $X \in U(2)$, Λ is a Hermitian operator ($\Lambda = \Lambda^*$) with $0 \leq \Lambda \leq I$. We can rewrite the boundary ∂B in the form

$$\partial B = \{Z = X\Lambda : \det(I - \Lambda) = 0, 0 \leq \Lambda \leq I\}.$$

Let us consider the structure of the boundary at points $z \in \partial B \setminus U$. After diagonalization of the matrix Λ we represent the matrix Z in the form

$$Z = XV \begin{pmatrix} 1 & 0 \\ 0 & \lambda \end{pmatrix} V^*, \tag{4}$$

where $V \in U(2)$, $0 \leq \lambda < 1$. The matrix V in this representation is defined up to multiplication from the right by a diagonal unitary matrix. The matrix X parametrizes points of the distinguished boundary U and the set $U(2)/\mathrm{diag}\, U(2)$ which parametrizes classes of matrices V is a 2-dimensional sphere S^2. At any point $Z \in \partial B \setminus U$ given by (4) we have a complex disc consisting of points

$$XV \begin{pmatrix} 1 & 0 \\ 0 & \alpha \end{pmatrix} V^*, \quad |\alpha| < 1, \quad \alpha \in \mathbb{C},$$

lying entirely on $\partial B \setminus U$. This disc is an analogue of a complex light ray in the future tube. Denote by ρ the real function

$$\rho(Z) = \rho(Z, Z^*) = -\det(I - Z^*Z), \quad Z \in \mathbb{C}[2 \times 2].$$

This is a local defining function of B at points of $\partial B \setminus U$ which means that any point $Z_0 \in \partial B \setminus U$ has a neighborhood Ω such that $B \cap \Omega = \{Z : \rho(Z) < 0\}$ and

$d\rho(Z) \neq 0$ for $Z \in \Omega$. The last inequality follows from the explicit formula for the differentials of ρ:

$$\partial\rho(Z) = -\partial(\det Z) \det Z^* + \mathrm{Tr}(dZ \cdot Z^*),$$

$$d\rho(Z) = -\bar{\partial}(\det Z^*) \det Z - \partial(\det Z) \det Z^* + \mathrm{Tr}(dZ \cdot Z^*) + \mathrm{Tr}(Z \cdot dZ^*).$$

These expressions are derived using the following identity for ρ

$$\rho(Z) = -1 - \det(ZZ^*) + \mathrm{Tr}(ZZ^*).$$

In particular, for $Z = Z_0 = \begin{pmatrix} 1 & 0 \\ 0 & \lambda \end{pmatrix}$ we obtain that

$$\partial(\det Z) = \lambda \, dz_{11} + dz_{22}, \quad \bar{\partial}(\det Z^*) = \lambda \, d\bar{z}_{11} + d\bar{z}_{22},$$

$$d\rho(Z) = (1 - \lambda^2)(dz_{11} + d\bar{z}_{11}) \neq 0 \quad \text{for } \lambda < 1.$$

It follows using the homogeneity of $\partial B \backslash U$ that $d\rho(Z) \neq 0$ for any $Z \in \partial B \backslash U$.

The Levi form of ρ in points $Z \in \partial B \backslash U$ is computed as follows

$$\mathscr{L}_\rho = \partial\bar{\partial}\rho(Z) = -\partial(\det Z) \wedge \bar{\partial}(\det Z^*) + \mathrm{Tr}(dZ \wedge dZ^*).$$

So, in particular, at the points $Z_0 = \begin{pmatrix} 1 & 0 \\ 0 & \lambda \end{pmatrix}$, $0 \leq \lambda < 1$ it is equal to

$$\mathscr{L}_\rho = (1 - \lambda^2) \, dz_{11} \wedge d\bar{z}_{11} + dz_{12} \wedge d\bar{z}_{12} + dz_{21} \wedge d\bar{z}_{21}$$

$$- \lambda(dz_{11} \wedge d\bar{z}_{22} + dz_{22} \wedge d\bar{z}_{11}).$$

The complex tangent space at a point $Z \in \partial B \backslash U$ is given by the equation $\left(\dfrac{\partial\rho}{\partial Z}(Z), W - Z \right) = 0$ where $(A, C) = \mathrm{Tr}(AC')$ is a complex linear inner product in the space of matrices. At the point Z_0 the complex tangent space is given by $w_{11} = 1$. The restriction of the Levi form to this space has the form

$$\mathscr{L}_\rho(W - Z_0, \overline{W} - \overline{Z}_0) = |w_{12}|^2 + |w_{21}|^2,$$

so it has one positive eigenvalue 1 of multiplicity 2, and one zero eigenvalue. Because of the homogeneity, the same assertion is true at any point of $\partial B \backslash U$.

Analytic automorphisms of the generalized unit disc are given by the mappings (cf. Siegel (1949), Hua (1958), Piatetski-Shapiro (1961)):

$$Z \rightarrow (AZ + B)(CZ + D)^{-1}, \quad Z \rightarrow {}'Z,$$

where the block 4×4-matrix $M = \begin{pmatrix} A & B \\ C & D \end{pmatrix}$ belongs to the unitary group $U(2, 2)$, i.e.

$$M^*JM = J \quad \text{where} \quad J = \begin{pmatrix} -I & 0 \\ 0 & I \end{pmatrix}.$$

The generalized unit disc B_m is defined analogously as

$$B_m = \{Z \in \mathbb{C}[m \times m] : ZZ^* < I\}.$$

For any m, B_m is biholomorphic to a tube domain over a cone in $\mathbb{C}^{m^2}$ (cf. Sect. 5.1), however for $m > 2$ the domains B_m and $\tau^+(m^2 - 1)$ are not biholomorphically equivalent.

2.3. A Realization of the Future Tube as the Lie Ball. The existence of the biholomorphic equivalence between the future tube $\tau^+(3)$ and the generalized unit disc (a classical Cartan domain of the Ist type, cf. Sect. 5.1) is, as was noted above, a low-dimensional effect. Analogously, for $n = 2$ the future tube $\tau^+(2)$ can be realized as a classical domain of the IIIrd type given by the set of symmetric matrices belonging to the generalized unit disc B_2, i.e. $\{Z \in B_2 : {}'Z = Z\}$, which is also a low-dimensional phenomenon. In this Section we shall construct a bounded realization of the future tube $\tau^+(n)$ for any n as a classical domain of the IVth type called the Lie ball. The biholomorphic mapping of $\tau^+(n)$ onto the Lie ball is a composition of two mappings.

The first mapping is a realization of $\tau^+ = \tau^+(n)$ as a domain on a complex quadric in $\mathbb{CP}^{n+2}$. Let us introduce the new variables

$$z_1 = \frac{s_1}{s_0}, \ldots, z_n = \frac{s_n}{s_0}, \quad z_0 = \frac{s_{n+1}}{s_0}.$$

In these variables the domain $\tau = \{z \in \mathbb{C}^{n+1} : y^2 > 0\}$ will transform to the domain

$$\mathscr{D}' = \{s \in \mathbb{C}^{n+3} : -|s_0|^2 - \cdots - |s_n|^2 + |s_{n+1}|^2 + 2\,\mathrm{Re}(\overline{s_0}s_{n+2}) > 0,$$
$$- s_0^2 - \cdots - s_n^2 + s_{n+1}^2 + 2s_0 s_{n+2} = 0\}$$

(note that $s_0 \neq 0$ for $s \in \mathscr{D}'$ so we can divide out s_0). Changing the variables s_0, $s_1, \ldots, s_{n+2}$ to the variables $t_0 = s_0 - s_{n+2}, t_1 = s_1, \ldots, t_{n+2} = s_{n+2}$ we can write $\mathscr{D}'$ in the form

$$\mathscr{D} = \{t \in \mathbb{C}^{n+3} : -|t_0|^2 - \cdots - |t_n|^2 + |t_{n+1}|^2 + |t_{n+2}|^2 > 0,$$
$$- t_0^2 - \cdots - t_n^2 + t_{n+1}^2 + t_{n+2}^2 = 0\}.$$

The domain $\mathscr{D}$ is a section of the domain $\widetilde{\mathscr{D}} = \{t \in \mathbb{C}^{n+3} : |t_0|^2 + \cdots + |t_n|^2 < |t_{n+1}|^2 + |t_{n+2}|^2\}$ by the complex quadric $\{t_0^2 + \cdots + t_n^2 = t_{n+1}^2 + t_{n+2}^2\}$. The domains $\mathscr{D}$ and $\widetilde{\mathscr{D}}$ are given by homogeneous relations so it's more natural to consider them as domains in $\mathbb{CP}^{n+2}$. Note that the Levi form of the domain $\widetilde{\mathscr{D}}$ being restricted to the complex tangent space of $\partial\widetilde{\mathscr{D}}$ at a point t with $t_{n+2} \neq 0$ has one negative and $n + 1$ positive eigenvalues. The domain $\mathscr{D}$ has two components distinguished by the sign of $\mathrm{Im}\dfrac{t_{n+1}}{t_{n+2}}$. The future tube τ^+ is biholomorphic to the domain $\mathscr{D}_+$ on the quadric in $\mathbb{CP}^{n+2}$ given in homogeneous coordinates as follows

$$\left\{ [t_0, t_1, \ldots, t_{n+2}] : |t_0|^2 + \cdots + |t_n|^2 < |t_{n+1}|^2 + |t_{n+2}|^2, \right.$$
$$\left. t_0^2 + \cdots + t_n^2 = t_{n+1}^2 + t_{n+2}^2, \ \mathrm{Im}\frac{t_{n+1}}{t_{n+2}} > 0 \right\}.$$

This representation of τ^+ as a domain on the quadric in $\mathbb{C}\mathbb{P}^{n+2}$ is closely related to the Penrose representation considered in the following Section.

The second mapping is a realization of $\mathcal{D}_+$ as the Lie ball and is given by the formula

$$w_0 = \frac{t_0}{t_{n+1} + it_{n+2}}, \ldots, \quad w_n = \frac{t_n}{t_{n+1} + it_{n+2}}.$$

Under this mapping the domain $\mathcal{D}_+$ transforms biholomorphically onto the domain

$$B_L = \{w \in \mathbb{C}^{n+1} : |w_0^2 + \cdots + w_n^2|^2 + 1 > 2|w_0|^2 + \cdots + 2|w_n|^2,$$

$$|w_0^2 + \cdots + w_n^2|^2 < 1\}$$

called *the classical domain of the IVth type* or *the Lie ball*. (In Hua's (1958) book this domain is called the Lie sphere. We prefer to call it the Lie ball reserving the name "Lie sphere" for the distinguished boundary of B_L).

The composed mapping of τ^+ onto B_L is given by the formula

$$w_0 = i\frac{1 + z^2}{(z + \mathbf{i})^2}, \quad w_1 = i\frac{2z_1}{(z + \mathbf{i})^2}, \ldots, w_n = i\frac{2z_n}{(z + \mathbf{i})^2}$$

where $\mathbf{i} = (i, 0, \ldots, 0)$. In particular, points of the form $z = (iy_0, 0, \ldots, 0)$ transform to points $w = \left(-i\frac{1 - y_0}{1 + y_0}, 0, \ldots, 0\right)$.

The distinguished boundary of the future tube transforms into the set $S_L = \{|w_0|^2 + \cdots + |w_n|^2 = 1, |w_0^2 + \cdots + w_n^2| = 1\}$. Let us consider this set in more detail. Set $w = u + iv, (z, \omega) = z_0\omega_0 + z_1\omega_1 + \cdots + z_n\omega_n$. Then the intersection of S_L with the complex sphere $\Sigma_1 = \{w : (w, w) = 1\}$ is given by the equations $|u|^2 = |v|^2 + 1, (u, v) = 0, |u|^2 + |v|^2 = 1$. It follows that $v = 0$; hence Σ_1 intersects S_L in the n-dimensional real sphere $\{u \in \mathbb{R}^{n+1} : |u|^2 = 1\}$. So the set S_L can be written as

$$S_L = \{w = e^{i\theta}u : |u|^2 = 1\}.$$

This set is called the distinguished boundary of B_L or the *Lie sphere*.

Consider now the smooth part of the boundary of B_L

$$\partial B_L \backslash S_L = \{|(w, w)|^2 + 1 = 2|w|^2, |(w, w)| < 1\}.$$

The complex sphere $\Sigma_\lambda = \{w : (w, w) = \lambda\}$, $|\lambda| < 1$, intersects $\partial B_L \backslash S_L$ in the set

$$\left\{w = u + iv : |u| = \frac{|1 + \lambda|}{2}, |v| = \frac{|1 - \lambda|}{2}, (u, v) = \frac{\text{Im } \lambda}{2}\right\}$$

which coincides with the product of spheres $S^n \times S^{n-1}$. This defines a fibration of $\partial B_L \backslash S_L$ by $(2n - 1)$-dimensional real submanifolds parametrized by points of the disc $\{\lambda \in \mathbb{C} : |\lambda| < 1\}$.

The local defining function of B_L at points of $\partial B_L \backslash S_L$ is given by

$$\rho_L(w) = 2|w|^2 - |(w, w)|^2 - 1.$$

Its differentials have the form

$$\partial \rho_L(w) = 2(\overline{w}, dw) = -2\overline{w}^2(w, dw),$$

$$\overline{\partial} \rho_L(w) = 2(w, d\overline{w}) - 2w^2(\overline{w}, d\overline{w})$$

whence the Levi form is computed as follows

$$\mathcal{L}_L = \partial \overline{\partial} \rho_L(w) = 2\, dw \wedge d\overline{w} - 4(w, dw) \wedge (\overline{w}, d\overline{w}).$$

The restriction of the Levi form to the complex tangent space $T_w^c(\partial B_L \backslash S_L)$ has the following properties: it is positively defined on vectors belonging to $T_w^c(\partial B_L \cap \Sigma_\lambda)$ where Σ_λ is a complex sphere through the point w, and equals to zero in the transversal direction (defined by the projection of the vector field $\partial/\partial \lambda$ on $T_w^c(\partial B_L)$).

Analytic automorphisms of the Lie ball B_L are given by the following transformations (cf. Hua (1958))

$$w \to \left\{ \left[\left(\frac{(w, w) + 1}{2}, i\frac{(w, w) - 1}{2} \right) A + wC \right] \binom{1}{i} \right\}^{-1}$$

$$\cdot \left[\left(\frac{(w, w) + 1}{2}, i\frac{(w, w) - 1}{2} \right) B + wD \right],$$

where A, B, C, D are respectively real $2 \times 2, 2 \times (n + 1), (n + 1) \times 2, (n + 1) \times (n + 1)$ matrices subject to the condition

$$'MJM = J,$$

where M, J are the block $(n + 3) \times (n + 3)$ matrices

$$M = \begin{pmatrix} A & B \\ C & D \end{pmatrix}, \quad J = \begin{pmatrix} I_2 & 0 \\ 0 & I_{n+1} \end{pmatrix}.$$

§ 3. Penrose Representation and Some Physical Applications

3.1. Penrose Representation and Twistor Transform. Denote by J an Hermitian 4×4-matrix having the eigenvalues $(+1, +1, -1, -1)$ and consider the set Ω_J of block 4×2-matrices

$$P = \begin{pmatrix} Z_1 \\ Z_2 \end{pmatrix}$$

of the form

$$\Omega_J = \{P : P^*JP > 0\}.$$

We introduce an equivalence relation in Ω_J by setting two matrices P and P' of the above type equivalent if there exists a non-degenerate matrix R such that $Z_1' = Z_1 R, Z_2' = Z_2 R$. The quotient of Ω_J with respect to this equivalence relation is denoted by $\mathcal{D}_J$.

The domain $\mathcal{D}_J$ can be identified with the Grassmann manifold of 2-dimensional (complex) subspaces in $\mathbb{C}^4$ which are positive with respect to J, i.e. $\mathbf{z}J\mathbf{z}^* > 0$ for any non-zero vector $\mathbf{z}$ of the considered 2-subspace (in other words, the restriction of the Hermitian form corresponding to J to the 2-subspace is positive definite). To prove this assertion, consider a vector $\mathbf{z} \in \mathbb{C}^4$ as a pair of two vectors $\mathbf{z} = (\omega, \pi)$ where $\omega, \pi \in \mathbb{C}^2$ and assign to a matrix $P \in \Omega_J$ a 2-subspace p in $\mathbb{C}^4$ by the equations

$$\pi Z_1 = \omega Z_2.$$

It is clear that a matrix $P' \in \Omega_J$ equivalent to P defines the same 2-subspace. Thus, we have assigned to an arbitrary element of $\mathcal{D}_J$ a 2-subspace in $\mathbb{C}^4$. We show that this subspace is positive with respect to J. The notion of positivity is invariant with respect to unitary transformations of $\mathbb{C}^4$ so we can assume that the matrix J has the diagonal form

$$J = \begin{pmatrix} -I & 0 \\ 0 & I \end{pmatrix}.$$

Then the condition $P^*JP > 0$ is reduced to $Z_2^*Z_2 > Z_1^*Z_1$ so the matrix Z_2 is non-degenerate. Hence we can identify the domain $\mathcal{D}_J$ with the set of matrices $P = \begin{pmatrix} Z \\ I \end{pmatrix}$ such that $Z^*Z < I$, i.e. with the generalized unit disc B. The positivity condition for the corresponding subspace

$$p = \{(\omega, \pi) : \pi Z = \omega\}$$

can be written as $\pi\pi^* > \omega\omega^*$ for $0 \neq (\omega, \pi) \in p$ which is equivalent to the inequality $\pi(I - ZZ^*)\pi^* > 0$. We have thus defined a correspondence between $\mathcal{D}_J$ and the Grassmann manifold of positive 2-subspaces in $\mathbb{C}^4$. It is easy to show that it is a one-to-one correspondence.

The space $\mathbb{C}^4$ with the Hermitian form $\Phi(\mathbf{z}) = |z_1|^2 + |z_2|^2 - |z_3|^2 - |z_4|^2$, $\mathbf{z} = (z_1, z_2, z_3, z_4) \in \mathbb{C}^4$, given by the matrix J, is called the *twistor space* and denoted by $\mathbb{T}$. A twistor $\mathbf{z} \in \mathbb{T}$ is called positive (respectively, negative, null) if $\Phi(\mathbf{z}) = \mathbf{z}J\mathbf{z}^* > 0$ (respectively, $\Phi(\mathbf{z}) < 0$, $\Phi(\mathbf{z}) = 0$). The corresponding subspaces of $\mathbb{T}$ are denoted by $\mathbb{T}^+$, $\mathbb{T}^-$, $\mathbb{N}$ respectively. We have shown above that the domain $\mathcal{D}_J$ which can be identified with the generalized unit disc B coincides with the Grassmann manifold $G_2(\mathbb{T}^+)$ of 2-subspaces in $\mathbb{T}^+$. So, B is identified with $G_2(\mathbb{T}^+)$.

If we take another matrix representation of Φ (or J), namely

$$J = \begin{pmatrix} 0 & iI \\ -iI & 0 \end{pmatrix},$$

we obtain a realization of $\mathcal{D}_J$ as the generalized upper halfplane H, so $G_2(\mathbb{T}^+)$ is identified also with H. The representation of the future tube $\tau^+ = \tau^+(3)$ which is biholomorphic to H, as the Grassmann manifold $G_2(\mathbb{T}^+)$ will be called the *Penrose representation*. The correspondence $\tau^+ \mapsto G_2(\mathbb{T}^+)$ is extended to the distinguished boundary M of τ^+ and to the whole space $\mathbb{C}^4$. If we identify M

with the complexified Minkowski space $\mathbb{C}M$, we obtain the embeddings $M \hookrightarrow G_2(\mathbb{N})$, $\mathbb{C}M \hookrightarrow G_2(\mathbb{T})$. The space $G_2(\mathbb{N})$ is the twistor model of the Minkowski space; $G_2(\mathbb{T})$ is the twistor model of the complexified Minkowski space.

Using these embeddings we can transform relativistic (conformally invariant) fields on the Minkowski space to the twistor space $\mathbb{T}$. This transformation is called the twistor (or Penrose) transform. Under this transform conformally invariant objects on M correspond to complex analytic objects on $\mathbb{T}$ such as holomorphic bundles, cohomologies with coefficients in these bundles and so on (cf. Twistors and Gauge Fields (1983) and references therein).

We note in conclusion that the constructed Penrose representation $\tau^+ \mapsto G_2(\mathbb{T}^+)$ is closely related to the realization of τ^+ as the domain $\mathcal{D}_+$ on the complex quadric in $\mathbb{C}P^5$ defined in this Section. To see this it is sufficient to represent $G_2(\mathbb{T})$ as a complex quadric in $\mathbb{C}P^5$ (cf. e.g. Chern (1956)).

3.2. Conformal Compactification of the Minkowski Space. The twistor model $G_2(\mathbb{N})$ of the Minkowski space constructed in the last Section is a compact space so it defines through the embedding $M \hookrightarrow G_2(\mathbb{N})$, a natural compactification $\mathbb{M}$ of the Minkowski space M. Using the correspondence between $G_2(\mathbb{T}^+)$ and the generalized unit disc B (cf. last Section) which can be extended to a homeomorphism of the distinguished boundaries $U \mapsto G_2(\mathbb{N})$, we can identify $\mathbb{M}$ with U and study the compactification $\mathbb{M}$ through the embedding $M \to U$ constructed in Sect. 2.1. The compactification $\mathbb{M}$ coincides with the *conformal compactification of Minkowski* space known in quantum field theory (cf. Penrose (1980, 1967), Uhlmann (1963)). It has the following properties. The "points at infinity" of $\mathbb{M}$ correspond to the points of the set $U_0 = \{X \in U : \det(I + X) = 0\}$ (cf. Sect. 2.1). We may represent elements of U in the form $U \ni X = e^{i\varphi/2}u$, where $0 \le \varphi \le 2\pi$, $u \in \mathrm{SU}(2)$, so

$$u = \begin{pmatrix} \alpha & \beta \\ -\bar{\beta} & \bar{\alpha} \end{pmatrix}, \quad |\alpha|^2 + |\beta|^2 = 1, \quad \alpha, \beta \in \mathbb{C}^2.$$

(This representation will be uniquely defined if we identify the pairs $(\varphi = 0, u)$ and $(\varphi = 2\pi, -u)$ for any $u \in \mathrm{SU}(2)$). The set U_0 in this parametrization is equal to $\{(\varphi, u) \in U : \mathrm{Re}\,\alpha + \cos\varphi/2 = 0\}$. Thus, topologically U_0 is the torus $S^2 \times S^1$, with one of the equators (corresponding to $\varphi = 0$ and $\varphi = 2\pi$) shrunken to a point.

Let us consider the topology of $\mathbb{M}$ in a neighborhood of the points at infinity using formula (3) from Sect. 2.1. Denote the points of M by $x = (x_0, \mathbf{x}) = (x_0, x_1, x_2, x_3)$ and consider the limits of various straight lines in $\mathbb{M}$. It follows from (3) that the limits in $\mathbb{M}$ of all "time" lines $x^0 = x_0^0 + t$, $\mathbf{x} = \mathbf{x}^0$ (where x^0 is a fixed point of M) and all "space" lines $x^0 = x_0^0$, $\mathbf{x} = \mathbf{x}^0 + \alpha t$ (where $\alpha = (\alpha_1, \alpha_2, \alpha_3)$ is a fixed point of the sphere $|\alpha|^2 = 1$) for $t \to \pm\infty$ coincide with each other and are equal to the unique point at infinity of $\mathbb{M}$ corresponding via (3) to the matrix $X = -I$. This point is denoted by I_0 and called the *spacetime infinity*. From the other side, the limits of the "light" line $\mathbf{x} = \alpha(x_0 - r)$ (where $|\alpha|^2 = 1$, r is a fixed real number) for $x_0 \to \pm\infty$ coincide and are equal to the point at infinity of $\mathbb{M}$ corresponding via (3) to the matrix

$$
X = \begin{bmatrix} \dfrac{r + i\alpha_3}{-r - i} & \dfrac{i\alpha_1 + \alpha_2}{-r - i} \\[3mm] \dfrac{i\alpha_1 - \alpha_2}{-r - i} & \dfrac{r - i\alpha_3}{-r - i} \end{bmatrix} \tag{5}
$$

The set of points of $\mathbb{M}$ corresponding to matrices of the form (5) is called the *light infinity* and denoted by $\mathfrak{J}$. Generalizing the last assertion we can prove that the limits in $\mathbb{M}$ of a "light" line $\mathbf{x} = \mathbf{x}^0 + \alpha x_0$ with $(x^0, \alpha) + r = 0$ (where $|\alpha|^2 = 1$, $r \in \mathbb{R}$, α and r are fixed) for $x_0 \to \pm\infty$ coincide and are equal to the point (5) at infinity. The limits in $\mathbb{M}$ of all non-light lines are equal to the spacetime infinity I_0. So the set of points at infinity of $\mathbb{M}$ is parametrized, according to (5), by pairs (r, α) where $\alpha \in S^2$, $-\infty \leq r \leq \infty$ and all points of the form $(\pm\infty, \alpha)$ are identified. $\Big($This parametrization is related to the parametrization of U_0 defined above through the change of variables $e^{i\varphi} = \dfrac{r - i}{r + i}\Big)$.

One can imagine the set of points at infinity of $\mathbb{M}$ as a "spinning top" with the equator shrunken to a point and identified with the vertices, and the upper and bottom cones identified along opposite generators (cf. Fig. 2). This interpretation was proposed by Penrose.

It is also possible to describe neighborhoods of points at infinity. Consider first the spacetime infinity I_0. Introduce the sets

$$
U_r^I = \overset{+}{U}_r^I \cup \bar{U}_r^I, \quad r \in \mathbb{R},
$$

where

$$
\overset{+}{U}_r^I = \{x : (x_0 - r)^2 > |\mathbf{x}|^2, x_0 > r\},
$$
$$
\bar{U}_r^I = \{x : (x_0 + r)^2 > |\mathbf{x}|^2, x_0 < -r\}.
$$

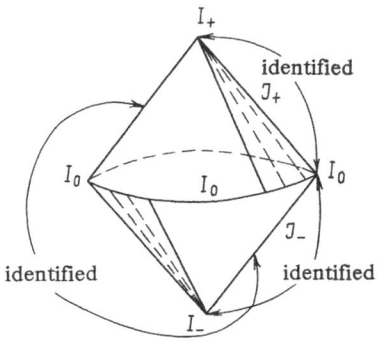

Fig. 2

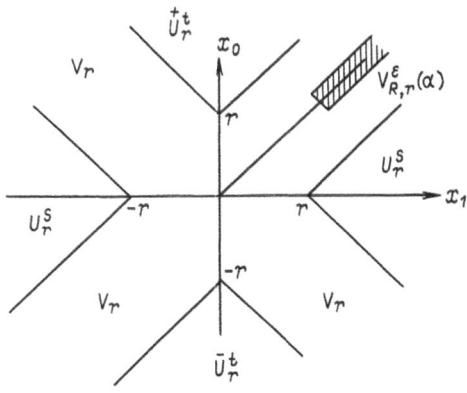

Fig. 3

In other words, $\overset{+}{U}{}^t_r$ is the interior of the future light cone with vertex at the point $(r, 0, 0, 0)$, $\bar{U}^t_r$ is the interior of the past light cone with vertex at point $(-r, 0, 0, 0)$ (cf. Fig. 3). Denote by U^s_r the complement of the set $\overset{+}{U}{}^t_{-r} \cup \bar{U}^t_{-r}$. The set U^s_r can be obtained by rotation of the cone $\{(x_0, x_1, 0, 0) : (x_1 - r)^2 > x_0^2, x_1 > r\}$ around the axis (x_0) in M. Finally, put $U_r = U^t_r \cup U^s_r$ and denote by V_r the complement of U_r in M. Then the completions $\tilde{U}_r$ of sets U_r in the topology of M (i.e. $\tilde{U}_r$ is the union of $\bar{U}_r$ and limit points at infinity of $\bar{U}_r$ in M) form, for $r \to +\infty$, a fundamental system of neighborhoods of the spacetime infinity I_0. Neighborhoods of a point of the light infinity $\mathfrak{J}$ with parameters $r = 0$, $\alpha = \alpha^0$ can be described as follows. Consider the subset $V^\varepsilon_r(\alpha^0)$ of the set V_r filled out by light lines $\mathbf{x} = \alpha(x_0 - s)$ with $|\alpha - \alpha^0| < \varepsilon$, $|s| < r$ and denote by $V^\varepsilon_{R,r}(\alpha^0)$ the intersection of $V^\varepsilon_r(\alpha^0)$ with the exterior of the ball: $\{|x| \le R\}$. Then the completions $\widetilde{V^\varepsilon_{R,r}}(\alpha^0)$ of sets $V^\varepsilon_{R,r}(\alpha^0)$ in M form for $R \to +\infty$, $r \to +0$, $\varepsilon \to +0$ a fundamental system of neighborhoods of the point $(0, \alpha^0) \in \mathfrak{J}$. Neighborhoods of the other points of $\mathfrak{J}$ can be described in an analogous way.

§4. Holomorphic Non-straightening

4.1. Holomorphic Non-straightening. In a neighborhood of any point $\zeta \in S$, the future tube $\tau^+ = \tau^+(n)$ looks locally like the product of a strictly pseudoconvex domain in $\mathbb{C}^n$ and a complex line. More precisely, we can find a neighborhood U of ζ and a diffeomorphism φ of this neighborhood onto an open subset V in $\mathbb{C}^{n+1}$ mapping $\tau^+ \cap V$ onto $(\mathbb{C}^1 \times \mathscr{D}') \cap V$ where $\mathscr{D}'$ is a strictly pseudoconvex domain in $\mathbb{C}^n$. Indeed, this diffeomorhism is given by the formula $w = \varphi(\zeta)$ where $w_0 = \zeta_0$, $w_1 = \zeta_1/\eta_0, \ldots, w_n = \zeta_n/\eta_0$. The domain $\mathscr{D}'$ has the form

$$\mathscr{D}' = \{w' = (w_1, \ldots, w_n) : (\operatorname{Im} w_1)^2 + \cdots + (\operatorname{Im} w_n)^2 < 1\}$$

which is a convex and strictly pseudoconvex domain (note that $\mathscr{D}'$ is not strictly convex because the tangent space at any point of $\partial\mathscr{D}'$ sticks to $\partial\mathscr{D}'$ along an n-dimensional real plane).

The constructed local diffeomorphism φ "straightens" the hypersurface S along complex light rays lying on S. However, there is no biholomorphism with the same property. Namely, we have the following.

Theorem 1 (Sergeev (1983, 1986), Sergeev-Vladimirov (1986)). *The hypersurface S cannot be biholomorphically straightened along complex light rays in a neighborhood of any of its points.*

This theorem is proved by checking the necessary condition for biholomorphic straightening found by Freeman (1970, 1977). In fact, the assertion of the theorem remains true if we weaken the definition of the straightening biholomorphism φ assuming only that φ is defined and holomorphic in a one-sided neighborhood $\tau^+ \cap U$ and smooth up to $U \cap \partial\tau^+$ (Sh. Tsyganov) or even that φ is a CR-diffeomorphism in a neighborhood of ζ in S (S. Pinchuk (1990), Sh. Tsyganov).

In Khenkin-Sergeev (1980) a notion of strictly pseudoconvex polyhedra was introduced unifying the notions of strictly pseudoconvex domains and that of analytic polyhedra. A domain Ω in $\mathbb{C}^m$ is called a *strictly pseudoconvex polyhedron* if there exist a domain $\Omega' \supset \bar{\Omega}$, holomorphic mappings χ^α, $\alpha = 1, \ldots, N$, of Ω' onto domains $\Omega'_\alpha \subset \mathbb{C}^{m_\alpha}$ with $m_\alpha \leq m$ and smooth strictly pseudoconvex domains Ω_α, $\bar{\Omega}_\alpha \subset \Omega'_\alpha$, such that Ω has the form

$$\Omega = \{z \in \Omega' : \chi^\alpha(\zeta) \in \Omega_\alpha, \alpha = 1, \ldots, N\}.$$

Thus, Ω is the intersection of the preimages of domains Ω_α with respect to the mappings χ^α. The boundary of Ω consists of smooth pieces $S_\alpha = \{\zeta \in \bar{\Omega} : \chi^\alpha(\zeta) \in \partial\Omega_\alpha\}$, $\alpha = 1, \ldots, N$ and each of these pieces is fibered by complex submanifolds of the form $(\chi^\alpha)^{-1}(\omega)$, $\omega \in \partial\Omega_\alpha$. It is evident that the map χ^α defines a biholomorphic straightening of the hypersurface S_α along these complex submanifolds in a neighborhood of any point on S_α. Moreover, if a polyhedron Ω is nondegenerate (cf. Khenkin-Sergeev, op. cit.), i.e. some conditions of general position type are satisfied on edges

$$S_A = S_{\alpha_1} \cap \cdots \cap S_{\alpha_s}, \quad A = \{\alpha_1, \ldots, \alpha_s\}, \quad 1 \leq \alpha_1 < \cdots < \alpha_s \leq N,$$

then also these edges S_A can be biholomorphically straightened in a neighborhood of any of their points along complex submanifolds of S_A. Conversely, any pseudoconvex domain (with piecewise smooth boundary with general position conditions satisfied on edges) which can be locally biholomorphically straightened in the above sense is locally a strictly pseudoconvex polyhedron. Hence Theorem 1 asserts that the future tube τ^+ gives an example of a pseudoconvex domain which is not (even locally) a strictly pseudoconvex polyhedron. However, it can be approximated up to the 2nd order by strictly pseudoconvex polyhedra as is shown in the next section.

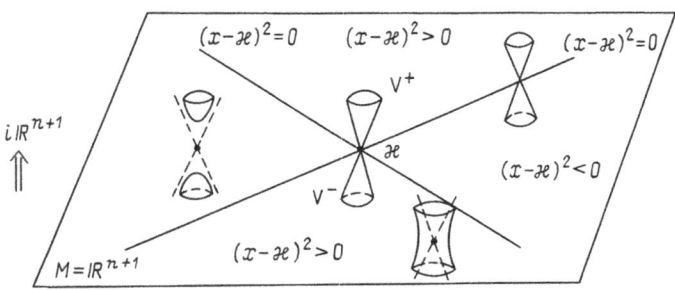

Fig. 4

4.2. Approximation by Strictly Pseudoconvex Polyhedra. Fix a point $\kappa \in \mathbb{R}^{n+1}$ and consider the domain

$$\mathscr{D}_\kappa = \{z \in \mathbb{C}^{n+1} : |z_0 - \kappa_0|^2 > |z' - \kappa'|^2\}$$

where $z' = (z_1, \ldots, z_n)$. The domain $\mathscr{D}_\kappa$ has the following properties. For any $x = \operatorname{Re} z$ belonging to the cone with vertex at κ: $\{x \in \mathbb{R}^{n+1} : (z - \kappa)^2 = 0\}$, the section $\mathscr{D}_\kappa \cap \{z : \operatorname{Re} z = x\}$ of $\mathscr{D}_\kappa$ with the fixed x is the interior of the light cone $\{y \in \mathbb{R}^{n+1} : y^2 > 0\}$. For other x the section of $\mathscr{D}_\kappa$ with fixed x coincides with the interior of the hyperboloid $\{z : (x - \kappa)^2 + y^2 > 0\}$ which has one cavity for $(x - \kappa)^2 > 0$ and two cavities for $(x - \kappa)^2 < 0$ (cf. Fig. 4). Note that $\mathscr{D}_\kappa$ is invariant under the subgroup of the Poincaré group in M fixing the point κ and sections of $\mathscr{D}_\kappa$ with fixed x are invariant under the action of the Lorentz group on these sections.

The holomorphic mapping

$$z \to \chi^\kappa(z) = (\chi_1^\kappa(z), \ldots, \chi_n^\kappa(z)), \quad \chi_j^\kappa(z) = \frac{z_j - \kappa_j}{z_0 - \kappa_0},$$

transforms $\mathscr{D}_\kappa$ onto the ball $\{|\chi_1^\kappa|^2 + \cdots + |\chi_n^\kappa|^2 < 1\}$. So $\mathscr{D}_\kappa$ is the preimage of the ball under the map χ^κ, however this map degenerates on the boundary of $\mathscr{D}_\kappa$ at the point κ ($\partial\mathscr{D}_\kappa$ also degenerates at this point). Let us extend the definition of a strictly pseudoconvex polyhedron Ω given above by allowing the maps χ^α to degenerate on $\partial\Omega$. In this case we shall say that Ω is a *strictly pseudoconvex polyhedron with singularities*. Thus, $\mathscr{D}_\kappa$ is a strictly pseudoconvex polyhedron with singularities.

As was noted above, for $x = \kappa$ the section of $\mathscr{D}_\kappa$ with fixed x coincides with the section of τ^+ for $y_0 > 0$. We can assert more than that. Namely, denote by S_κ the smooth hypersurface $\partial\mathscr{D}_\kappa \cap \{y_0 > 0\}$. Then S_κ coincides with S to the 1st order at any point $z = x + iy \in S$, i.e.

$$T_z S_x = T_z S, \quad T_z^c S_x = T_z^c S.$$

The Levi forms of $\mathscr{D}_\kappa$ and τ^+ also coincide at these points (the Levi form of $\mathscr{D}_\kappa$ is computed using the defining function $r_\kappa(z) = -\frac{1}{2}|z_0 - \kappa_0|^2 + \frac{1}{2}|z' - \kappa'|^2$).

§ 5. Generalizations

5.1. Tube Cones. A *tube cone* or a *Siegel domain of the 1st kind* is a domain of the form

$$T^C = \{z = x + iy \in \mathbb{C}^m : y \in C\} = \mathbb{R}^m + iC$$

where C is an open cone in $\mathbb{R}^m$ with vertex at the origin. According to *Bochner's Tube Theorem* (cf. Vladimirov (1964)), any function holomorphic in T^C can be holomorphically extended to the tube cone $T^{\mathrm{ch}\, C}$ where ch C is the convex hull of C. Hence, it is natural to suppose that the cone C is convex. We shall also assume that the cone C is *proper*, i.e. its closure $\bar{C}$ does not contain a whole line (cf. the motivation of this condition in the note to Theorem 3 from Chap. 2, Sect. 1.2).

Besides the future cone, we have the following examples of convex proper cones:

1) *The octant* $\mathbb{R}^m_+ = \{y \in \mathbb{R}^m : y_1 > 0, \ldots, y_m > 0\}$. The tube cone $T_+ = T^{\mathbb{R}^m_+}$ is biholomorphic to the polydisc $\{z \in \mathbb{C}^m : |z_1| < 1, \ldots, |z_m| < 1\}$.

2) *The cone* $\mathcal{H}_l \subset \mathbb{R}^m$ *with* $m = l^2$ consisting of all complex positive definite Hermitian $l \times l$-matrices. For $l = 2$ the tube cone $T^{\mathcal{H}_2}$ coincides with the generalized upper halfplane H (cf. Sect. 2.1). For any l the tube cone $T^{\mathcal{H}_l}$ is biholomorphic to the generalized unit disc B_l (cf. Sect. 2.2) which is a particular case of a *classical Cartan domain of the Ist type* (Cartan (1935), Siegel (1949), Piatetski-Shapiro (1961)) consisting of complex $p \times q$-matrices Z, $p \geq q \geq 1$, subject to the condition $ZZ^* < I$. This domain is biholomorphic to a tube cone only for $p = q$.

3) *The cone* $\mathcal{P}_l \subset \mathbb{R}^m$ *with* $m = \dfrac{l(l+1)}{2}$ consisting of all real positive definite symmetric $l \times l$-matrices. For $l = 2$ the tube cone $T^{\mathcal{P}_2}$ is biholomorphic to the future tube $\tau^+(2)$. For any l the tube cone $T^{\mathcal{P}_l}$ is biholomorphic to the *classical Cartan domain of the IIIrd type* consisting of complex $l \times l$-matrices Z such that $ZZ^* < I$ and $'Z = Z$.

4) The cone $Q_l \subset \mathbb{R}^m$ with $m = 2l^2 - l$ consisting of all quaternion positive definite quaternion-Hermitian $l \times l$-matrices. The tube cone T^{Q_l} is biholomorphic to *the classical Cartan domain of the IInd type* consisting of complex $p \times p$-matrices Z such that $ZZ^* < I$, $'Z = -Z$ with $p = 2l$.

To characterize the common properties of these cones including the light cone let us give the following definitions. For a cone C we call the cone $C^* = \{\eta \in \mathbb{R}^m : (\eta, y) \geq 0, \forall y \in C\}$ the *dual cone*. A cone C is self-dual if $C^* = \bar{C}$. A cone C is called *homogeneous* if the group of linear automorphisms of C (i.e. linear non-degenerate transformations of $\mathbb{R}^m$ mapping C into itself) acts transitively on C, i.e. for any y, $y' \in C$ there exists an automorphism of C mapping y to y'. All the cones listed above are self-dual and homogeneous; such cones are also called *domains of positivity* (Koecher (1957), Rothaus (1960)). It turns out that almost the only examples of self-dual homogeneous cones are the ones listed above. More precisely, any self-dual homogeneous cone C (which is convex and proper) can be represented as the direct sum of light cones $V^+(n)$, cones of type 2)–4) and an exceptional cone in 27-dimensional space which can be realized in the space of matrices over the Cayley numbers (cf. Vinberg (1963)). Tube cones T^C over domains of positivity can be realized as the direct sums of classical Cartan domains of the types I–IV (domains of the IVth type were introduced in Sect. 2.3) and an exceptional domain in 27-dimensional space. So they form a subclass of bounded symmetric domains in $\mathbb{C}^m$ (cf. Helgason (1978)) which can be realized as tube cones and for this reason they are called *bounded symmetric domains of tube type* (arbitrary bounded symmetric domains in $\mathbb{C}^m$ can be realized as Siegel domains of the IInd kind, cf. below).

Arbitrary tube cones have the following general properties. Any tube cone is biholomorphic to a bounded domain because it can be mapped by a non-degenerate linear transformation into the tube cone T_+ biholomorphic to the polydisc. Analytic automorphisms of a tube cone T^C continuous in the closure of T^C have the form $z \rightarrow Az + b$ where A is an affine transformation of the cone C onto itself, $b \in \mathbb{R}^m$.

A further generalization of tube cones is connected with the notion of *Siegel domains of the IInd kind*. Recall (cf. Piatetski-Shapiro (1961) that a Siegel domain of the IInd kind is a domain in $\mathbb{C}^{k+m}$ of the type

$$\{(z, w) \in \mathbb{C}^k \times \mathbb{C}^m : \mathrm{Im}\, z - F(w, w) \in C\}$$

where $F : \mathbb{C}^m \times \mathbb{C}^m \rightarrow \mathbb{C}^k$ is a sesquilinear non-degenerate form with values in $\mathbb{C}^k$ which is C-Hermitian in the sense that $F(w, w) \in \bar{C}$ for any $w \in \mathbb{C}^m$ and $F(w, w) = 0$ only when $w = 0$. Tube cones (Siegel domains of the Ist kind) correspond to the case $m = 0$, $F = 0$. The other extreme case is $k = 1$, $C = \mathbb{R}_+$. In this case the Siegel domain coincides with the unbounded realization of the ball in $\mathbb{C}^{m+1}$. We have restricted ourselves here to the case of tube cones.

Another generalization of tube cones is considered in the next section.

5.2. Tuboids. Let us call a profile $\bigwedge = \bigwedge(\Omega)$ over an open set Ω in $\mathbb{R}^m$ a domain in $\mathbb{C}^m$ of the form

$$\bigwedge = \{z = x + iy \in \mathbb{C}^m : x \in \Omega, y \in \bigwedge_x\}$$

where the fiber $\bigwedge_x$ for any $x \in \Omega$ is an open proper cone in $\mathbb{R}^m$. We call the fiber convex hull ch $\bigwedge$ of a profile Λ the profile having the fibers (ch $\bigwedge)_x$ equal to the convex hull of $\bigwedge_x$ for any $x \in \Omega$. A profile $\bigwedge'$ is compact in a profile $\bigwedge$, $\bigwedge'(\Omega) \Subset \bigwedge(\Omega)$, if $\bigwedge'_x \Subset \bigwedge_x$ for any $x \in \Omega$. This means that $\overline{\bigwedge'_x} \subset \bigwedge_x \cup \{0\}$ for any $x \in \Omega$. A *Tuboid* $\mathscr{D} = \mathscr{D}(\bigwedge)$ with profile $\bigwedge = \bigwedge(\Omega)$ is a domain $\mathscr{D} \subset \bigwedge$ of the form

$$\mathscr{D} = \{z = x + iy \in \mathbb{C}^m : x \in \Omega, y \in \mathscr{D}_x\}$$

with the following property: for any profile $\bigwedge' \Subset \bigwedge$ there exists a complex neighborhood $\tilde{\Omega}$ of the set Ω such that $\tilde{\Omega} \cap \bigwedge' \subset \mathscr{D}$ (cf. Bros (1976), Bros, Iagolnitzer (1974–75, 1976)). In other words, the set $\mathscr{D}$ near $x \in \Omega$ looks

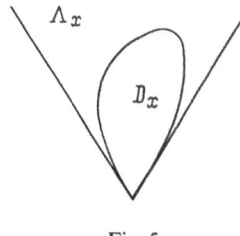

Fig. 5

"asymptotically" like the tube cone over $\bigwedge_x$ (cf. Fig. 5). If, in particular, $\mathscr{D} = \Omega + iC_R$ where $C_R = C \cap B(0, R)$ is the intersection of C with the ball $B(0, R) = \{|y| < R\}$ in $\mathbb{R}^m$ we shall call the tuboid $\mathscr{D}$ a *local tube* over Ω.

Many of the results valid for tube cones can be extended to tuboids. Fourier transform, which is crucial for complex analysis in tube cones, has been generalized to tuboids in the form of the local generalized Fourier transform now also called the FBI-transform due to the names of its inventors (and not to the known agency) the French mathematical physicists J. Bros and D. Iagolnitzer (Bros-Iagolnitzer (1974–75, 1976)).

We give two more examples. The first of them is an analogue of Bochner's Tube Theorem. It asserts that any function holomorphic in a tuboid $\mathscr{D}(A)$ can be holomorphically extended to a tuboid $\mathscr{D}'(\bigwedge')$ with profile $\bigwedge' = \mathrm{ch}\, A$ (the tuboid $\mathscr{D}'$ depends only on the tuboid $\mathscr{D}(\bigwedge)$ and not on the function). This result can be considered as a microlocal version of Bochner's Theorem (cf. a local variant of Bochner's Theorem in Komatsu (1972)). The second result can be considered as a microlocal version of Grauert's theorem on the holomorphic convexity of totally-real sets: for any tuboid $\mathscr{D} = \mathscr{D}(\bigwedge)$ there exists a tuboid $\mathscr{D}'(\bigwedge) \subset \mathscr{D}$ with the same profile which is a domain of holomorphy.

Bibliographical Notes

This chapter is of introductory character, so the exposition in the first four sections is rather detailed and many of the omitted proofs can be obtained by the reader using the given results. The material of the first two sections is known in general though it is hard to give the appropriate references for Sect. 1 and Sect. 2.1, 2.2. The transformations of Sect. 2.3 and further information on classical domains of the IVth type can be found in (Siegel (1949), Hua (1958), Piatetski-Shapiro (1961)). The twistor theory which is barely touched in Sect. 3.1 is studied in the collection of original papers of R. Penrose and his collaborators (Twistors and Gauge Fields (1983)) and in the books (Atiyah (1979), Manin (1984)) (cf. also the review article Sergeev (1991)). The exposition of Sect. 3.2 is based on (Sergeev-Vladimirov (1986)). Section 4 contains the results of (Sergeev

(1983, 1986) and Sergeev-Vladimirov (1986)). The last section is of expository character. Cf. further information on tube cones in (Vladimirov (1964, 1979)). Classical Cartan domains and corresponding homogeneous tube cones are considered in (Cartan (1935), Siegel (1949), Hua (1958), Piatetski-Shapiro (1961), Koraṅyi-Wolf (1965), Wolf (1972)). In Sect. 5.2 we formulate some of the results of (Bros (1976), Bros-Iagolnitzer (1974–75, 1976)).

Chapter 2
Boundary Properties of Holomorphic Functions

Throughout this chapter, C will denote a convex open proper cone in $\mathbb{R}^m$.

§ 1. Boundary Values in L^p and $\mathcal{H}_s$

1.1. The Spaces $H^p(T^C)$. We define $H^p(T^C)$, $0 < p \le \infty$, as the space consisting of all functions $f \in \mathcal{O}(T^C)$ having the finite norm

$$\|f\|_{H^p} = \sup_{y \in C} \left[\int_{\mathbb{R}^m} |f(x + iy)|^p \, dx \right]^{1/p} \quad \text{for } 0 < p < \infty,$$

$$\|f\|_{H^\infty} = \sup_{y \in T^C} |f(z)| \quad \text{for } p = \infty.$$

The spaces $H^p(T^C)$ are Banach spaces for all p, $1 \le p \le \infty$.

Theorem 1 (Stein-Weiss-Weiss (1964), Stein-Weiss (1971)). *Any function* $f \in H^p(T^C)$, $1 \le p < \infty$, *has boundary values in L^p for $y \to 0$, $y \in C$.*

If $C = \mathbb{R}^m_+$ we can assert moreover that for any function $f \in H^p(T_+)$ with $0 < p < \infty$ there exists almost everywhere on $\mathbb{R}^m$ (with respect to Lebesgue measure on $\mathbb{R}^m$) the limit

$$f(x + iy) \to f(x) \quad \text{for} \quad y \to 0, \quad y \in \mathbb{R}^m_+,$$

so that

$$\|f(x + iy) - f(x)\|_{L^p} \to 0 \quad \text{for} \quad y \to 0, \quad y \in \mathbb{R}^m_+.$$

This result can be extended to general tube cones only in the following weak form.

Theorem 2 (Stein-Weiss-Weiss (1964), Stein-Weiss (1971)). *Let $f \in H^p(T^C)$ with $0 < p < \infty$. Then for any compact subcone $C' \Subset C$ (cf. Chap. 1, Sect. 5.2) there exists the limit*

$$f(x + iy) \to f(x) \quad \text{for} \quad y \to 0, y \in C'.$$

for almost all $x \in \mathbb{R}^m$ and

$$\|f(x + iy) - f(x)\|_{L^p} \to 0, \quad \text{for } y \to 0, \, y \in C'.$$

1.2. The Spaces $H^{(s)}(C)$. Denote by $\mathscr{L}_s^2$, $s \in \mathbb{R}$, the Hilbert space of functions g on $\mathbb{R}^m$ having the finite norm

$$\|g\|_{(s)} = \left[\int_{\mathbb{R}^m} |g(\xi)|^2 (1 + |\xi|)^s \, d\xi \right]^{1/2},$$

and let the *space* $\mathscr{H}_s$ consist of all functions f which are the Fourier transforms of functions $g \in \mathscr{L}_s^2 : f = F[g]$. The space $\mathscr{H}_s$ is provided with the norm $\|f\|_{\mathscr{H}_s} = \|g\|_{(s)}$.

We define next the Banach *space $H^{(s)}(C)$* (cf. Vladimirov (1979)) consisting of all functions f holomorphic in T^C and having the finite norm

$$\|f\|^{(s)} = \sup_{y \in C} \|f(x + iy)\|_{\mathscr{H}_s}.$$

For $s = 0$ the space $H^{(0)}(C)$ coincides with $H^2(T^C)$.

Theorem 3. *Any function $f \in H^{(s)}(C)$ has a boundary value $f(x)$ in $\mathscr{H}_s$,*

$$f(x + iy) \to f(x) \text{ in } \mathscr{H}_s \quad \text{for } y \to 0, \, y \in C.$$

The function f belongs to $H^{(s)}(C)$ if and only if its spectral function $g = F^{-1}[f(x)]$ belongs to $\mathscr{L}_s^2$ with its support $\operatorname{supp} g \subset C^*$ where C^* is the cone dual to C (cf. Chap. 1, Sect. 5.1). Moreover, $\|f\|^{(s)} = \|f(x)\|_{\mathscr{H}_s} = \|g\|_{(s)}$ and the Laplace transform $g \to L[g] = f$ is an isomorphism between the space of functions in $\mathscr{L}_s^2$ with supports in C^* and the space $H^{(s)}(C)$.

This theorem is proved in Vladimirov (1979) (cf. also Tillmann (1961)). If a cone C is not proper and $f \in H^{(s)}(C)$ then it follows that $f \equiv 0$. This motivates our requirement that the cone C be proper.

§2. Boundary Values in Spaces of Distributions and Hyperfunctions

2.1. The Space $H(C)$. Let us define the algebra $H(C)$ of slowly increasing (or temperate) functions $f \in \mathcal{O}(T^C)$ as those satisfying the following growth condition

$$|f(z)| \le M(1 + |z|^2)^{\alpha/2} [1 + \Delta^{-\beta}(y)], \quad \zeta \in T^C,$$

for some constants $M > 0$, α, $\beta \ge 0$, where $\Delta(y)$ is the distance from y to the boundary of the cone C (cf. Vladimirov (1979)). The topology in $H(C)$ is defined by the system of seminorms

$$\|f\|^{(\alpha, \beta)} = \sup_{z \in T^C} |f(z)|/(1 + |z|^2)^{\alpha/2}(1 + \Delta^{-\beta}(y)), \quad \alpha, \beta \ge 0.$$

Theorem 4 (Vladimirov (1960, 1961, 1979)). *The Laplace transform* $g \to L[g] = f$ *is an isomorphism between the space* $\mathscr{S}'(C^*)$ *of tempered distributions with supports in* C^* *and the space* $H(C)$. *Any function* $f \in H(C)$ *has a boundary value* $f(x)$ *in* $\mathscr{S}'$ *for* $y \to 0$, $y \in C$.

The existence of boundary values of functions from $H(C)$ was proved also in Tillmann (1961). Generalizing this theorem to the case of non-convex tube cones we can obtain a variant of Bochner's Tube Theorem (with estimates) for the space $H(C)$.

2.2. Boundary Values in the Sense of Hyperfunctions. Recall first the definition of hyperfunctions. Let Ω be a domain in $\mathbb{R}^m$ and $\tilde{\Omega}$ – a complex neighborhood thereof in $\mathbb{C}^m$, $\tilde{\Omega} \cap \mathbb{R}^m = \Omega$, which is a domain of holomorphy. The *space of hyperfunctions* $\mathscr{B}(\Omega)$ is by definition the cohomology group $H^{m-1}(\tilde{\Omega} \backslash \Omega, \mathcal{O})$ with coefficients in the sheaf $\mathcal{O}$ of holomorphic functions (this definition does not depend on the choice of the complex neighborhood $\tilde{\Omega}$) (Sato (1959–60)). Considering the covering of $\tilde{\Omega} \backslash \Omega$ by the domains of holomorphy

$$\tilde{\Omega}_j^\pm = \{z = x + iy \in \tilde{\Omega} : \pm y_j > 0\}, \quad j = 1, \ldots, m,$$

we obtain, using the Leray Theorem, that any hyperfunction in Ω is given by a collection of 2^m functions f_ε holomorphic in the domains

$$\tilde{\Omega}_\varepsilon = \{z \in \tilde{\Omega} : \varepsilon_j y_j > 0, j = 1, \ldots, m\}, \quad \varepsilon = (\varepsilon_1, \ldots, \varepsilon_m), \quad \varepsilon_j = \pm 1$$

(cf. Fig. 6). (The collection $\{f_\varepsilon\}$ is defined up to the addition of an $(m-2)$-coboundary of the covering $\{\tilde{\Omega}_j^\pm\}$ with coefficients in $\mathcal{O}$). For us it is more convenient to use another representation of hyperfunctions which we shall obtain using the following covering of $\tilde{\Omega} \backslash \Omega$. Consider half spaces in $\mathbb{C}^m$ of the form

$$E_j = \{z = x + iy : (y, e_j) > 0\}, \quad j = 0, \ldots, m,$$

where $e_0, \ldots, e_m$ are unit vectors in $\mathbb{R}^m$ such that $\bigcup_j E_j = \mathbb{C}^m \backslash \mathbb{R}^m$. Then the domains $\tilde{\Omega}_j = \tilde{\Omega} \cap E_j$, $j = 0, \ldots, m$, form a covering of $\tilde{\Omega} \backslash \Omega$ by domains of

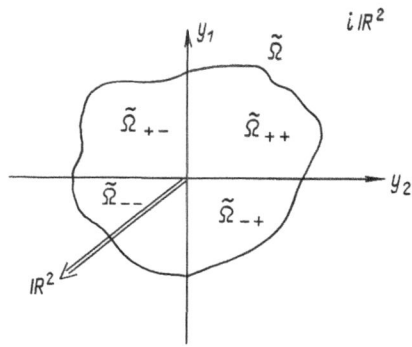

Fig. 6

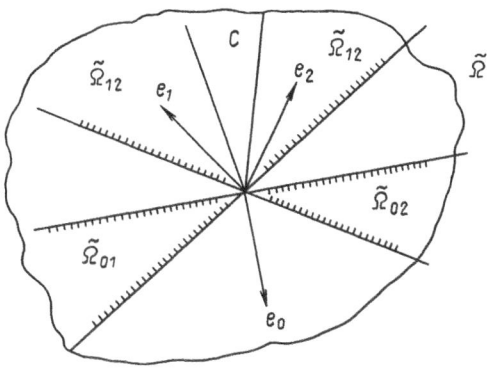

Fig. 7

holomorphy. Using the Leray Theorem for this covering we can represent a hyperfunction on Ω by a collection of $m + 1$ functions $f_{j_1 \ldots j_m}$, $0 \le j_1 < j_2 < \cdots < j_m \le m$, holomorphic in the domains $\tilde{\Omega}_{j_1 \ldots j_m} = \tilde{\Omega}_{j_1} \cap \cdots \cap \tilde{\Omega}_{j_m}$ (and defined up to the addition of an $(m - 2)$-coboundary of the covering $\{\tilde{\Omega}_j\}$ with coefficients in $\mathcal{O}$) (cf. Fig. 7).

Take now a function $f \in \mathcal{O}(T^C \cap \tilde{\Omega})$ and choose vectors $e_0, \ldots, e_m$ so that $\tilde{\Omega}_{1 \ldots m} \subset T^C \cap \tilde{\Omega}$. Assign to the function f a collection of functions $f_{j_1 \ldots j_m}$ where $f_{1 \ldots m} = \mathrm{sgn}(e_1 \wedge \cdots \wedge e_m)f$ (here $\mathrm{sgn}(e_1 \wedge \cdots \wedge e_m)$ is the orientation of the polyvector $(e_1 \wedge \cdots \wedge e_m)$ and the other components $f_{j_1 \ldots j_m}$ are set equal to zero. This collection defines a hyperfunction on Ω which is called the *boundary value of f* and denoted by bv f. The boundary value bv f does not depend on the choice of vectors $e_0, \ldots, e_m$ satisfying the above hypothesis nor on the complex neighborhood $\tilde{\Omega}$. Thus bv f can be computed using any domain of the form $T^{C'} \cap \tilde{\Omega}$ where $C' \subset C$, $\tilde{\Omega}' \subset \tilde{\Omega}$ and $\tilde{\Omega}'$ is a domain of holomorphy such that $\tilde{\Omega}' \cap \mathbb{R}^m = \Omega$.

2.3. Distributional and Hyperfunctional Boundary Values in Tuboids. Let D be a tuboid with profile $\bigwedge$ over a domain $\Omega \subset \mathbb{R}^m$ (cf. Chap. 1, Sect. 5.2). We can assume (cf. loc. cit.) that the profile $\bigwedge$ is fiber-convex and D is a domain of holomorphy. Let $f \in \mathcal{O}(D)$. Then for any point $x \in \Omega$ and a neighborhood $U = U(x)$ of x, we can take a local tube of the form $\mathcal{D}_U = U + iC_R$ so that $\mathcal{D}_U \subset D$ and define the boundary value $\mathrm{bv}_U f \in \mathcal{B}(U)$ as in the Sect. 2.2. These boundary values coincide on the intersection of neighborhoods $U \cap U'$ and thus define the unique hyperfunction bv $f \in \mathcal{B}(\Omega)$ which is called the *boundary value* of f (cf. Zharinov (1983)).

A function $f \in \mathcal{O}(D)$ is called a locally *slowly growing* (or *tempered*) *function*, $f \in H_{\mathrm{loc}}(\bigwedge)$, if for any point $x \in \Omega$ there exists a neighborhood U of x and a local tube $U + iC_R \subset D$ where the following estimate is satisfied: $|f(z)| \le M/|y|^N$ for some constants $M, N > 0$. A function f of such type has by Theorem 4 a boundary value on U in the sense of distributions. This boundary value coincides with the boundary value $bv_U f$ in the sense of hyperfunctions (Martineau (1964)). Thus, any function $f \in H_{loc}(\bigwedge)$ has a boundary value on Ω in the sense of distributions (in $\mathcal{D}'(\Omega)$) and this boundary value coincides with bv f. For functions of the class $H_{loc}(\bigwedge)$ we have the microlocal analogue of Bochner's Tube Theorem (with estimates) from Chap. 1, Sect. 5.2. There is also the following interesting property: if a function $f \in \mathcal{O}(D)$ is locally tempered in a tuboid $D' \subset D$ over the same domain Ω then it is locally tempered also in the tuboid D.

§3. Boundary Values of Bounded Holomorphic Functions

3.1. Auxiliary Results. To study the boundary properties of holomorphic functions in tube cones T^C the following two assertions formulated as lemmas may be useful.

Lemma 1. *A tube cone T^C is biholomorphically equivalent to a bounded domain $\mathcal{D}$ contained in the polydisc. We can choose the biholomorphism in such a way that it maps the distinguished boundary of T^C into the distinguished boundary of the polydisc.*

To prove this lemma it is sufficient to consider a homogeneous linear transformation taking the cone C into the octant $\mathbb{R}^m_+$. Its complexification maps T^C into T_+ and, combined with the biholomorphism of T_+ onto the polydisc, gives the required map. This lemma often allows one to restrict the proofs of statements for general tube cones to the case of T_+.

To formulate the second assertion we introduce the following definition. An oriented C^1-smooth hypersurface M in $\mathbb{R}^m$ will be called *C-spacelike* if the normal to M at any of its points belongs to C (cf. Fig. 8). Using the next lemma we can sometimes restrict the study of the boundary properties of holomorphic functions in T^C to the case of strictly pseudoconvex domains.

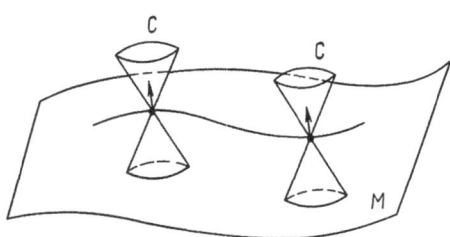

Fig. 8

Lemma 2. *Let M be a C-spacelike hypersurface in $\mathbb{R}^m$ of class C^2. There exists a C^2-smooth strictly pseudoconvex domain $\mathscr{D} \subset \mathbb{C}^m$ such that $\mathscr{D}$ contains T^C in a neighborhood of $\mathbb{R}^m$ and $T_x M \subset T_x^c(\partial \mathscr{D})$ for any point $x \in M$.*

The last assertion means that M is an integral submanifold of $\mathscr{D}$. We give an idea of the proof. Let us assume for simplicity that T^C is the future tube τ^+ in $\mathbb{C}^{n+1}$ (cf. Chap. 1, Sect. 1.1) and the hypersurface M is given by the equation $x_0 = s(x') = s(x_1, \ldots, x_n)$. If s is of class C^3 we can define a domain $\mathscr{D}$ as follows

$$\mathscr{D} = \left\{ z = x + iy \in \mathbb{C}^{n+1} : y_0 > \sum_{j=1}^{n} \frac{\partial s(x')}{\partial x_j} y_j + \sum_{j=0}^{n} y_j^2 \right\}.$$

Then in some neighborhood U of $\mathbb{R}^{n+1}$ in $\mathbb{C}^{n+1}$ we have: 1) $\mathscr{D} \cap U \supset \tau^+ \cap U$; 2) $\partial \mathscr{D} \cap U$ is C^2-smooth and strictly pseudoconvex; 3) M is an integral submanifold of $\partial \mathscr{D}$ (cf. Fig. 9). To prove the lemma when M is of class C^2 we note that the defining function of $\partial \mathscr{D}$ has 2nd derivatives at points where $y = 0$ so we can approximate this function by a C^2-smooth function defining the required domain.

We introduce now the *algebra* $A(T^C)$. Consider the one-point compactification $\tilde{\mathbb{C}}^m = \mathbb{C}^m \cup \{\infty\}$ (where a base of neighbourhoods of $\infty \in \tilde{\mathbb{C}}^m$ is given by the exteriors of balls: $\{|z| > R\}$) and denote by $\tilde{T}^C$ the closure of T^C in $\tilde{\mathbb{C}}^m$ so that $\tilde{T}^C = \bar{T}^C \cup \{\infty\}$. The algebra $A(T^C)$ consists of all functions holomorphic in T^C and continuous in $\tilde{T}^C$. It follows from Lemma 1 that the Shilov boundary of $A(T^C)$ coincides with $\tilde{\mathbb{R}}^m = \mathbb{R}^m \cup \{\infty\}$. There is an even stronger assertion.

Theorem 5 (Vladimirov (1979)). *If the boundary value of $f \in H(C)$ is bounded, i.e. $|f(x)| \leq M$ for almost all $x \in \mathbb{R}^m$ then $|f(z)| \leq M$ for all $z \in T^C$.*

3.2. Fatou and Lindelöf Theorems. For functions $f \in H^\infty(T^C)$ we have the following analogue of the classical *Fatou Theorem*: for almost all $x^0 \in \mathbb{R}^m$ there exists a limit of $f(z)$ when $z \to x^0$ in the restricted admissible way (in the sense of the definition given in Chap. 4, Sect. 2.3). This assertion follows from the corresponding result for T_+ (cf. Zygmund (1958), Stein-Weiss (1971)) using Lemma 1. A stronger assertion is also true. We say that a function $f \in \mathcal{O}(T^C)$ is *bounded* at $x^0 \in \mathbb{R}^m$ *in the restricted sense* if it is bounded in some approach set $\Gamma_C^\alpha(x^0)$ (cf. Chap. 4, Sect. 2.3). Let $\Omega + iC_R$ be a local tube over an open set $\Omega \subset \mathbb{R}^m$ (cf. Chap. 1, Sect. 5.2).

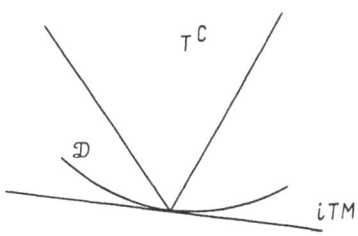

Fig. 9

Theorem 6 (Zygmund (1958), Stein-Weiss (1971)). *If a function* $f \in \mathcal{O}(\Omega + iC_R)$ *is bounded in the restricted sense at each point* $x \in \Omega$ *then* f *has a restricted admissible limit almost everywhere on* Ω.

This result can be further strengthened using the following Theorem of Drozhzhinov-Zavialov (1982) and Khurumov (1983).

Theorem 7. *Let* $f \in \mathcal{O}(\Omega + iC_R)$ *where* Ω *is open in* $\mathbb{R}^m$ *and let* $x \in \Omega$. *If* f *is bounded on some smooth totally real m-dimensional submanifold*

$$M \subset (\Omega + iC_R) \cup \{x\}$$

going through x *then* f *is bounded in the restricted sense at* x.

Hence, if a function $f \in \mathcal{O}(\Omega + iC_R)$ is bounded on some smooth $(m + 1)$-dimensional submanifold of $\Omega + iC_R$ with edge Ω then it has a restricted admissible limit almost everywhere on Ω. What is proved by Drozhzhinov-Zavialov is in fact the following assertion implying Theorem 7. Let $\bigwedge$ be a profile over Ω (Chap. 1, Sect. 5.2) and

$$\lambda = \{z = x + iy \in \mathbb{C}^m : x \in \Omega,\, y \in \lambda_x \text{ where } \lambda_x = te_x, t > 0, e_x \in \bigwedge_x\}$$

be a one-dimensional smooth profile over Ω contained in $\bigwedge$. If a function f is holomorphic in a tuboid $\mathscr{D} = \mathscr{D}(\bigwedge)$ and bounded on $\lambda \cap \mathscr{D}$ then it is bounded in some tuboid $\mathscr{D}' = \mathscr{D}'(\bigwedge)$ with the same profile $\bigwedge$ over Ω (cf. Fig. 10). Note that, generally speaking, f is not bounded in the tuboid $\mathscr{D}$.

A result close to Theorem 7 (namely, a variant of the "two-constants" theorem for our situation) was proved in Jöricke (1982). A nice and short proof of Theorem 7 was proposed by Gonchar. It is based on his "boundary" variant of the theorem on separate analyticity (Gonchar (1985)).

The classical *Lindelöf Theorem* does not have a direct extension to tube cones (Chirka (1973)). In fact, from the above formulation of the Fatou Theorem for tube cones we could expect that the following extension of the Lindelöf Theorem is true. Let a function $f \in \mathcal{O}(\Omega + iC_R)$, where $\Omega + iC_R$ is a local tube over an open set $\Omega \subset \mathbb{R}^m$, be bounded in the restricted sense at $x^0 \in \Omega$. Suppose that f

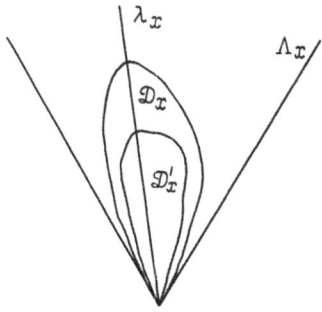

Fig. 10

has a limit along a continuous curve $\Gamma_0(t)$ lying in $\Omega + iC_R$ for $0 \leq t < 1$ and approaching x^0 for $t \to 1$. Then f has the same limit along any continuous curve $\Gamma(t)$ such that $\Gamma(t) \to x^0$ for $t \to 1$ lying in $\Omega + iC'_R$ for $0 \leq t < 1$, where C' is a compact subcone of C. Unfortunately, this assertion is not true. To see this, it is sufficient to consider the tube cone T_+ in $\mathbb{C}^2$ and the function $f(z_0, z_1) = z_1/z_0$ holomorphic in T_+ and bounded at the origin in the restricted sense. However, f has different limits along distinct rays at the origin (it is not difficult to change this example in such a way that f would be bounded at the origin).

The correct extension of the Lindelöf Theorem asserts that f has the same limit only for curves $\Gamma(t)$ "tangential" to $\Gamma_0(t)$ for $t \to 1$. More precisely, let $\Gamma(t)$ be a continuous curve in $\Omega + iC_R$ of the same type as before with the endpoint x^0. Denote by $\lambda_y(x^0)$, $y \in C_R$, the complex ray

$$\lambda_y(x^0) = \{z = x^0 + \alpha y : \alpha \in \mathbb{C}, \operatorname{Im} \alpha > 0\}$$

at x^0 with the direction y. Denote by $\gamma(t)$ the orthogonal projection of $\Gamma(t)$ onto $\lambda_y(x^0)$. We shall say that $\Gamma(t) \to x^0$ alongside $\lambda_y(x^0)$ if

$$\frac{|\Gamma(t) - \gamma(t)|}{|\operatorname{Im} \gamma(t)|} \to 0 \qquad \text{for } t \to 1.$$

Theorem 8 (Chirka (1973), Sergeev (1989)). *Let f be holomorphic in a local tube $\Omega + iC_R$ over an open set $\Omega \subset \mathbb{R}^m$ and bounded in the restricted sense at $x^0 \in \Omega$. Suppose that f has a limit along a continuous curve $\Gamma_0(t)$, $0 \leq t < 1$, in $\Omega + iC_R$ such that $\Gamma_0(t) \to x^0$ alongside some complex ray $\lambda_y(x^0)$, $y \in C_R$. Then f has the same limit along any continuous curve $\Gamma(t)$, $0 \leq t < 1$, in $\Omega + iC_R$ such that $\Gamma(t) \to x^0$ alongside $\lambda_y(x^0)$.*

Again, according to Drozhzhinov-Zavialov and Khurumov (op. cit.) it is sufficient to require f to be bounded on some smooth totally-real n-dimensional submanifold going through the point x^0.

3.3. Uniqueness Theorems. For functions bounded and holomorphic in tube cones we have the following well-known uniqueness theorem.

Theorem 9 (Zygmund (1958), Stein-Weiss (1971)). *If a function $f \in H^\infty(T^C)$ has restricted admissible limit 0 on a set $E \subset \mathbb{R}^m$ of positive measure then $f \equiv 0$.*

A set $E \subset \mathbb{R}^m$ is called a *uniqueness set* for the algebra $A(T^C)$ (or a *determining set* in the terminology of Rudin (1969)) if for any function $f \in A(T^C)$ the equality $f(x) = 0$, $x \in E$, implies that $f \equiv 0$. According to Theorem 9 any set E of positive Lebesgue measure on $\mathbb{R}^m$ is a uniqueness set. On the other hand, not every set of (even infinite) $(m-1)$-dimensional Hausdorff measure on $\mathbb{R}^m$ is a uniqueness set (cf. Sect. 5.1).

Rudin (1969), Sect. 5.1, gives an example of a compact uniqueness set for $A(T_+)$ having finite linear measure. By the same methods as in Rudin (op. cit.) or using Lemma 1 it is easy to construct examples of compact uniqueness sets for $A(T^C)$ having finite linear measure. Note that a compact set of linear measure zero on $\mathbb{R}^m$ cannot be a uniqueness set for $A(T^C)$ (cf. Sect. 5.1).

§4. Inner Functions and Holomorphic Mappings

4.1. Rational Inner Functions. A function $f \in H^\infty(T^C)$ is *inner* if its limit boundary values on $\mathbb{R}^m$ (which exist almost everywhere on $\mathbb{R}^m$ by Theorem 6, Sect. 3.2) have modulus one almost everywhere on $\mathbb{R}^m$. The functions

$$\frac{(z, q) - \alpha}{(\bar{z}, q) - \bar{\alpha}} \quad \text{where } q \in V^+, \quad \alpha \in \mathbb{C}, \, \text{Im } \alpha > 0,$$

provide examples of inner functions in the future tube τ^+ (cf. Vladimirov (1983b)). Other examples can be generated by applying to these functions automorphisms of τ^+ and taking the product of different functions. In the case of the tube cone T_+ which is biholomorphic to the polydisc, inner functions are given by Blaschke products (cf. Rudin (1969)). For the generalized unit disc B_m rational inner functions are given by the function det Z and functions obtained from this one by composing with automorphisms of B_m and taking products.

A complete description of rational inner functions in the polydisc was given in Rudin (1969). This result was extended also to general bounded symmetric domains in Korányi-Vagi (1979). To formulate this extension we need to introduce some notation. Let $\mathscr{D}$ be a bounded symmetric domain (cf. Helgason (1978)). Denote by N the invariant norm on $\mathscr{D}$. The general definition of N is given in Korányi-Vagi (op. cit). We note here that for the generalized unit disc and classical domains of the IIIrd kind N is given by the determinant; for classical domains of the IInd kind – by the Pfaffian and for the Lie spheres – by (z, z). Set $r(z) = \dfrac{\text{grad } N(z)}{N(z)}$, $z \in \mathscr{D}$, and denote by $\tilde{Q}(z)$ the polynomial obtained from a polynomial $Q(z)$ by the conjugation of coefficients.

Theorem 10 (Korányi-Vagi (1979)). *Let $\mathscr{D}$ be a bounded symmetric domain of tube type (Chap. 1, Sect. 5.1) and f a rational inner function on $\mathscr{D}$. Then f has the form*

$$f(z) = M(z) \frac{\tilde{Q}(r(z))}{Q(z)}, \quad z \in \mathscr{D},$$

where $Q(z)$ is a polynomial having no zeros in $\mathscr{D}$, and $M(z)$ is a homogeneous polynomial having modulus one on the distinguished boundary of $\mathscr{D}$. If $Q(0) = 1$ then M and Q are uniquely defined.

If $\mathscr{D}$ is one of the classical domains then $M(z)$ coincides, up to a constant with modulus 1, with a power of the norm $N(z)$. The proof of Theorem 10, as in Rudin (1969), is based on the following Lemma which is interesting in itself.

Lemma 3 (Korányi-Vagi (1979)). *Let E be a compact uniqueness set for polynomials lying on the distinguished boundary of $\mathscr{D}$ and let $f \in \mathcal{O}(\mathscr{D})$. If for each $x \in E$ the function $f_x(\lambda) = f(x\lambda)$, $|\lambda| < 1$, is a rational function of degree k (i.e. the maximum of the degrees of the numerator and denominator is equal to k) which is* *continuous for* $|\lambda| \leq 1$ *then* $f = P/Q$ *where* P, Q *are polynomials without common divisors and the degree of* f *is* k. *If, in addition, the function* f_x *is inner for each* $x \in E$ *then* Q *has no zeros in* $\mathscr{D} \cup E$ *and* $P(x) = M(x)\overline{Q}(x)$ *for* $x \in E$ *where* M *is a homogeneous polynomial having modulus one on* E.

Note that all inner functions in $A(\mathscr{D})$ are rational; the same is true for inner functions meromorphic in a neighborhood of $\overline{\mathscr{D}}$ (Korányi-Vagi (op. cit.)).

4.2. General Inner Functions. Infinite Blaschke products provide examples of nonrational inner functions in the tube cone T_+. The analogues of the Blaschke products in the future tube τ^+ are given by the functions

$$\prod_{k=1}^{\infty} \left[\frac{(z, q_k) - \alpha_k}{(z, q_k) - \overline{\alpha}_k} \cdot \frac{\overline{\alpha}_k - i}{\alpha_k - i} \cdot \left| \frac{\alpha_k - i}{\overline{\alpha}_k - i} \right| \right]^{n_k}$$

where α_k are complex numbers with $\operatorname{Im} \alpha_k > 0$, $\alpha_k \neq \alpha_l$ for $k \neq l$; $q_k = (1, q_k') \in V^+$ and n_k are positive integers. This infinite product converges uniformly on compacta in τ^+ if and only if the following series is finite

$$\sum_{k=1}^{\infty} n_k \frac{\operatorname{Im} \alpha_k}{1 + |\alpha_k|^2} < \infty.$$

This is proved in Vladimirov (1983b) where Blaschke products for the generalized unit disc are also described.

The general results of Aleksandrov and Löw on inner functions (cf. Aleksandrov (1984, 1983), Löw (1984) and Aleksandrov's article in this volume) are true also for inner functions in tube cones. We give here several of their results.

Theorem 11. *Let* φ *be a positive lower semicontinuous function on* $\mathbb{R}^m$ *with* $\varphi \in L^p(\mathbb{R}^m)$, $1 \leq p \leq \infty$. *Then there exists a function* $f \in H^p(T^C)$ *such that* $|f(x)| = \varphi(x)$ *almost everywhere on* $\mathbb{R}^m$.

Theorem 12. *Let* φ *be a positive lower semicontinuous function on* $\mathbb{R}^m$ *with* $\varphi \in L^1(\mathbb{R}^m)$. *Then for any* $\varepsilon > 0$ *there exists a function* $f \in A(T^C)$ *such that* $|f(x)| \leq \varphi(x)$ *almost everywhere on* $\mathbb{R}^m$ *and the measure of the set* $\{x \in \mathbb{R}^m : |f(x)| \neq \varphi(x)\}$ *is less than* ε.

It follows that inner functions are dense in the unit ball of $H^{\infty}(T^C)$ in the topology of uniform convergence on compacta in T^C. Also, for any function f belonging to the unit ball of $A(T^C)$ there exists an inner function in T^C having the same zeros as f.

4.3. Holomorphic Mappings. Let $\mathscr{D} = T^C$ and $\mathscr{D}' = T^{C'}$ be tube cones in $\mathbb{C}^m$ with $m \geq 2$. The domains $\mathscr{D}$ and $\mathscr{D}'$ are biholomorphically equivalent if and only if the cones C and C' are affinely equivalent; any biholomorphic mapping $F : \mathscr{D} \to \mathscr{D}'$ is necessarily rational (cf. Matsushima (1972), Murakami (1972), Yang (1982)). The last assertion is also true for proper mappings F (recall that

F is proper if the preimage of any compact subset of $\mathscr{D}'$ under this mapping is compact in $\mathscr{D}$).

Theorem 13. *Let* $F : \mathscr{D} \to \mathscr{D}'$ *be a proper holomorphic mapping. Then F is rational. If the cones C and C' are irreducible domains of positivity* (Chap. 1, Sect. 5.1) *then F is biholomorphic.*

This theorem was proved by Khenkin and Tumanov (cf. Khenkin-Tumanov (1983)) using a result of Bell (1982) (it is also true for Siegel domains of the IInd kind). In the case of the polydisc it was proved in Rudin (1969).

Let $\mathscr{D}$ denote now a bounded symmetric domain. A holomorphic mapping $F : \mathscr{D} \to \mathscr{D}$ is called *inner* if its boundary value F^* on the distinguished boundary S of $\mathscr{D}$ has the property: $F^*(x) \in S$ for almost all $x \in S$. The importance of inner mappings is due to the following theorem of Korányi-Vagi (1976).

Theorem 14. *Let T be a linear isometry of the space $H^p(\mathscr{D})$, $0 < p < \infty$, $p \neq 2$, into itself and let $g = T(1)$. Then, there exists an inner mapping $F : \mathscr{D} \to \mathscr{D}$ such that $Tf = g(f \circ F)$ for any $f \in H^p(\mathscr{D})$. Moreover*

$$\int_S (h \circ F^*) |g^*|^p \, d\mu = \int_S h \, d\mu$$

for any $h \in L^\infty(S)$ where μ is the invariant measure on S. Conversely, if F is an inner mapping and the last equality is true for some function $g \in H^p(\mathscr{D})$ and for any continuous function h on S then the operator $Tf := g(f \circ F)$ defines an isometry of $H^p(\mathscr{D})$. The isometry T maps $H^p(\mathscr{D})$ onto $H^p(\mathscr{D})$ if and only if F is an isomorphism of $\mathscr{D}$ and the function g is given by

$$g(z) = \alpha \left(\frac{\mathscr{K}_u^2(z)}{\mathscr{K}_u(u)} \right)^{1/p}$$

where $\mathscr{K}_u(z) = \mathscr{K}(u, z)$ is the Cauchy kernel of $\mathscr{D}$, $\alpha \in \mathbb{C}$, $|\alpha| = 1$ and $u = F^{-1}(0)$.

§ 5. Interpolation Sets

5.1. Properties of Interpolation Sets. Let K be a compact subset of the distinguished boundary $\tilde{\mathbb{R}}^m$ of $\tilde{T}^C$ (cf. Sect. 3.1). We call K a *zero set* for the algebra $A(T^C)$ if there exists a function $f \in A(T^C)$ equal to zero on K such that $f(z) \neq 0$ on $\tilde{T}^C \setminus K$. K is called a *peak set* for $A(T^C)$ if there exists a *peak function* $f \in A(T^C)$ equal to 1 on K such that $|f(z)| < 1$ for $z \in \tilde{T}^C \setminus K$. We call K an *interpolation set* for $A(T^C)$ if any continuous function on K can be extended to a function in $A(T^C)$. Finally, K is called a *peak interpolation set* for $A(T^C)$ if for any continuous function $g \not\equiv 0$ on K there exists a function $f \in A(T^C)$ such that $f(x) = g(x)$ for all $x \in K$ and $|f(z)| < \sup_K |g(x)|$ for any $z \in \tilde{T}^C \setminus K$.

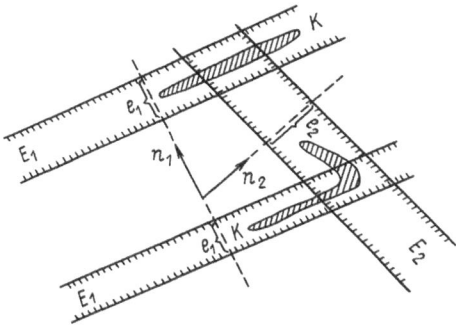

Fig. 11

Theorem 15. *All the properties of a compact $K \subset \tilde{\mathbb{R}}^m$ listed above are equivalent to each other and to the following Bishop's property: for any finite Borel measure μ on $\tilde{\mathbb{R}}^m$ orthogonal to $A(T^C)$ (i.e. $\int f \, d\mu = 0$ for any $f \in A(T^C)$), we have $\int_K |d\mu| = 0$.*

This theorem follows from the corresponding theorem of Rudin (1969) for T_+ using Lemma 1. We shall give one more result on interpolation sets, well known for the tube cone T_+. Let us say that a compact K is of *zero width* with respect to a set N consisting of unit vectors in $\mathbb{R}^m$ if for any $\varepsilon > 0$ there exist a collection of vectors $\{n_i\} \subset N$ and a collection of Borel subsets $\{e_i\}$ of the real line $\mathbb{R}$ such that $\sum |e_i| < \varepsilon$ (where $|e_i|$ is the Lebesgue measure of e_i) and K is contained in the union $\bigcup \tilde{E}_i$ where $E_i = \{x \in \mathbb{R}^m : (x, n_i) \in e_i\}$ (cf. Fig. 11).

Theorem 16. *Let N be a compact set of unit C-like vectors (i.e. vectors belonging to C) in $\mathbb{R}^m$. If a compact set K has zero width with respect to N then it is an interpolation set for $A(T^C)$.*

This theorem follows from the *Forelli Theorem* for T_+ (cf. Rudin (1969)) using Lemma 1. As a corollary of this theorem we obtain that compact sets K of linear measure zero are interpolation sets.

5.2. Interpolation Manifolds. A C^1-smooth submanifold M in $\mathbb{R}^m$ is called an *interpolation manifold* if any compact $K \subset \tilde{M}$ is an interpolation set for $A(T^C)$. It follows from Theorem 16 that any C^1-smooth C-spacelike curve (i.e. a curve such that its tangent vector at any of its points lies outside $\overline{C} \cup (-\overline{C})$) is an interpolation manifold (the smoothness condition here can be weakened, cf. Rudin (1969, 1971)). This result can be extended to submanifolds of $\mathbb{R}^m$ of arbitrary dimension $\leq (m - 1)$. We formulate here (for the sake of simplicity) this extension for the case of hypersurfaces in $\mathbb{R}^m$.

Theorem 17. *All C^1-smooth C-spacelike hypersurfaces in $\mathbb{R}^m$ are interpolation manifolds.*

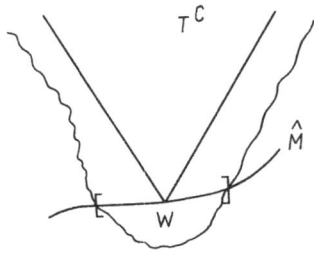

Fig. 12

This theorem follows from the corresponding assertion for smooth strictly pseudoconvex domains (cf. Khenkin-Tumanov (1976), Nagel (1976), Rudin (1978)) using Lemma 2. A similar proof was proposed by Saerens (1984). There is a simple proof of the theorem in the real analytic case proposed by Burns-Stout (1976). Let M be a real analytic C-spacelike hypersurface in $\mathbb{R}^m$ and $\hat{M}$ its complexification. By the hypothesis on M there exists a neighborhood W of M in $\hat{M}$ such that $W \cap \bar{T}^C = M$ (cf. Fig. 12). Let $\hat{f}$ be a holomorphic function on W equal to a given real analytic function f on M. Since W is a complex submanifold in a neighborhood of $\bar{T}^C$ in $\mathbb{C}^m$, by Cartan's Theorem $\hat{f}$ can be extended to a function holomorphic in a neighborhood of $\bar{T}^C$. We have proved that any real analytic function f on M can be extended to a function holomorphic in a neighborhood of $\bar{T}^C$. Using Theorem 15 it is easy to prove now that M is an interpolation manifold. Conversely, if M is a real-analytic interpolation submanifold of $\mathbb{R}^m$ then it has no tangent C-like vectors. Indeed, assume the opposite and consider an arc of a real-analytic C-like curve on M. Then (by Theorem 15) there exists a function $f \in A(T^C)$ which is equal to zero on this arc and $f(z) \neq 0$ at other points of $\tilde{T}^C$. But the complexification of the curve (because it has C-like tangent vectors) has non-void intersection with T^C. Thus the zero set of f intersects T^C. Contradiction.

A partial converse to Theorem 17 can be proved also in the smooth case. Namely, if a C^2-smooth hypersurface M on $\mathbb{R}^m$ is an interpolation manifold for $A(T^C)$ then it has no C-like tangent vectors (Saerens (1984)).

A result combining the Forelli Theorem with Theorem 17 was proved by Labonde (1985).

Bibliographical Notes

Boundary values in the space $H^2(T^C)$ were studied by Bochner (1944), and further results on boundary values in Hardy spaces $H^p(T^C)$ were given in Stein-Weiss (1971). The assertions on the spaces $H^{(s)}(C)$ and $H(C)$ given in Sect. 1.2, 2.1 are contained in Vladimirov (1979) (where also more general spaces $H_a^{(s)}(C)$, and $H_a(C)$ with exponential scale of type a at infinity along the imaginary space are considered; we restricted ourselves for the sake of simplicity to the case $a = 0$). For further information on hyperfunctions and their boundary values cf. Schapira (1970), Morimoto (1973), Sato-Kawai-Kashiwara (1973). Hyperfunctional boundary values in tuboids were considered in Zharinov (1983). Fatou and Lindelöf theorems were considered in many papers (cf. Chirka (1973), Chirka-Khenkin (1975), other references are given in the Notes to Chap. 4). New variants of Fatou and Lindelöf theorems formulated in Sect. 3.2 were proved in Drozhzhinov-Zav'ialov (1982) and Khurumov (1983). The results of Sect. 3.3, 4.1 are parallel to those for the polydisc (Rudin (1969)). The general properties of inner functions were studied in Aleksandrov (1984, 1983), Löw (1984), Rudin (1980). The results of Sect. 5.1 are analogous to those for the polydisc (Rudin (1969)). For the description of interpolation manifolds in Sect. 5.2 see Burns-Stout (1976), Stout (1981), Saerens (1984), Sergeev-Vladimirov (1985), and Sergeev (1989).

Chapter 3
"Edge-of-the-Wedge" Theorem and Related Problems

§ 1. "Edge-of-the-Wedge" Theorem

1.1. Theorem of Bogolubov. This theorem was announced by Bogolubov at the International Conference in Seattle (September 1956) (the first detailed proof was published in Bogolubov-Medvedev-Polivanov (1958)). We formulate it in the form convenient for our goals. Denote temporarily a cone C by C^+, the opposite cone $-C$ by C^- and by Ω a domain in $\mathbb{R}^m$.

Theorem of Bogolubov. *Let $f_\pm \in \mathcal{O}(\mathcal{D}_\pm)$ be functions of locally slow growth (cf. Chap. 2, Sect. 2.3) in local tubes $\mathcal{D}_\pm = \Omega + iC_R^\pm$. Suppose that their boundary values (in the distributional sense) coincide on Ω. Then there exists a complex neighborhood $\tilde{\Omega}$ of Ω (cf. Fig. 13) and a function f which is holomorphic and has locally slow growth in $\tilde{\Omega} \cup \mathcal{D}_+ \cup \mathcal{D}_-$ equal to $f_\pm$ on $\mathcal{D}_\pm$.*

Note that the neighborhood $\tilde{\Omega}$ does not depend on f and is described rather explicitly (cf. Vladimirov (1964)). The "Edge-of-the-Wedge" Theorem of Bogolubov has generated many generalizations, first in quantum field theory and then in several complex variables. Now (along with different variants and generalizations) it constitutes, in fact, a separate chapter in the theory of functions of several complex variables. We wish to emphasize, in particular, its close relation to the local Bochner Tube Theorem (cf. Chap. 1, Sect. 5.2) and the *theorem on separate analyticity* (Siciak (1969), Zakharyuta (1976)).

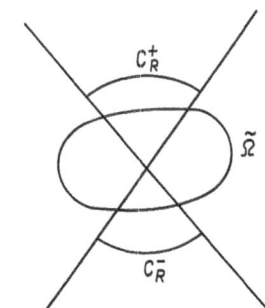

Fig. 13

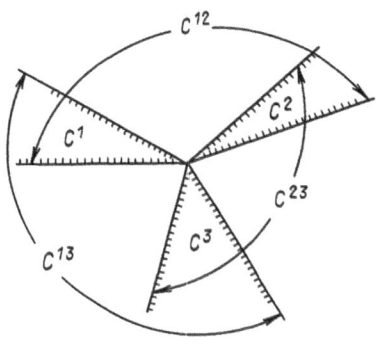

Fig. 14

1.2. Theorem of Martineau. We formulate here one of the generalizations of Bogolubov's Theorem proved by Martineau (1970).

Theorem of Martineau. *Let $f_k \in \mathcal{O}(\mathcal{D}_k)$ be functions of locally slow growth given in local tubes $\mathcal{D}_k = \Omega + iC_R^k$, $k = 1, \ldots, N$, over a domain Ω in $\mathbb{R}^m$. Suppose that the boundary values $f_k(x)$ of $f_k(z)$ on Ω (in the distributional sense) satisfy the following condition*

$$\sum_{k=1}^{N} f_k(x) = 0, \quad x \in \Omega.$$

Then there exists a complex neighborhood $\tilde{\Omega}$ of Ω and functions $f_{jk}, j, k = 1, \ldots, N$, which are holomorphic and of locally slow growth in the domains

$$\mathcal{D}_{jk} = (\Omega + iC_R^{jk}) \cap \tilde{\Omega},$$

where $C^{jk} = \mathrm{ch}(C^j \cup C^k)$ is the convex hull of the cones C^j and C^k (cf. Fig. 14), and satisfy the following conditions:

1) $f_{jk} = -f_{kj}, j, k = 1, \ldots, N$;

2) $f_k(z) = \sum_{j=1}^{N} f_{kj}(z), z \in \mathcal{D}_k \cap \tilde{\Omega}.$

Other generalizations of the "edge-of-the-wedge" theorem will be given in Sect. 3.1.

§ 2. "C-convex Hull" Theorem

2.1. "C-convex Hull" Theorem. Consider again Bogolubov's Theorem which we reformulate in another form. Consider a "unified" function f in the domain $\mathscr{D} = \mathscr{D}_+ \cup \mathscr{D}_- \cup \Omega$ which is holomorphic in $\mathscr{D}_+ \cup \mathscr{D}_-$ and belongs to the space $\mathscr{D}'(\Omega)$ on Ω (in other words, the boundary values $f_\pm(x)$ of f on $\mathbb{R}^m$ from $\mathscr{D}_+$ and $\mathscr{D}_-$ exist in the distributuional sense and coincide on Ω). Bogolubov's Theorem provides a holomorphic extension of any such function along "imaginary directions" into the domain $\tilde{\mathscr{D}} = \mathscr{D}_+ \cup \mathscr{D}_- \cup \tilde{\Omega}$. From this point of view, Bogolubov's Theorem gives an estimate of the holomorphic hull of $\mathscr{D}$ with respect to functions of locally slow growth near $\mathbb{R}^m$. It appears that sometimes we can considerably improve this estimate using the extension along "real directions".

Namely, denote temporarily a cone $C^+ \cup C^-$ by C and call a C^1-smooth curve in $\mathbb{R}^m$ *C-like* if its tangent vectors at each of its points x belong to the cone $x + C$. The *C-convex hull* $B_C(\Omega)$ of Ω is the convex hull of Ω with respect to C-like curves, more precisely, the smallest open neighborhood of Ω in $\mathbb{R}^m$ satisfying the following condition: along with any arc $[x', x'']$ of a C-like curve it contains also the "diamond" $(x' + C^+) \cap (x'' + C^-)$ (we suppose that the parameter on a curve is chosen in such a way that a tangent vector at an arbitrary point x "points to the future", i.e. belongs to $x + C^+$) (cf. Fig. 15).

"C-convex Hull" Theorem. *Let a function f be holomorphic in a domain of the form $\mathscr{D}_+ \cup \mathscr{D}_- \cup \tilde{\Omega}$ where $\mathscr{D}_\pm = \Omega + iC_R^\mp$ and $\tilde{\Omega}$ is a complex neighborhood of Ω. Then f is extends to a holomorphic function in a domain*

$$\mathscr{D}_+ \cup \mathscr{D}_- \cup \widetilde{B_C(\Omega)}$$

where $\widetilde{B_C(\Omega)}$ is a complex neighborhood of the C-convex Hull $B_C(\Omega)$ of Ω.

This theorem was proved by Vladimirov (1960, 1961). Other proofs and extensions for the case of the light cone $C = V$ can be found in Vladimirov

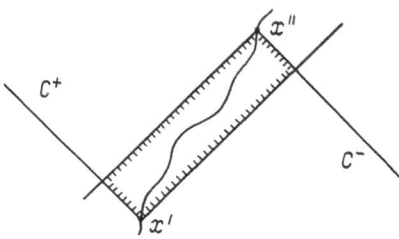

Fig. 15

(1964), Borchers (1961), Araki (1963); variants of this theorem for classes of ultradistributions and hyperfunctions were obtained by Beurling (1972) and Morimoto (1973). From the "C-convex Hull" Theorem we can deduce an interesting quasianalytic property of distributions. Namely, denote by $L(C)$ the class of distributions $f \in \mathscr{S}'(\mathbb{R}^m)$ represented as the jump $f(x) = f_+(x) - f_-(x)$ of boundary values of functions $f_\pm \in H(C^\pm)$. This class consists precisely of functions whose Fourier transforms vanish outside the cone $C^* = (C^+)^* \cup (C^-)^*$ (cf. Vladimirov (1964)).

Theorem 1 (Vladimirov (1964)). *If a function $f \in L(C)$ vanishes on an open set $\Omega \subset \mathbb{R}^m$ then it vanishes also on its C-convex hull $B_C(\Omega)$.*

2.2. Holomorphic Hulls and Dyson Domains. It is natural to consider, in connection with the "Edge-of-the-Wedge" and "C-convex Hull" theorems, the problem of describing holomorphic hulls of domains of the form $\mathscr{D}_+ \cup \mathscr{D}_- \cup \Omega$. This problem is not solved in general but there is one particular case, important for physical applications, when it is possible to obtain a simple description of the above holomorphic hull. Namely, consider domains of the form $\mathscr{D} = \mathscr{D}(\Omega) = \tau^+ \cup \tau^- \cup \Omega$ where the domain Ω lies between two spacelike hypersurfaces (cf. Chap. 2, Sect. 3.1). For any domain $\mathscr{D}$ of this type we can construct its holomorphic hull $\widetilde{\mathscr{D}}(\Omega)$ in $\mathbb{C}^{n+1}$ in the following way. We call a complex hyperboloid $\{z \in \mathbb{C}^{n+1} : (z - u)^2 = \lambda^2\}$ where $u \in \mathbb{R}^{n+1}$, $\lambda \in \mathbb{R}$, *admissible* for Ω if its real section does not intersect Ω (cf. Fig. 16). Denote by $\widetilde{\mathscr{D}} = \widetilde{\mathscr{D}}(\Omega)$ the domain in $\mathbb{C}^{n+1}$ obtained by deleting all complex hyperboloids admissible for Ω. Then $\widetilde{\mathscr{D}}$ is a domain of holomorphy which is called the *Dyson domain* associated with Ω. We show that $\widetilde{\mathscr{D}} \supset \mathscr{D}$. It is sufficient to prove that $\widetilde{\mathscr{D}} \supset \tau = \tau^+ \cup \tau^-$. If this is not so then $(z - u)^2 = \lambda^2$ for $z \in \tau$ and some $u \in \mathbb{R}^{n+1}$, $\lambda \in \mathbb{R}$. This equation is equivalent to two equations $(x - u)^2 = y^2 + \lambda^2$, $(x - u) \cdot y = 0$. It follows from the first equation that $(x - u)^2 > 0$ which contradicts the second equation. Thus $\widetilde{\mathscr{D}}$ is a domain of holomorphy (in fact, a polynomially convex domain) containing $\mathscr{D}$. The natural question is whether it coincides with the holomorphic hull of $\mathscr{D}$. The positive answer to this question follows from a theorem proved in Vladimirov

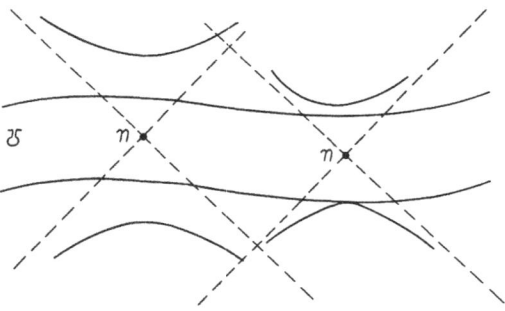

Fig. 16

(1964), §33, with the help of the Jost-Lehmann-Dyson integral representation (cf. Chap. 4, Sect. 3.1) and a theorem of Pflug (1974) (cf. also Bros-Messiah-Stora (1961)). The Jost-Lehmann-Dyson representation allows one also to describe the holomorphic hull of domains $\mathscr{D} = T^{C^+} \cup T^{C^-} \cup \Omega$ where $C^+ = C$, $C^- = -C$ and $\Omega = (C^+ + a) \cup (C^- - b)$ where $a, b \in C^+$ (or, more generally, Ω is an n-separated set in the sense of Vladimirov-Zharinov (1970)). In this case the holomorphic hull $\widetilde{\mathscr{D}}$ is described as above by using admissible complex hyperplanes defined in analogy with admissible hyperboloids (cf. Vladimirov-Zharinov, op. cit.)

There is one more interesting result connected with Dyson domains. It is the "*Finite Covariance*" *Theorem* proved by Bogolubov-Vladimirov (1958) for so called 1-point functions. Let f be a holomorphic function in the Dyson domain $\tau^+(3) \cup \tau^-(3) \cup \tilde{J}$ associated with the domain $J = \{\xi \in \mathbb{R}^4 : x^2 < 0\}$ and let $f \in H(V^+ \cup V^-)$. Then $f(z) = \sum_{v=1}^{k} \mathscr{P}_v(z) f_v(z^2)$ where $\mathscr{P}_v$ are polynomials, $f_v(\zeta)$–functions of a single complex variable ζ holomorphic and of slow growth on the complex plane $\mathbb{C}^1$ slit along the positive real half-line. This theorem was extended in Bros-Epstein-Glaser (1967) and Bogolubov-Vladimirov (1971) to so called N-point functions f, when the tube $\tau^\pm$ is replaced by the direct product $\tau_N^\pm = \tau^\pm \times \cdots \times \tau^\pm$ (N times) in $\mathbb{C}^{4N}$, assuming that the "*extended future tube conjecture*" is true. This conjecture asserts that the extended future tube τ_N' (to be defined) is a domain of holomorphy. The domain τ_N' consists of points in $\mathbb{C}^{4N}$ which can be represented in the form $(\bigwedge z^1, \ldots, \bigwedge z^N)$ where $(z^1, \ldots, z^N) \in \tau_N^+$, $\bigwedge$ is a transformation from $L_+(\mathbb{C})$, the proper complex Lorentz group (or the component of the identity of the complex Lorentz group). The extended future tube conjecture still remains unproved for $N \geq 3$ (cf. review articles of Vladimirov (1970, 1982, 1983a)). The compact version of this conjecture (where τ^+ is replaced by the generalized unit disc B_2 and the Lorentz group $L_+(\mathbb{C})$ – by the group $SL(2, \mathbb{C}) \times SL(2, \mathbb{C})$) is proved in Heinzner-Sergeev (1991).

§3. Analytic Representations

3.1. Decomposition of Hyperfunctions in Tuboids. Extensions of the "Edge-of-the-Wedge" Theorem.

A *decomposition theorem* was already formulated in Sect. 2.1: a function $f \in \mathscr{S}'(\mathbb{R}^m)$ can be represented as the jump $f(x) = f_+(x) - f_-(x)$ of boundary values of functions $f_\pm \in H(C^\pm)$ if its Fourier transform vanishes outside $(C^+)^* \cup (C^-)^*$ (this assertion can be extended also to the Riemann-Hilbert problem in T^{C^+}, cf. Vladimirov (1965)). We give here some generalizations of this result.

We define the *microlocal singular support* $SS(f)$ (the *singular spectrum* in the terminology of Sato-Kawai-Kashiwara (1973)) of a hyperfunction $f \in \mathscr{B}(\Omega)$ where Ω is open in $\mathbb{R}^m$ as the complement of the set of points $(x, \sigma) \in \Omega \times S^{m-1}$

having the following property. A point $(x, \sigma) \notin SS(f)$ if there exists a neighborhood U of x such that for some collection of local tubes $\mathscr{D}_v = U + iC_R^v$, $v = 1, \ldots, k$, such that $\sigma \notin \bigcup_{v=1}^{k} (C^v)^*$ (cf. Fig. 17) there exist functions $f_v \in \mathcal{O}(\mathscr{D}_v)$ such that we have the representation

$$f = \sum_{v=1}^{k} \text{bv } f_v.$$

Let now $\mathscr{D}$ be a tuboid with profile $\bigwedge$ over Ω. Denote by $\bigwedge^*$ the profile dual to $\bigwedge$, i.e. $\bigwedge_x^* := (\bigwedge_x)^*$ for $x \in \Omega$ and define pr $\bigwedge$ to be the subset of $\Omega \times S^{m-1}$ of the form

$$\text{pr} \bigwedge = \{(x, \sigma) : x \in \Omega, \sigma \in \bigwedge_x \cap S^{m-1}\}.$$

Theorem 2 (Sato-Kawai-Kashiwara (1973), Zharinov (1983), Morimoto (1973)). *A hyperfunction $f \in \mathscr{B}(\Omega)$ can be represented as the sum of functions $f_v \in \mathcal{O}(\mathscr{D}_v)$ holomorphic in tuboids $\mathscr{D}_v = \mathscr{D}_v(\bigwedge_v)$, $v = 1, \ldots, k$, over Ω*

$$f = \sum_{v=1}^{k} \text{bv } f_v$$

if and only if its microlocal singular support $SS(f)$ is contained in pr $\bigwedge_1^* \cup \cdots \cup$ pr $\bigwedge_k^*$.

The space of hyperfunctions on Ω having microlocal singular support contained in the projection pr $\bigwedge$ of a profile $\bigwedge = \bigwedge(\Omega)$ is denoted by $\mathscr{B}(\Omega, \bigwedge)$.

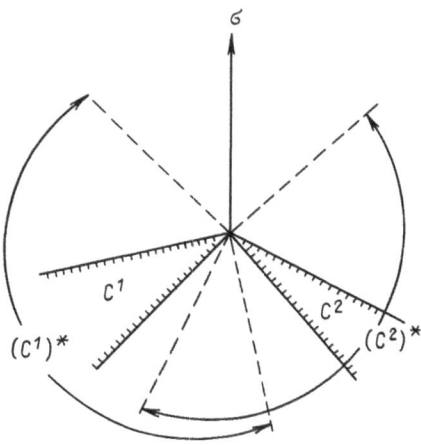

Fig. 17

The microlocal variants of Bochner's Tube Theorem (Chap. 1, Sect. 5.2) and of Bogolubov's "Edge-of-the-Wedge" Theorem (Sect. 1.1) follow immediately from Theorem 2. We formulate now a microlocal version of *Martineau's Theorem* (Sect. 1.2).

Theorem 3 (Zharinov (1983), Morimoto (1973)). *Let f_v, $v = 1, \ldots, k$, be functions holomorphic in tuboids $\mathscr{D}_v = \mathscr{D}_v(\bigwedge_v)$ over Ω. If*

$$f = \sum_{v=1}^{k} \text{bv } f_v = 0$$

on Ω then there exist functions $f_{\mu v} = -f_{v\mu}$ holomorphic in some tuboids $\mathscr{D}_{\mu v} = \mathscr{D}_{v\mu}$ with profiles $\bigwedge_{\mu v} = \text{ch}(\bigwedge_\mu \cup \bigwedge_v)$ such that

$$f_v(z) = \sum_{\mu=1}^{k} f_{\mu v}(z), \quad z \in \mathscr{D}_v \cap \mathscr{D}'_v, \quad v = 1, \ldots, k,$$

where $\mathscr{D}'_v = \bigcap_{\mu=1}^{k} \mathscr{D}_{\mu v}$.

There is a more general formulation of the "Edge-of-the-Wedge" Theorem due to Zharinov (1980, 1983) which implies the Theorems 2 and 3. Namely, let $\bigwedge$ be a fiber convex profile over an open set Ω in $\mathbb{R}^m$ (Chap. 1, Sect. 5.2). Denote by $\mathcal{O}(\Omega, \bigwedge) = \lim \mathcal{O}(\mathscr{D}(\bigwedge))$ the inductive limit of the spaces $\mathcal{O}(\mathscr{D}(\bigwedge))$ with respect to all tuboids $\vec{\mathscr{D}}(\bigwedge)$ with profile $\bigwedge$ over Ω. In other words, $\mathcal{O}(\Omega, \bigwedge)$ consists of functions "holomorphic in directions from $\bigwedge$". In Chap. 2, Sect. 2.2 we defined the boundary value map bv : $\mathcal{O}(\Omega, \Lambda) \to \mathscr{B}(\Omega)$ assigning to a function $f \in \mathcal{O}(\mathscr{D}(\bigwedge))$ a hyperfunction bv f. By Theorem 2 we have bv $\mathcal{O}(\Omega, \bigwedge) = \mathscr{B}(\Omega, \bigwedge^*)$. Consider now a more general situation. Let $\bigwedge_1, \ldots, \bigwedge_N$ be a collection of fiber convex and fiber proper profiles over Ω. Denote by $\bigwedge_{v_1 \ldots v_p}$, $1 \le p \le N$, the profile ch $(\bigwedge_{v_1} \cup \cdots \cup \bigwedge_{v_1})$, $1 \le v_1, \ldots, v_p \le N$. Introduce the space $\mathcal{O}_p(\Omega, \{\bigwedge_v\})$ of p-chains with respect to $\{\bigwedge_v\}$ consisting of collections $f = \{f_{v_1 \ldots v_p}\}$ of functions $f_{v_1 \ldots v_p} \in \mathcal{O}(\Omega, \bigwedge_{v_1 \ldots v_p})$ skew-symmetric with respect to permutations of the indices $v_1, \ldots, v_p$. Define now the boundary operator $\delta_p : \mathcal{O}_p(\Omega, \{\bigwedge_v\}) \to \mathcal{O}_{p-1}(\Omega, \{\Lambda_v\})$ by the formula

$$(\delta_p f)_{v_1 \ldots v_{p-1}} = \sum_{v=1}^{N} f_{vv_1 \ldots v_{p-1}} \quad \text{for} \quad f = \{f_{v_1 \ldots v_p}\} \in \mathcal{O}_p(\Omega, \{\bigwedge_v\}).$$

Using, as above, the boundary value map bv, we introduce the spaces of p-chains $A_p(\Omega, \{\bigwedge_v\}) = \text{bv } \mathcal{O}_p(\Omega, \{\bigwedge_v\})$ and extend in the natural way the action of δ_p to these spaces. Then the following *generalized "Edge-of-the-Wedge" Theorem* is true.

Theorem 4 (Zharinov (1983)). *The homology sequence*

$$0 \leftarrow \mathscr{B}\left(\Omega, \bigcup_{v=1}^{N} \bigwedge_v^*\right) \xleftarrow{\delta_1} A_1(\Omega, \{\bigwedge_v\}) \xleftarrow{\delta_2} \cdots \xleftarrow{\delta_N} A_N(\Omega, \{\bigwedge_v\}) \leftarrow 0$$

is exact.

Theorem 2 is equivalent to the exactness of this sequence in the term $\mathscr{B}$, Theorem 3 – in the term A_1.

3.2. Generalized Fourier and Radon Transforms. The microlocal singular support of a distribution can be defined as for hyperfunctions, however we prefer to give here another equivalent definition in terms of the generalized Fourier transform which has proven to be important for applications in several complex variables.

Let $f \in \mathscr{S}'(\mathbb{R}^m)$. *The generalized Fourier transform* or *FBI-transform* of f is defined at a point $x^0 \in \mathbb{R}^m$ by the formula (Bros-Iagolnitzer (1974–75))

$$F_{x^0}(\xi, \xi^0) = (2\pi)^{-m/2} \int_{\mathbb{R}^m} f(x) e^{-i(\xi, x) - \xi^0 |x - x^0|^2} \, dx$$

where $\xi \in \mathbb{R}^m$, $\xi^0 > 0$. The function $F_{x^0}(\xi, \xi^0)$ is defined for $\xi^0 > 0$ and satisfies the estimate

$$|F_{x^0}(\xi, \xi^0)| < M[(\xi^0)^2 + |\xi|^2]^{\alpha/2}/(\xi^0)^\beta$$

for some constants $M > 0$, α, $\beta \geq 0$. Its boundary value in $\mathscr{S}'$ for $\xi^0 \to 0$ coincides with the usual Fourier transform of f. (We can generalize this definition replacing the function $|x - x^0|^2$ in the exponent by a function of more general form, cf. Bros-Iagolnitzer, op. cit). The FBI-transform has many properties of the usual Fourier transform, in particular there is an inversion formula for this transform.

We define now the *microlocal singular support* $SS(f)$ of a distribution $f \in \mathscr{S}'(\mathbb{R}^m)$ (the *essential support* in the terminology of Bros-Iagolnitzer, op. cit.) as the complement of the set of points $(x, \sigma) \in \mathbb{R}^m \times S^{m-1}$ having the following property. A point $(x, \sigma) \notin SS(f)$ if there exists a conical neighborhood C of σ such that for some $\lambda, \alpha, \beta, \gamma > 0$ the following estuimate is true

$$|F_x(\xi, \xi^0)| < M \frac{[(\xi^0)^2 + |\xi|]^{\alpha/2}}{|\xi|^\beta} e^{-\gamma \xi^0}$$

for $\xi \in C$, $0 < \xi^0 \leq \lambda |\xi|$. Setting here $\xi^0 = \lambda |\xi|$ we obtain that a point $(x, \sigma) \notin SS(f)$ if $F_x(\xi, \lambda |\xi|)$ decreases exponentially for $|\xi| \to \infty$ in a conical neighborhood of σ. (This definition can be extended to functions of the class $\mathscr{D}'(\Omega)$, $\Omega \subset \mathbb{R}^m$, using a "cut-function", i.e. a C^∞-function with compact support which is real-analytic in a neighborhood of the considered point x).

Thus defined, the microlocal singular support of a distribution coincides with its microlocal singular support in the sense of the boundary value mapping defined in Sect. 3.1, and with its *analytic wave front set* in the sense of Hörmander (1971) (The equivalence of these three definitions is proved in Bony (1976)). The decomposition theorems for hyperfunctions given in Sect. 3.1 have their analogues for distributions and can be reformulated in terms of the FBI-transform (cf. Bros-Iagolnitzer, op. cit, and Zharinov (1983)).

With the generalized Fourier transform, as in the case of the usual Fourier transform, is closely connected the *generalized Radon transform*. Just as the usual Radon transform is based on the decomposition of the δ-function through "plane waves", the generalized Radon transform is based on the decomposition of the δ-function through "curvilinear waves"

$$\delta(x) = \frac{(m-1)!}{(-2\pi i)^m} \int_{S^{m-1}} \frac{[1 + (x, \xi)] \omega'(\xi)}{[(x, \xi) + i|x|^2 + i0]^m}$$

where the kernel $[(x, \xi) + i|x|^2 + i0]^{-m}$ is defined as the limit in the sense of distributions of the functions

$$[(x + iy, \xi + i\eta) + i(x + iy, x + iy)]^{-m}$$

as $y \to 0, \eta \to 0$ and

$$\omega'(\xi) = \sum_{j=1}^{m} (-1)^{j+1} d\xi_1 \wedge \cdots \wedge \widehat{d\xi_j} \wedge \cdots \wedge d\xi_m.$$

This formula was proved in Bony (1976); a similar formula was proved earlier in Sato-Kawai-Kashiwara (1973), p. 473. It is possible to reformulate the definition of the microlocal singular support and the decomposition theorems of Sect. 3.1 using the generalized Radon transform (cf. Kataoka (1981)).

3.3. Factorization of Hyperfunctions. The multidimensional *factorization problem* (cf. Sergeev (1978)) is a multiplicative analogue of the decomposition problem for hyperfunctions considered in Sect. 3.1. The results of Sect. 3.1 can be partially extended to this problem. Let U be a convex open subset of $\mathbb{R}^m$ and $\tilde{U}$ a complex convex neighborhood thereof in $\mathbb{C}^n$. The space of *multiplicative hyperfunctions* $\mathscr{B}^*(U)$ (cf. Sergeev (1975)) is by definition the cohomology group $H^{m-1}(\tilde{U} \setminus U, \mathcal{O}^*)$ with coefficients in the sheaf $\mathcal{O}^*$ of holomorphic functions without zeros. (This definition does not depend on the choice of $\tilde{U}$ and defines a sheaf $\mathscr{B}^*$ of multiplicative hyperfunctions on $\mathbb{R}^m$. The space $\mathscr{B}^*(\Omega)$ for any open subset Ω of $\mathbb{R}^m$ consists of sections of the sheaf $\mathscr{B}^*$ over Ω).

Using, as in Chap. 2, Sect. 2.2, different special coverings of $\tilde{U} \setminus U$, we can represent a multiplicative hyperfunction by a collection of 2^m functions $f_\varepsilon \in \mathcal{O}^*(\tilde{U}_\varepsilon)$ or by a collection of $m + 1$ functions $f_{j_1 \ldots j_m} \in \mathcal{O}^*(\tilde{U}_{j_1 \ldots j_m})$ (defined up to multiplication by an $(m - 2)$-coboundary). This allows us to define the boundary value map $bv^*(f)$ for functions from $\mathcal{O}^*(T^C \cap \tilde{U})$. The exponential map $\mathcal{O} \to \mathcal{O}^*$ induces an exponential map $\mathscr{B} \to \mathscr{B}^*$ which can be included in the exact sequence

$$0 \to \mathbb{Z} \to \mathscr{B}(U) \to \mathscr{B}^*(U) \to 1$$

for convex U. Hence the assertions of Sect. 3.1 being of microlocal character can be extended to the multiplicative case using the above exact sequence and its analogues.

Bibliographical Notes

Problems related to Bogolubov's "Edge-of-the-Wedge" Theorem are considered, for example, in the books and review articles: Vladimirov (1965, 1969, 1983a, 1971, 1982), Zharinov (1983), Morimoto (1973). On the "C-convex Hull" Theorem cf. Vladimirov (1964, 1983a, 1971, 1982), Morimoto (1973). In the exposition of Sect. 3.1 we followed Zharinov (1983). The generalized Fourier transform and its properties were studied in Bros-Iagolnitzer (1974–75), the generalized Radon transform – in Kataoka (1981). The multiplicative theory of hyperfunctions was considered in Sergeev (1975).

Chapter 4
Integral Representations

§ 1. Cauchy-Bochner Integral Representation

1.1. Cauchy-Bochner Integral in Tube Cones. The *Cauchy kernel of a tube cone T^C* is defined as the Laplace transform of the characteristic function θ_{C^*} of the dual cone C^*, i.e.

$$\mathscr{K}_C(z) = \int_{C^*} e^{i(z,\xi)} \, d\xi, \quad z \in T^C.$$

The Cauchy kernel is evidently holomorphic in T^C. There is an other representation for this kernel (cf. Vladimirov (1979))

$$\mathscr{K}_C(z) = i^m \Gamma(m) \int_{\mathrm{pr}\, C^*} \frac{d\sigma}{(\sigma, z)^m}, \quad z \in T^C, \tag{1}$$

where $\mathrm{pr}\, C^* = C^* \cap S^{m-1}$ (cf. Chap. 3, Sect. 3.1). As both sides of (1) are holomorphic in T^C, to prove (1) it is sufficient to prove it, say, for $z = iy$, $y \in C$. In this case we have

$$\mathscr{K}_C(iy) = \int_{C^*} e^{-(y,\xi)} \, d\xi = \int_{\mathrm{pr}\, C^*} d\sigma \int_0^\infty e^{-\rho(y,\sigma)} \rho^{m-1} \, d\rho$$

$$= \int_{\mathrm{pr}\, C^*} \frac{d\sigma}{(y,\sigma)^m} \int_0^\infty e^{-u} u^{m-1} \, du = i^m \Gamma(m) \int_{\mathrm{pr}\, C^*} \frac{d\sigma}{(iy,\sigma)^m},$$

q.e.d.

It follows from (1) that the Cauchy kernel $\mathscr{K}_C(z)$ is in fact holomorphic in a larger domain containing T^C and T^{-C}, namely in the domain

$$\mathbb{C}^m \setminus \bigcup_{\sigma \in \mathrm{pr}\, C^*} \{z \in \mathbb{C}^m : (z, \sigma) = 0\}.$$

This domain contains, besides T^C and T^{-C}, also real points belonging to the cones C and $-C$ in $\mathbb{R}^m$.

In the case of the future tube $\tau^+ = \tau^+(n)$ the Cauchy kernel has the form (cf. Bochner (1944), Vladimirov (1979))

$$\mathcal{K}_{V^+}(z) = \frac{2^n \pi^{(n-1)/2} \Gamma\left(\dfrac{n+1}{2}\right)}{(-z^2)^{(n+1)/2}}, \quad z^2 = z_0^2 - z_1^2 - \cdots - z_n^2. \tag{2}$$

In the case of the tube cone T_+ over the octant $\mathbb{R}^m_+$ the Cauchy kernel is the direct product of the usual Cauchy kernels

$$\mathcal{K}_{\mathbb{R}^m_+}(z) = \frac{i^m}{z_1 \ldots z_m}.$$

The Cauchy kernel for a general tube cone T^C satisfies the following estimate

$$|D^\alpha \mathcal{K}_C(z)| \leq \frac{M_\alpha}{\varDelta^{m+|\alpha|}(y)}, \quad z \in T^C$$

where $\varDelta(y)$ is the distance from y to the boundary of the cone C

$$\varDelta(y) = \inf_{\sigma \in \mathrm{pr}\, C^*} (\sigma, y).$$

In the norm of the space $\mathcal{H}_s$ (cf. Chap. 2, Sect. 1.2) the Cauchy kernel is estimated by

$$\|D^\alpha \mathcal{K}_C(z)\|_{\mathcal{H}_s} \leq M_{\alpha,s} \frac{1 + \Delta^{-s}(y)}{\varDelta^{m/2+|\alpha|}(y)}, \quad z \in T^C$$

with the usual multi-index notations. It follows from the last estimate that $\mathcal{K}_C(z)$ has a boundary value, as $y \to 0$, $y \in C$, in each space $\mathcal{H}_s$ with $s < -m/2$, which coincides with the Fourier transform of the characteristic function θ_{C^*}. Using the representation (1) we obtain the following formula for $\mathcal{K}_C(x)$

$$\mathcal{K}_C(x) = \pi(-i)^{m-1} \int_{\mathrm{pr}\, C^*} \delta^{(m-1)}((x,\sigma))\, d\sigma - (-i)^m \int_{\mathrm{pr}\, C^*} \mathscr{P}^{(m-1)}\left(\frac{1}{(x,\sigma)}\right) d\sigma \tag{3}$$

where $\mathscr{P}\left(\dfrac{1}{\lambda}\right)$ is the principal value in the Cauchy sense (cf. Vladimirov (1979)).

The importance of the Cauchy kernel is explained, in particular, by the following theorem.

Theorem 1. *A function f is in $H^{(s)}(C)$ (cf. Chap. 2, Sect. 1.2) if and only if it is given by the Cauchy-Bochner integral representation*

$$f(z) = \frac{1}{(2\pi)^m} \int_{\mathbb{R}^m} \mathcal{K}_C(z - t) f(t)\, dt = (2\pi)^{-m} \mathcal{K}_C * f(z), \quad z \in T^C \tag{4}$$

where $f(t)$ is the boundary value of f in $\mathcal{H}_s$.

This theorem was proved for $s = 0$ in Bochner (1944), the general case was considered in Vladimirov (1969a, 1979).

1.2. Cauchy-Bochner Integral for Classical Domains. We give here explicit formulas for the Cauchy kernel of some classical domains. In the case of the cone $\mathscr{H}_l$ of positive definite Hermitian $l \times l$-matrices (Chap. 1, Sect. 5.1) the Cauchy kernel has the form (Bochner (1944))

$$\mathscr{K}_{\mathscr{H}_l}(Z) = \pi^{l(l-1)/2} i^{l^2} \frac{1! \dots (l-1)!}{(\det Z)^l}$$

The Cauchy-Bochner integral in the representation (4) in this case is taken over the space of Hermitian $l \times l$-matrices. Analogous explicit formulas are known for the cones $\mathscr{P}_l$ and Q_l from Chap. 1, Sect. 5.1.

We present also Cauchy-Bochner representations for the generalized unit disc (Chap. 1, Sect. 2.2) and the Lie ball (Chap. 1, Sect. 2.3). In the case of the generalized unit disc B_m the Cauchy-Bochner representation for holomorphic functions has the form (Bochner (1944), Hua (1958))

$$f(Z) = \frac{1! \cdot 2! \cdot \dots \cdot (m-1)!}{(2\pi)^{m(m+1)/2}} \int_{U_m} \frac{f(X)\omega(X)}{[\det(I - ZX^*)]^m}, \quad Z \in B_m$$

where $\omega(X)$ is the volume form on the space of unitary $m \times m$-matrices U_m, $\omega(X) = [\operatorname{Tr}(dX \wedge dX^*)]^{\wedge m}$, i.e. the m-th exterior power of the form $\operatorname{Tr}(dX \wedge dX^*)$.

For the Lie ball this representation takes the form (Hua (1958))

$$f(Z) = \frac{\Gamma\left(\dfrac{n+1}{2}\right)}{2\pi^{(n+3)/2}} \int_{S_L} \frac{f(u)e^{-i(n+1)\theta}\omega(u)}{[(u-z, u-z)]^{(n+1)/2}}$$

$$= i\frac{\Gamma\left(\dfrac{n+1}{2}\right)}{2\pi^{(n+3)/2}} \int_0^\pi d\theta \int_{|x|=1} \frac{f(x, \theta)\omega(x)}{[(xe^{i\theta} - z, xe^{i\theta} - z)]^{(n+1)/2}}, \quad z \in B_L$$

where $u = xe^{i\theta} \in S_L$, $\omega(u)$ and $\omega(x)$ are the volume forms on S_L and S^n respectively.

1.3. Hilbert Transform. The *Hilbert transform* of a function $f \in \mathscr{H}_s$ is a function of the form

$$(Hf)(x) = -\frac{2}{(2\pi)^m} \int_{\mathbb{R}^m} \operatorname{Im} \mathscr{K}_C(x - t)f(t) \, dt$$

$$= -\frac{2}{(2\pi)^m} \operatorname{Im} \mathscr{K}_C * f(x), \quad x \in \mathbb{R}^m.$$

(Here the integral should be considered as the convolution of the distributions $\operatorname{Im} \mathscr{K}_C(x)$ and $f(x)$) (cf. Vladimirov (1979)). The explicit expression for the kernel

Im $\mathcal{K}_C(x)$ is given by formula (3). The Hilbert transform of $f \in \mathcal{H}_s$ belongs again to $\mathcal{H}_s$ and its support is contained in $C^* \cup (-C^*)$. The function f is expressed through Hf by the same integral as above but with the "+" sign before it.

Theorem 2 (Vladimirov (1969a, 1979)). *If $f(x)$ is the boundary value of a function $f \in H^{(s)}(C)$ then its real and imaginary parts are connected by the Hilbert transform*

$$\text{Im } f(x) = \frac{2}{(2\pi)^m} \int_{\mathbb{R}^m} \text{Im } \mathcal{K}_C(x - t) \, \text{Re } f(t) \, dt = \frac{2}{(2\pi)^m} \text{Im } \mathcal{K}_C * \text{Re } f(x), \, x \in \mathbb{R}^m.$$

Such formulas are also called *dispersion relations*.

1.4. Estimates of the Cauchy-Bochner Integral. Denote by $K_C f(z)$ the *Cauchy-Bochner integral* given by the right hand side of (4) and consider it as an integral operator acting on functions $f(t)$ defined on $\mathbb{R}^m$. Then the following Theorem holds.

Theorem 3 (Vladimirov (1979)). *The Cauchy-Bochner integral $K_C f(z)$ of a function $f \in \mathcal{H}_s$ satisfies the following estimate in the $\mathcal{H}_s$-norm*

$$\|D^\alpha(K_C f)(z)\|_{\mathcal{H}_s} \le M_\alpha \frac{\|f\|_{\mathcal{H}_s}}{\Delta^{|\alpha|}(y)}, \quad z \in T^C.$$

In particular, the Cauchy-Bochner integral operator K_C is bounded as an operator $K_C : \mathcal{H}_s \to \mathcal{H}_s$.

This result is true for any tube cone T^C (with an open convex proper cone C). But if we pass from the Hilbert spaces $\mathcal{H}_s$ to the Banach spaces L_p (or the Lipschitz spaces Λ_α) the estimates will depend essentially upon the tube cone considered. To see this, we give some estimates for the Cauchy-Bochner integral in the future tube and compare them with the corresponding results for the tube cone $T_+ = T^{\mathbb{R}^m_+}$ over the octant. Denote temporarily the Cauchy-Bochner integral for the future tube $\tau^+ = \tau^+(n)$ in $\mathbb{C}^{n+1}$ by K and the Cauchy integral for T_+ in $\mathbb{C}^{n+1}$ – by K_0.

First, we consider estimates in L_p spaces with $1 < p < \infty$. For $p = 2$, as we know from Theorem 3, K and K_0 are bounded in $L_2(\mathbb{R}^{n+1})$. For $1 < p < \infty$ the same assertion is true for K_0 but not for K. It follows from a theorem of Fefferman (1970), that the Cauchy-Bochner integral K is unbounded in $L_p(\mathbb{R}^{n+1})$ for any $p \ne 2$, $1 < p < \infty$ (cf. Stein (1971)).

The difference between K and K_0 becomes perhaps even more clear if we consider the behavior of these operators on the space $L_c^\infty(\mathbb{R}^{n+1})$, $n \ge 2$, of essentially bounded functions on $\mathbb{R}^{n+1}$ with compact supports. But first we need to define the types of estimates we shall consider. Usually we prove estimates for $Kf(z) = Kf(x + iy)$ at points x of the distinguished boundary when y belongs to a compact subcone C of V^+. We call such estimates *conical*. More precisely, we say that a function $g(z)$ holomorphic in τ^+ satisfies some *estimate* $(*)$ at a point $x^0 \in \mathbb{R}^{n+1}$ if for any compact subcone C of V^+ and for any r, $0 < r \le r_0$, the estimate (∗) is satisfied for $z = x + iy \in \tau^+$ such that $|x - x^0| < r$, $|y| < r$, $y \in C$, with a constant depending on r and C. At points $z^0 = x^0 + iy^0$ of the smooth part S of the boundary $\partial \tau^+$ we consider estimates of another type. We say that a function g satisfies some *local estimate* (∗) at a point $z^0 \in S$ if for any r, $0 < r \leq r_0$, the estimate (∗) is satisfied for $z = x + iy \in \tau^+$ such that $|x - x^0| < r$, $|y - y^0| < r$, with a constant depending on r. Now we can formulate the estimates for K in $L_C^\infty(\mathbb{R}^{n+1})$.

Theorem 4. *The Cauchy-Bochner integral $Kf = \mathcal{K}_{V^+} f$ of a function $f \in L_C^\infty(\mathbb{R}^{n+1})$ given by formulas (4), (2) satisfies at any point $x^0 \in \mathbb{R}^{n+1}$ the following conical estimate*

$$|Kf(z)| \leq M \frac{\|f\|_{L^\infty}}{y_0^{(n-1)/2}}.\tag{5}$$

At points $z^0 \in S$ it satisfies the local estimate

$$|Kf(z)| \leq M \|f\|_{L^\infty} \cdot |\ln|y - y^0||.\tag{6}$$

These estimates are sharp.

The estimate (5) was proved in Jöricke (1983), estimate (6) – in Sergeev (1986). Estimates analogous to (5) were proved also for the Lipschitz spaces Λ_α and for some classical domains (cf. Jöricke, op. cit., and Mitchell-Sampson (1982)).

We see from (5) that $|Kf(z)|$ could grow like a power of y_0 when $z = x + iy \to \mathbb{R}^{n+1}$. For K_0 it is well known that $|K_0 f(z)|$ can grow only logarithmically when $z \to \mathbb{R}^{n+1}$.

1.5. Schwartz Representation. We call an open convex proper cone C *regular* if its Cauchy kernel $\mathcal{K}_C(z)$ is a divisor in the algebra $H(C)$, i.e. $1/\mathcal{K}_C(z) \in H(C)$. All cones from the examples given in Chap. 1, Sect. 5.1 are regular. It can be proved also that any open proper convex cone C in $\mathbb{R}^m$ with $m \leq 3$ is regular; for $m > 3$ this is not true (Danilov (1985)). The *Schwartz kernel of a tube cone* T^C with regular C is the function

$$\mathscr{S}_C(z, z^0) = \frac{2\mathcal{K}_C(z)\mathcal{K}_C(-\bar{z}^0)}{(2\pi)^m \mathcal{K}_C(z - \bar{z}^0)} - \mathscr{P}_C(x^0, y^0), \quad z, z^0 \in T^C \tag{7}$$

where $\mathscr{P}_C(x^0, y^0)$ is the Poisson kernel for T^C (cf. Sect. 2.1 below). For $z^0 = z$ the Schwartz kernel coincides with the Poisson kernel.

For the future tube $\tau^+ = \tau^+(n)$ the Schwartz kernel is given by (Vladimirov (1979))

$$\mathscr{S}_{V^+}(z, z^0) = \frac{\Gamma\left(\dfrac{n+1}{2}\right)[-(z - \bar{z}^0)^2]^{(n+1)/2}}{\pi^{(n+3)/2}(-z^2)^{(n+1)/2}[-(\bar{z}^0)^2]^{(n+1)/2}} - \mathscr{P}_{V^+}(x^0, y^0)$$

(The explicit formula for $\mathscr{P}_{V^+}$ is given in Sect. 2.1). In the case of the octant $C = \mathbb{R}_+^m$ the Schwartz kernel is given by

$$\mathscr{S}_{\mathbb{R}_+^m}(z, z^0) = \frac{2i^m}{(2\pi)^m}\left(\frac{1}{z_1} - \frac{1}{\bar{z}_1^0}\right) \cdot \ldots \cdot \left(\frac{1}{z_m} - \frac{1}{\bar{z}_m^0}\right) - \mathscr{P}_{\mathbb{R}_+^m}(x^0, y^0)$$

(for $\mathscr{P}_{\mathbb{R}_+^m}$ cf. Sect. 2.1). In particular, for $m = 1$ we have

$$\mathscr{S}_{\mathbb{R}_+}(z, z^0) = \frac{i}{\pi}\left(\frac{1}{z} - \frac{x^0}{|z^0|^2}\right)$$

and Re $\mathscr{S}_{\mathbb{R}_+}(z, z^0)$ coincides with the usual Poisson kernel $\mathscr{P}(x, y)$ on the plane.

Theorem 5 (Vladimirov (1979)). *Let a function $f \in H(C)$, for a regular cone C, satisfy the condition: $f(x)\mathscr{K}_C(x - \bar{z}^0) \in \mathscr{H}_s$ for some s and all $z^0 \in T^C$ (where $f(x)$ is the boundary value of f in $\mathscr{S}'$). Then f has the Schwarz representation*

$$f(z) = i\int_{\mathbb{R}^m} \mathscr{S}_C(z - t, z^0 - t)\,\mathrm{Im}\,f(t)\,dt + \mathrm{Re}\,f(z^0), \quad z, z^0 \in T^C$$

where the integral should be considered as the value of the distribution functional $\mathrm{Im}\,f(t)$ *on the function* $\mathscr{S}_C(z - t, z^0 - t)$.

§2. Poisson Integral Representation

2.1. Poisson Integral in Tube Cones. Let C be a convex open cone in $\mathbb{R}^m$. The *Poisson kernel of the tube cone* T^C is the function

$$\mathscr{P}_C(x, y) = \frac{|\mathscr{K}_C(x + iy)|^2}{(2\pi)^m \mathscr{K}_C(2iy)}, \quad x + iy \in T^C \tag{8}$$

where $\mathscr{K}_C(z)$ is the Cauchy kernel (cf. Sect. 1.1).

The Poisson kernel $\mathscr{P}_C(x, y)$ is non-negative in T^C and satisfies:

$$\int_{\mathbb{R}^m} \mathscr{P}_C(x, y)\,dx = 1, \quad y \in C, \quad \int_{|x| > \delta} \mathscr{P}_C(x, y)\,dx \to 0, \quad \text{as } y \to 0, \quad y \in C$$

for any $\delta > 0$; i.e. it has properties analogous to the ones for the usual Poisson kernel (cf. Vladimirov (1979), Stein-Weiss (1971)).

The Poisson kernel for the future tube $\tau^+ = \tau^+(n)$ is given by the formula (Vladimirov, op. cit.)

$$\mathscr{P}_{V_+}(x, y) = \frac{2^n \Gamma\left(\dfrac{n + 1}{2}\right)(y^2)^{(n+1)/2}}{\pi^{(n+3)/2}|(x + iy)^2|^{n+1}}, \quad x + iy \in \tau^+.$$

In the case of $C = \mathbb{R}_+^m$ the Poisson kernel is the product of the usual Poisson kernels

$$\mathscr{P}_{\mathbb{R}_+^m}(x, y) = \frac{y_1 \cdots \cdots y_m}{\pi^m |z_1|^2 \cdots \cdots |z_m|^2}, \quad x + iy \in T_+.$$

The Poisson kernel for a general tube cone satisfies the estimates (Vladimirov, op. cit.)

$$|D^\alpha \mathscr{P}_C(x, y)| \le M_\alpha \frac{|y|^m}{\Delta^{2m+|\alpha|}(y)}, \quad x + iy \in T^C$$

$$\|D^\alpha \mathscr{P}_C(x, y)\|_{\mathscr{H}_s} \le M_{\alpha, s, p} \frac{[1 + \Delta^{-s}(y)][1 + \Delta^{-p}(y)]}{\Delta^{m+|\alpha|}(y)} |y|^m, \quad x + iy \in T^C$$

for some constants M_α, $M_{\alpha, s, p} > 0$ and for all $s \ge 0$, $p > s + \dfrac{m}{2}$, using the usual multi-index notations ($\Delta(y)$ is the distance of y from ∂C).

Theorem 6. *A function f is in $H^{(s)}(C)$ if and only if it can be given by the Poisson integral representation*

$$f(z) = \int_{\mathbb{R}^m} \mathscr{P}_C(x - t, y)g(t)\, dt = \mathscr{P}_C * g(z), \quad z = x + iy \in T^C \tag{9}$$

where $g \in \mathscr{H}_s$ and the support of the Fourier transform $F^{-1}[g]$ belongs to C^*. If f is given by formula (9) then $g(t)$ coincides with the boundary value $f(t)$ of f in $\mathscr{H}_s$. Accordingly, $\operatorname{Re} f(z)$ (resp. $\operatorname{Im} f(z)$) is given by formula (9) with $g(t) = \operatorname{Re} f(t)$ (resp. $\operatorname{Im} f(t)$).

This theorem was proved in Vladimirov (1979), the case $s = 0$ – in Stein-Weiss (1971).

2.2. Poisson Integral in Classical Domains. We shall now give explicit formulas for the Poisson kernel of some classical domains. In the case of the cone $\mathscr{H}_l$ of positive definite Hermitian $l \times l$-matrices (Chap. 1, Sect. 5.1) the Poisson kernel has the form

$$\mathscr{P}_{\mathscr{H}_l}(X, Y) = c_l \left[\frac{\det Y}{|\det Z|^2} \right]^l, \quad Z = X + iY \in T^{\mathscr{H}_l}$$

where

$$c_l = \frac{2^l(-1)^l i^{l^2}}{\pi^{l-1}} \cdot \frac{l(l - 1)}{2} \cdot 1! \cdots (l - 1)!$$

The Poisson integral in the representation (9), in this case, is taken over the space of Hermitian $l \times l$-matrices. Analogous formulas are known for the cones $\mathscr{P}_l$ and Q_l from Chap. 1, Sect. 5.1.

In the case of the generalized unit disc B_n (Chap. 1, Sect. 2.2) the Poisson representation takes the form (Hua (1958))

$$f(Z) = \frac{1! \cdots (m - 1)!}{(2\pi)^{m(m+1)/2}} \int_{U_m} f(X) \frac{[\det(I - ZZ^*)]^m}{|\det(Z - X)|^{2m}} \omega(X), \quad Z \in B_m$$

where $\omega(X)$ is the volume form on U_m (cf. Sect. 1.2). For the Lie ball (Chap. 1, Sect. 2.3) this representation is given by (Hua (1958))

$$f(z) = \frac{\Gamma\left(\dfrac{n+1}{2}\right)}{2\pi^{(n+3)/2}} \int_{S_L} f(u) \frac{[1 + |(z, z)|^2 - 2|z|^2]^{(n+1)/2}}{|(u - z, u - z)|^{n+1}} \omega(u), \quad z \in B_L$$

where $\omega(u)$ is the volume form on the Lie sphere S_L (cf. Sect. 1.2).

2.3. Boundary Properties of the Poisson Integral. Denote by $P_C f(z)$ the *Poisson integral* given by the right hand side of (9) and also denote by $P_C[d\mu](z)$ the Poisson integral of a complex-valued Borel measure μ on $\mathbb{R}^m$ given by the same formula (9). We formulate in the next theorem some boundary properties and estimates for the Poisson integral in different spaces.

Theorem 7. 1) *If $f \in \mathcal{H}_s$ then*

$$\|P_C(x + iy)\|_{\mathcal{H}_s} \le \|f\|_{\mathcal{H}_s}, \quad y \in C.$$

Moreover,

$$P_C f(x + iy) \to f(x) \quad \text{in } \mathcal{H}_s \text{ for } y \to 0, \ y \in C.$$

2) *If $f \in L^p(\mathbb{R}^m)$, $1 \le p < \infty$, then*

$$\|P_C f(x + iy)\|_{L^p} \le \|f\|_{L^p} \quad \text{for any } y \in C$$

and

$$P_C f(x + iy) \to f(x) \quad \text{in } L^p \text{ for } y \to 0, \ y \in C.$$

3) *If $f \in L^\infty(\mathbb{R}^m)$ then*

$$|P_C f(z)| \le \|f\|_{L^\infty} \quad \text{for } z \in T^C$$

and

$$P_C f(x + iy) \to f(x) \quad \text{for } y \to 0, \ y \in C$$

in the weak* topology of the space L^∞. If, in addition, the function f is continuous at a point x then

$$P_C f(x + iy) \to f(x) \quad \text{for } y \to 0, \ y \in C$$

4) *For a finite Borel measure μ on $\mathbb{R}^m$*

$$\|P_C[d\mu](x + iy)\|_{L^1} \le \|\mu\| = \int_{\mathbb{R}^m} |d\mu|.$$

Moreover,

$$P_C[d\mu](x + iy) \to \mu(x) \quad \text{for } y \to 0, \ y \in C$$

in the weak* topology of the measure space.

The assertions of this Theorem follow from the basic properties of the Poisson kernel given in Sect. 2.1. Assertion 1) was proved in Vladimirov (1979); assertion 2) – in Korányi (1965), Stein-Weiss (1971); assertion 3) – in Vladimirov (op. cit.), Korányi (op. cit.); assertion 4) – in Korányi (op. cit.).

The problem of pointwise convergence of Poisson integrals is much more complicated. Let us first define the sets of approach. We say that $z = x + iy \in T^C$ tends to x^0 in the *restricted admissible sense* if $z \to x^0$ staying inside the set

$$\Gamma_{C'}^{\alpha}(x^0) = \{x + iy \in T^C : |x - x^0| < \alpha|y|, y \in C' \Subset C\}$$

for some constant $\alpha > 0$ and some compact subcone C' in C.

Most results on the pointwise convergence of Poisson integrals have been proved in the homogeneous case so we restrict our attention now to tube cones T^C corresponding to the classical domains (cf. Chap. 1, Sect. 5.1).

Theorem 8 (Stein-Weiss (1969), Weiss (1972), Stein-Weiss (1971)). *If $f \in L^p(\mathbb{R}^m)$, $1 \le p \le \infty$, then its Poisson integral $P_C f(z)$ converges to $f(x^0)$ for almost all $x^0 \in \mathbb{R}^m$ when $z \to x^0$ in the restricted admissible sense. If μ is a finite Borel measure on $\mathbb{R}^m$ such that its absolutely continuous part (with respect to dx) is equal to $f(x)$ then $P_C[d\mu][z]$ converges to $f(x^0)$ for almost all $x^0 \in \mathbb{R}^m$ when $z \to x^0$ in the restricted admissible sense.*

A further generalization of the notion of restricted admissible convergence was proposed by Korányi (1969, 1972). His definition uses, in full strength, the theory of semisimple Lie groups so for the precise formulations we should require much background material. To avoid this, we prefer to give here only a sketch of his results referring for the details to the papers of Korányi (1969, 1972, 1976, 1979) and Stein (1983).

For a bounded symmetric domain it is possible to construct several (in general) different compactifications. To each of these compactifications corresponds its own distinguished boundary and Poisson integral and the notion of the restricted admissible limit can be introduced at points of the distinguished boundaries of these compactifications in an invariant way using approaching sets of the type $\Gamma_{C'}^{\alpha}(x^0)$. Theorem 8 remains true for any compactification of a bounded symmetric domain as was proved by Stein (1983). The notion of the restricted admissible limit can be formulated also for the other boundary points of these compactifications outside their distinguished boundaries. An analogue of Theorem 8 for these points was proved for functions of class L^p, $1 \le p \le \infty$, in Korányi (1979), Stein (1983).

The notion of the restricted admissible limit is a natural extension of the notion of nontangential limit to domains of tube type. But for some tube cones we can assert the existence of a limit in a stronger sense. In particular, if $C = \mathbb{R}^m_+$ we can replace the approach sets $\Gamma_C^{\alpha}(x^0)$ by the approach sets

$$\Gamma^{\alpha}(x^0) = \{x + iy \in T_+ : |x - x^0| < \alpha|y|, y \in C\}, \quad \alpha > 0.$$

Then for any function $f \in L^p(\mathbb{R}^m)$, $p > 1$, its Poisson integral $P_{\mathbb{R}^m_+} f(z)$ converges to $f(x^0)$ almost everywhere on $\mathbb{R}^m$ when $z \to x^0$ satying within some approaching set $\Gamma^\alpha(x^0)$ (cf. Stein-Weiss (1971)). However, if we translate this notion of unrestricted limit directly to general tube cones then the last assertion fails. The counterexample given in Stein-Weiss (1969) shows that for the future tube $\tau^+(n)$, $n \geq 2$, this unrestricted limit does not exist in any L^p, $1 \leq p \leq \infty$. The correct extension of the restricted admissible limit to homogeneous tube cones was found in Korányi (1969, 1972). This notion, called the *admissible* (or *semi restricted admissible*) limit, can be defined, as in the restricted case, for each of the compactifications of a bounded symmetric domain in an invariant way and for all boundary points. It was shown in Lindahl (1972) (cf. also Knapp-Williamson (1971)) that Theorem 8 remains valid for any compactification of a bounded symmetric domain of tube type for functions $f \in L^p$ when $z \to x^0$ in the admissible way and $p_0 < p \leq \infty$ where p_0 depends upon the domain and, in principle, can be arbitrarily large (however, there are no counterexamples with $p > 1$). An analogue of this result for the other points of the boundaries was proved in Korányi (1979), Stein (1983).

Let us illustrate now the notion of the admissible limit, considering the case of the future tube $\tau^+(2)$ in $\mathbb{C}^3$ (cf. Korányi (1976)). For the standard compactification of $\tau^+(2)$ analogous to the one considered in Chap. 1, Sect. 3.2 we obtain an admissible limit which coincides with the restricted admissible limit at points of the distinguished boundary $\mathbb{R}^3$. At other points of the boundary of $\tau^+(2)$ the admissible limit is nontangential in some directions while in other directions (e.g. on the plane (y_1, y_2)) contact of the 1st order with the boundary is allowed. Another (so called, maximal) compactification of $\tau^+(2)$ provides us with another notion of admissible limit at points of $\mathbb{R}^3$ for which tangential approach (of any order) is allowed along almost all real light rays.

2.4. Pluriharmonic Functions. Denote by $RP(T^C)$ the space of *pluriharmonic functions*, i.e. functions on T^C which are the real parts of holomorphic functions. It follows from Theorem 6 (Sect. 2.1) that the Poisson integral $P_C g(z)$ of a real function $g \in \mathcal{H}_s$ belongs to $RP(T^C)$ if and only if the Fourier transform $F^{-1}[g]$ vanishes outside $C^* \cup (-C^*)$. The discrete analogue of this assertion for the generalized unit disc B_2 was proved in Vladimirov (1974) for slowly growing functions; its analogue for bounded symmetric domains and L^2 functions is contained in Schmid (1969).

For functions of the class $RP(T^C)$ we have the following generalization of Rudin's "Correction" Theorem (Rudin (1969)).

Theorem 9. *Let g be a lower semicontinuous positive function on $\mathbb{R}^m$, $g \in L^1(\mathbb{R}^m)$. Then there exists a positive singular (with respect to Lebesgue measure) measure σ on $\mathbb{R}^m$ such that the Poisson integral*

$$P_C[g \, dx - d\sigma]$$

belongs to $RP(T^C)$.

This theorem was proved in a more general situation in Alexandrov (1984). The following *"Localization" Theorem of Rudin* is also related to the class $RP(T^C)$.

Theorem 10. *Let Ω be an open subset of $\mathbb{R}^m$. There exists an open set $\mathscr{D} = T^{C^+} \cup T^{C^-} \cup \tilde{\Omega}$ where $C^+ = C$, $C^- = -C$, $\tilde{\Omega}$ is a complex neighborhood of Ω, having the following property. If the Poisson integral $P_C[d\mu]$ of a measure μ on $\mathbb{R}^m$ belongs to $RP(T^{C\pm})$ and the support of μ does not intersect Ω then $P_C[d\mu]$ belongs to $RP(\mathscr{D})$ and vanishes on Ω.*

This theorem was proved in Rudin (1970) for the polydisc but its proof, based on the "Edge-of-the-Wedge" Theorem, is valid for general tube cones. Note that the theorem is *not* true for arbitrary Borel measures (a counterexample is given in Rudin (1969), Sect. 2.3).

For other results on pluriharmonic functions cf. Vladimirov (1979), Stoll (1974) and Sect. 4.2, Chap. 4.

2.5. Functions given by Poisson Integrals. In the case when the cone C is the octant $\mathbb{R}^m_+$ the class of functions given by Poisson integrals coincides with the class of m-harmonic functions, i.e. functions which are harmonic with respect to each of the variables separately. The class of real m-harmonic functions strictly contains the class $RP(T^{\mathbb{R}^m_+})$ (cf. Rudin (1969)). What is the characterization of functions given by Poisson integrals in the case of a general tube cone? We consider first this question for the generalized unit disc B_m (Chap. 1, Sect. 2.2). Let us introduce a matrix operator Δ_Z whose components are differential operators of the 2nd order

$$(\Delta_Z)_{ij} = \sum_{r,s,t=1}^{m} \left(\delta_{ir} - \sum_{\alpha=1}^{m} z_{i\alpha}\bar{z}_{r\alpha} \right) \left(\delta_{st} - \sum_{\beta=1}^{m} \bar{z}_{\beta s} z_{\beta t} \right) \frac{\partial^2}{\partial \bar{z}_{rs} \partial z_{jt}} \quad 1 \le i, j \le m.$$

This operator can be written symbolically in the form (cf. Hua (1958))

$$\Delta_Z = (I - ZZ^*)\bar{\partial}_z \cdot (I - Z^*Z)^t \partial_z$$

where

$$Z = (z_{ij}), \quad \partial_z = \partial/\partial Z = (\partial/\partial z_{ij}), \quad 1 \le i, j \le m.$$

The trace $\mathrm{Tr}\,\Delta_Z$ of Δ_Z is the *invariant* (with respect to automorphisms of B_m) Laplacian of B_m so that functions u given by Poisson integrals in B_m are harmonic with respect to $\mathrm{Tr}\,\Delta_Z$

$$(\mathrm{Tr}\,\Delta_Z)u = 0.$$

This result was proved by Hua (1958) who noted, moreover, that functions u given by Poisson integrals in B_m satisfy in fact the system of differential equations

$$\Delta_Z u = 0$$

(this assertion was proved in Hua (op. cit.) for classical Cartan domains of the 1st type). E.M. Stein has conjectured that the equations found by Hua Loo-keng completely characterize functions given by Poisson integrals. This conjecture was proved for bounded symmetric domains $\mathscr{D}$ of tube type in Johnson-Korányi (1980), Berline-Vergne (1981) (cf. also Lasalle (1984a, b), Johnson (1984a, b)) (partial results in the same direction were proved in Korányi-Malliavin (1975), Johnson (1978)). We formulate here the result of Berline-Vergne (op. cit.). They constructed a system of differential operators of the 2nd order called Hua operators, which coincides with $\varDelta_z$ in the case of the generalized unit disc, and proved the following assertion. A function F in a domain $\mathscr{D}$ is the Poisson integral of some hyperfunction over the distinguished boundary (Shilov boundary) of $\mathscr{D}$ if and only if it satisfies the *Hua equations*. It is interesting to compare this result with the *theorem of Fürstenberg* (1963) which asserts that any bounded function in a bounded symmetric domain $\mathscr{D}$ which is annihilated by all invariant differential operators without a constant term in $\mathscr{D}$ is in fact the Poisson integral of some bounded function over the distinguished boundary of $\mathscr{D}$. Here we take the Poisson integral and the distinguished boundary with respect to the maximal compactification of $\mathscr{D}$ mentioned before in Sect. 2.3.

§ 3. Other Integral Representations

3.1. Bergman Representation. The Bergman representation for classical domains and Siegel domains was constructed in Hua (1958), Rothaus (1960), Gindikin (1964), (for the general properties of the Bergman representation cf., e.g. Fuks (1963)). In the case of the future tube $\tau^+ = \tau^+(n)$, $n \geq 2$, it has the form (cf. Sergeev (1985))

$$f(z) = \left(\frac{i}{2}\right)^{n+1} \frac{(n+1)!}{2(2\pi)^{n+1}} \int_{\tau^+} \frac{f(w)\, dw \wedge d\overline{w}}{\left[\left(\dfrac{z-\overline{w}}{2i}\right)^2\right]^{n+1}}, \quad z \in \tau^+ \tag{10}$$

for functions $f \in L^2(\tau^+) \cap \mathcal{O}(\tau^+)$. We see that, in contrast with the Cauchy and Poisson integral representations considered above, the Bergman representation involves the integration over the entire domain τ^+. The *Bergman operator* $Kf(z)$ given by the right hand side of (10) is an orthogonal projector of $L^2(\tau^+)$ onto the space $L^2(\tau^+) \cap \mathcal{O}(\tau^+)$. We can estimate the operator Kf on functions $f \in L_c^\infty(\tau^+)$ vanishing outside some ball $\{|z| < R\}$ as follows (cf. the definition of conical and local estimates in Sect. 1.4).

Theorem 11 (Sergeev (1985)). *The Bergman operator $Kf(z)$, $f \in L_c^\infty(\tau^+)$, given by the right hand side of (10), has the following conical estimate at any point $x^0 \in \mathbb{R}^{n+1}$*

$$|Kf(z)| \leq C \frac{\|f\|_{L^\infty}}{y_0^{(n-1)/2}}.$$

At points $z^0 = x^0 + iy^0 \in S$ we have the following local estimate

$$|Kf(z)| \le C\|f\|_{L^\infty} \cdot |\ln|y - y^0||.$$

Very little is known about estimates of the Bergman operator in L^p norms with $1 < p < \infty$, $p \ne 2$. D. Bekollé (1984) has considered this operator in the case of the future tube $\tau^+(2)$ in $\mathbb{C}^3$. He proved that it is bounded in L^p for p "close" to 2 and is unbounded for p "close" to 1 and infinity. Still some gap in between these two subsets of the p-axis remains where it is unknown whether the operator is bounded.

3.2. Cauchy-Fantappiè Type Representations. We begin with a general scheme for the construction of *Cauchy-Fantappiè type* integral *representations* for holomorphic and smooth functions in smooth domains. After that we shall show how these representations are modified when applied to the future tube and Dyson domains. A detailed exposition of Cauchy-Fantappiè integral representations and further references can be found in Aizenberg-Yuzhakov (1979), Chirka-Khenkin (1975), Khenkin-Leiterer (1984), Leray (1959) and this series, vol. 7, part II.

Let $\mathscr{D}$ be a C^1-smooth domain in $\mathbb{C}^m$ and $\varphi(\zeta, z)$ a C^1-smooth function on $\partial\mathscr{D} \times \bar{\mathscr{D}}$ which is holomorphic in $z \in \mathscr{D}$. Such a function φ is called a *barrier function* if $\varphi(\zeta, \zeta) = 0$ and $\varphi(\zeta, z) \ne 0$ for $(\zeta, z) \in \partial\mathscr{D} \times \mathscr{D}$. For example, the function $\sum_{i=1}^m \bar{\zeta}_i(\zeta_i - z_i)$ is a barrier function for the ball $\left\{\sum_{i=1}^m |z_i|^2 < 1\right\}$ in $\mathbb{C}^m$.

By the *Hefer representation* of a barrier function φ we mean the representation of φ in the form

$$\varphi(\zeta, z) = \sum_{i=1}^m P_i(\zeta, z)(\zeta_i - z), \quad (\zeta, z) \in \partial\mathscr{D} \times \bar{\mathscr{D}}$$

where P_i are C^1-smooth in their domain of definition and holomorphic in $z \in \mathscr{D}$. Denote by P the column vector ${}^t(P_1, \ldots, P_m)$ so that $\varphi(\zeta, z) = (P(\zeta, z), \zeta - z)$ and define the vector-function

$$w(\zeta, z) = \frac{P(\zeta, z)}{\varphi(\zeta, z)}, \quad (\zeta, z) \in \partial\mathscr{D} \times \mathscr{D}.$$

Consider also the universal *barrier function* (non-holomorphic in z) of *Martinelli-Bochner*

$$\varphi_0(\zeta, z) = |\zeta - z|^2 = (P^0(\zeta, z), \zeta - z), \quad P^0(\zeta, z) = {}^t(\bar{\zeta}_1 - \bar{z}_1, \ldots, \bar{\zeta}_m - \bar{z}_m),$$

and set

$$w_0(\zeta, z) = \frac{P^0(\zeta, z)}{\varphi_0(\zeta, z)}.$$

Denote, at last, by $w(\zeta, z, \lambda)$ the linear combination of the vector-functions w and w_0:

$$w(\zeta, z, \lambda) = (1 - \lambda)w_0(\zeta, z) + \lambda w(\zeta, z), \quad (\zeta, z) \in \partial\mathscr{D} \times \mathscr{D}, \quad 0 \le \lambda \le 1.$$

The *kernel of the Cauchy-Fantappiè representation* associated with a barrier function φ is given by the differential form

$$\Omega(\zeta, z, \lambda) = \det(w, dw, \ldots, dw) \wedge d\zeta, \quad (\zeta, z) \in \partial \mathscr{D} \times \mathscr{D}, \quad 0 \leq \lambda \leq 1$$

where $w = w(\zeta, z, \lambda)$, $d\zeta = d\zeta_1 \wedge \cdots \wedge d\zeta_m$. The differential d is taken with respect to the variables ζ, λ and the determinant is expanded with respect to columns replacing the multiplication by exterior multiplication.

Let us introduce the *space $E(\mathscr{D})$* consisting of functions $v \in C^1(\overline{\mathscr{D}})$ decreasing sufficiently rapidly at infinity along with their $\bar{\partial}$-derivatives (the rate of decay depends on the kernel of the representation). The Cauchy-Fantappiè integral representation (associated with a barrier function φ) for functions $v \in E(\mathscr{D})$ has the form

$$-(2\pi i)^m v(z) = \int_{\mathscr{D}} \bar{\partial} v \wedge \Omega_0 + \int_{\partial \mathscr{D} \times [0,1]} \bar{\partial} v \wedge \Omega - \int_{\partial \mathscr{D}} v \Omega_1, \quad z \in \mathscr{D} \quad (11)$$

where Ω_0 (respectively Ω_1) denotes the restriction of Ω to the set $\{\lambda = 0\}$ (respectively $\{\lambda = 1\}$). For holomorphic functions $v = f \in E(\mathscr{D})$ this representation takes the form

$$f(z) = \frac{1}{(2\pi i)^m} \int_{\partial \mathscr{D}} v \Omega_1, \quad z \in \mathscr{D}. \quad (12)$$

By slightly modifying this construction, let us show now how to obtain the Cauchy-Bochner integral representation in τ^+ in $\mathbb{C}^{n+1}$ with $n \geq 2$. We define first the natural barrier function for τ^+ using the convexity of τ^+. For $\zeta = \xi + i\eta \in S$ (cf. Chap. 1, Sect. 1.1), $z \in \tau^+$ we put

$$\varphi(\zeta, z) = 1 - \sum_{j=1}^{n} \frac{\eta_j}{\eta_0} \frac{z_j - \xi_j}{z_0 - \xi_0}. \quad (13)$$

Note that the equation $\varphi(\zeta, z) = 0$ (with ζ fixed) defines the complex tangent space $T_\zeta^c \tau^+$ at ζ (cf. Fig. 18) so the function φ is a barrier. Taking the limit in (13) for $\eta_0 \to 0$, $\eta_j/\eta_0 \to \sigma_j$ we obtain the function

$$\varphi_\sigma(\xi, z) = 1 - \sum_{j=1}^{n} \sigma_j \frac{z_j - \xi_j}{z_0 - \xi_0} \quad (14)$$

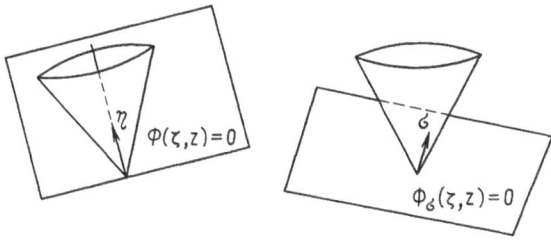

Fig. 18

where $\sigma = (\sigma_1, \ldots, \sigma_n)$, $|\sigma| = 1$. Note that for any σ, $|\sigma| \le 1$, the function $\varphi_\sigma(\xi, z)$ is a barrier function on $\mathbb{R}^{n+1} \times \tau^+$ (i.e. $\varphi_\sigma(\xi, z) \ne 0$ for $(\xi, z) \in \mathbb{R}^{n+1} \times \tau^+$). The equation $\varphi_\sigma(\xi, z) = 0$ (with ξ fixed) defines a support space of τ^+ at the point ξ (cf. Fig. 18). Thus we can say that the functions $\varphi(\zeta, z)$, $\varphi_\sigma(\xi, z)$ form a family of barrier functions for τ^+. We can consider this family as one barrier function (in the sense of the definition given in the beginning of this section) for the domain $\mathscr{D}$ obtained by applying a real monoidal transformation or σ-process to τ^+ along $\mathbb{R}^{n+1}$. The Cauchy-Fantappiè representation (12) for holomorphic functions associated with this barrier function coincides with the Cauchy-Bochner representation (1). For $E(\tau^+)$, in this case, we may take the space $E_n(\tau^+)$ of functions v decreasing at infinity along with their $\bar{\partial}$-derivatives faster than $1/|z|^n$, i.e. $|v(z)| \le C/|z|^{n+\varepsilon}$, $|\bar{\partial}v(z)| \le C/|z|^{n+\varepsilon}$ for $|z| > R$ and for some positive constants $C, \varepsilon > 0$.

Another natural integral representation on τ^+ can be obtained by considering the Cauchy-Fantappiè representation associated with the barrier function for τ^+ given by the Levi polynomial for τ^+

$$\varphi(\zeta, z) = i\eta \cdot (\zeta - z) - \frac{(\zeta - z)^2}{4}, \quad (\zeta, z) \in \partial\tau^+ \times \tau^+. \tag{16}$$

The integral operator $Kf(z)$ for $f \in L_c^\infty(\partial\tau^+)$ defined by the right hand side of (12) with the barrier function (16) satisfies the following estimates (cf. Sergeev (1986))

1) conical estimate at points $x^0 \in \mathbb{R}^{n+1}$:

$$|Kf(z)| \le C\|f\|_{L^\infty} \frac{|\ln y_0|}{y_0^{(n-1)/2}},$$

2) local estimate at points $z^0 = x^0 + iy^0 \in S$:

$$|Kf(z)| \le C\frac{\|f\|_{L^\infty}}{|y - y^0|}.$$

Another application of Cauchy-Fantappiè representations is considered in the next section.

3.3. The Jost-Lehmann-Dyson Representation.

Let us consider a Dyson domain (cf. Chap. 3, Sect. 2.2.) $\widetilde{\mathscr{D}} = \tau^+ \cup \tau^- \cup \widetilde{\Omega}$ associated with a domain Ω in $\mathbb{R}^{n+1}$ bounded by two spacelike hypersurfaces. Let Σ be a spacelike hypersurface in Ω. We shall construct an integral representation for holomorphic functions in $\widetilde{\mathscr{D}}$ using the Cauchy-Fantappiè representation. We define first a barrier function for $\widetilde{\mathscr{D}}$. Through any non-real point $\zeta \in \partial\widetilde{\mathscr{D}}$ there passes the unique admissible hyperboloid $(z - u)^2 = \lambda^2$ where $u \in \Sigma$; we denote by $u(\zeta)$, $\lambda(\zeta)$ the parameters of this hyperboloid. On the other hand, through any real point ξ of the set $\mathbb{R}^{n+1} \setminus \Omega$, which is the union of non-intersecting domains $\mathscr{D}_+$ and $\mathscr{D}_-$, there passes a whole family of complex hyperboloids $(z - u)^2 = \lambda^2$; their centers u fill up a domain Σ_ξ on the hypersurface Σ; the parameter λ of any such hyperboloid is uniquely determined by u. We define a barrier function φ_u in $\widetilde{\mathscr{D}}$ by the formula

$$\varphi_u(\zeta, z) = (\zeta - u)^2 - (z - u)^2 \quad \text{where} \quad \begin{cases} u = u(\zeta), & \zeta \in \partial\widetilde{\mathscr{D}} \setminus \mathbb{R}^{n+1}, \\ u \in \Sigma_\xi, & \zeta = \xi \in \mathbb{R}^{n+1} \setminus \Omega. \end{cases} \quad (17)$$

We obtain again a family φ_u of barrier functions which can be interpreted as in Sect. 3.2 as a single barrier function of a domain obtained from $\widetilde{\mathscr{D}}$ by application of a real monoidal transformation. The Cauchy-Fantappiè representation (12) for holomorphic functions associated with the barrier function (17) has the form

$$f(z) = \frac{1}{(2\pi i)^{n+1}} \int_{\mathscr{D}_+ \cup \mathscr{D}_-} \varepsilon(\xi)[f_+(\xi) - f_-(\xi)] \int_{\Sigma_\xi} \Omega_1, \quad z \in \widetilde{\mathscr{D}}, \quad (18)$$

where $\varepsilon(\xi) = \pm 1$ for $\xi \in \mathscr{D}_\pm$, $f_\pm(x)$ is the boundary value of $f(x + iy)$ as $y \to 0$, $y \in V^\pm$.

The integral representation (18) was obtained in Bros-Itzykson-Pham (1966) and, as was shown in that paper, coincides with the *Jost-Lehmann-Dyson representation* introduced in Jost-Lehmann (1957), Dyson (1958). It makes sense for functions $f \in H^{(s)}(V^+ \cup V^-) \cup \mathcal{O}(\widetilde{\mathscr{D}})$ (cf. Vladimirov (1964)). An integral representation of the Jost-Lehmann-Dyson type for domains $\Omega = (C^+ + a) \cup (C^- - b)$, $a, b \in C^+$, and for similar domains of more general type (cf. Chap. 3, Sect. 2.2) was obtained in Vladimirov-Zharinov (1970).

In the limit case when the domain Ω degenerates to a spacelike hypersurface Σ we obtain from (18) an integral representation for functions f holomorphic in the future tube τ^+. In this case Σ_ξ is the intersection of Σ with the light cone $\{(\xi - u)^2 \geq 0\}$ (cf. Fig. 19). In particular, if $\Sigma = \{u : u_0 = 0\}$ we can compute explicitly the kernel of the corresponding integral representation (18) (cf. Sergeev (1986))

$$f(z) = \frac{2^{2n} \pi^{(n-1)/2} \Gamma\left(\dfrac{n+1}{2}\right)}{(2\pi i)^{n+1}}$$

$$\times \int_{\mathscr{D}_+ \cup \mathscr{D}_-} \frac{\varepsilon(\xi)\xi_0^n(\xi_0 + z_0)f(\xi)\,d\xi}{\{[\xi_0^2 - z_0^2 - (\zeta' - z')^2]^2 - 4z_0^2(\xi' - z')^2\}^{(n+1)/2}}, \quad (19)$$

where $z \in \tau^+$, $(\xi' - z')^2 = (\xi_1 - z_1)^2 + \cdots + (\xi_n - z_n)^2$. Certain estimates for the integral operator defined by the right hand side of (19) in the space $L_c^\infty(\mathbb{R}^{n+1})$ were given in Sergeev (op. cit.).

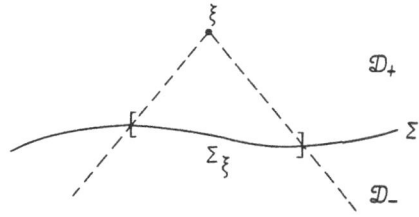

Fig. 19

3.4. Representations for Solutions of the $\bar{\partial}$-equation. The Cauchy-Fantappiè representation (11) can be used to represent smooth solutions of the $\bar{\partial}$-equation

$$\bar{\partial}v = u \tag{20}$$

in a domain $\mathscr{D}$ where u is a smooth $\bar{\partial}$-closed $(0, 1)$-form, with coefficients in $E(\mathscr{D})$. Indeed, the function

$$-(2\pi i)^{n+1}v(z) = \int_{\mathscr{D}} u \wedge \Omega_0 + \int_{\partial\mathscr{D}\times[0,1]} u \wedge \Omega, \quad \zeta \in \mathscr{D} \tag{21}$$

is a solution of (20) singled out by the property that the integral

$$\int_{\partial\mathscr{D}} v\Omega_1 = 0.$$

Using the formula (21) for the different barrier functions defined in Sect. 3.2, 3.3 we obtain different solutions of the $\bar{\partial}$-equation in τ^+. We shall give here some L^∞-estimates of these solutions assuming that the form u is bounded in τ^+ and has a finite support in the sense that it vanishes outside some ball $\{|z| < R\}$ (depending on u).

Consider first the Cauchy-Bochner solution, i.e. the solution of (20) given by formula (21) with the barrier function (13), (14). This solution satisfies the following estimate:

$$|v(z)| \le C\frac{\|u\|_{L^\infty}}{(\sqrt{y^2})^{n-1}} \tag{22}$$

for $z \in \tau^+$, $|z| < r$. This estimate was obtained in Sergeev (1986); the analogous estimate for the generalized unit disc B_m was given in Khenkin-Leiterer (1984), Probl. 4.7. It is not known whether this estimate is sharp. It is known (cf. Sergeev (1991)) that there exists a right hand side u of the $\bar{\partial}$-equation (20) in τ^+ in $\mathbb{C}^{n+1}$ with $n \ge 3$, which is bounded in τ^+ and has finite support, such that the following is true: any solution $v(z)$ of (20) annihilated by the Cauchy-Bochner integral operator (cf. Sect. 1.4) grows like a power $C/y_0^{(n-3)/2}$ for some sequence $z \to x^0 \in \mathbb{R}^{n+1}$. In particular, the rate of the growth is increasing with n.

For the solution given by formula (21) with the barrier function (16) given by the Levi polynomial of τ^+ there is the following estimate

1) conical estimate at points $x^0 \in \mathbb{R}^{n+1}$:

$$|v(z)| \le C\frac{\|u\|_\infty}{y_0^{(n-1)/2}};$$

2) local estimate at points $z^0 = x^0 + iy^0 \in S$:

$$|v(z)| \le C\|u\|_{L^\infty}|\ln|y - y_0||.$$

This estimate is proved in Polyakov (1985) (cf. also Sergeev (1986)). In the last paper some other solutions of (20) with estimates in τ^+ are also considered (e.g. the solution associated with the Jost-Lehmann-Dyson barrier function).

All the solutions of (20) in τ^+ in $\mathbb{C}^{n+1}$ so far constructed are estimated by terms growing like a power (increasing) when z approaches $\mathbb{R}^{n+1}$. In Khenkin-Sergeev (1980) it was conjectured that the $\bar{\partial}$-equation in τ^+ (for $n \geq 2$) has no uniformly bounded estimate, i.e. there exists a bounded smooth $(0, 1)$-form u in τ^+ having finite support such that any solution of (20), with the right hand side given by u, grows (like a power?) when z approaches the distinguished boundary. This conjecture is still unproved. It is proved only (cf. Sergeev (1988, 1991)) that the boundary τ^+ contains a Sibony-type compact set, i.e. a polynomially convex compact subset $X \subset \partial\tau^+$ having the following properties:

1) there exists a system of smooth strictly pseudoconvex neighborhoods

$$U_1 \supset U_2 \supset \cdots \supset U_k \supset \cdots \supset X$$

of the compact set X such that

$$X = \bigcap_{k=1}^{\infty} U_k;$$

2) there exists a system of right hand sides u_k of the $\bar{\partial}$-equations

$$\bar{\partial}v_k = u_k \tag{23}$$

in U_k such that u_k is a smooth $\bar{\partial}$-closed $(0, 1)$-form with finite support in $\mathbb{C}^{n+1}$ for $k = 1, 2, \ldots$ and the norms of u_k are uniformly bounded

$$\|u_k\|_{L^\infty(U_k)} \leq C < \infty, \quad k = 1, 2, \ldots,$$

where C does not depend on k;

3) for any system v_k of solutions of (23) we have

$$\|v_k\|_X \to \infty \quad \text{as} \quad k \to \infty,$$

i.e. the sequence of norms of v_k is unbounded. For a further discussion and the proof of this assertion, see Sergeev (op. cit.).

§4. Functions with Nonnegative Imaginary Part

4.1. Properties of Functions with Nonnegative Imaginary Part in Tube Cones.
Denote by $H_+(T^C)$ the *class of all holomorphic functions with nonnegative imaginary part in T^C*. Functions of this class satisfy the estimate (Vladimirov (1969b, 1979))

$$|f(z)| \leq M\frac{1 + |z|^2}{|y|}, \quad z \in T^{C'}$$

for any cone $C' \Subset C$ where the constant M depends on C'. In particular, $H_+(T^C) \subset H(C)$. The boundary value of the imaginary part $\mu(x) = \operatorname{Im} f(x)$ is a nonnegative measure of slow growth. The Poisson integral of this measure satisfies the estimate (Vladimirov (1978a))

$$P_C[d\mu](z) \leq \operatorname{Im} f(z), \quad z \in T^C.$$

We note that $\mu(x)$ is the boundary value of Im $f(z)$ not only in the space $\mathcal{S}'$ but also in the following "weak" sense

$$\int_{\mathbb{R}^m} \text{Im } f(x + iy^0) P_C(x, y) \varphi(x) \, dx \to \int_{\mathbb{R}^m} P_C(x, y) \varphi(x) \, d\mu(x)$$

as $y^0 \to 0$, $y^0 \in C' \Subset C$ for any bounded continuous function φ on $\mathbb{R}^m$.

For functions $f \in H_+(T^C)$ we have the following uniqueness theorem (Vladimirov (1979)): if $\mu(x) = \text{Im } f(x) = 0$ then $f(z) = (a, z) + b$ where $a \in C^*$, Im $b = 0$ (here C^* is the dual cone of C, cf. Chap. 1, Sect. 5.1).

Denote by $h(\text{Im } f; y)$ the *growth indicator* of a function $f \in H_+(T^C)$, i.e.

$$h(\text{Im } f; y) = \lim_{t \to \infty} \frac{\text{Im } f(z^0 + ity)}{t}, \quad z^0 \in T^C, y \in C.$$

This limit does not depend on z^0 and defines a nonnegative concave function homogeneous of degree 1 in C, moreover

$$h(\text{Im } f; y) \le \text{Im } f(z^0 + iy)$$

for $z^0 \in T^C$, $y \in C$ (cf. Vladimirov (op. cit.)).

4.2. Integral Representation. The following theorem is an extension to $H_+(T^C)$ of the well-known (in the one-dimensional case) *integral representation of Herglotz-Nevanlinna*.

Theorem 12 (Vladimirov (1978a, 1979)). *The following conditions for a function $f \in H_+(T^C)$, $\mu(x) = \text{Im } f(x)$, are equivalent:*
(i) *the Poisson integral $P_C[d\mu]$ is pluriharmonic in T^C;*
(ii) *the function Im $f(z)$ is represented by the Poisson formula*

$$\text{Im } f(z) = P_C[d\mu](z) + (a, y), \quad z \in T^C \tag{24}$$

for some $a \in C^*$;
(iii) *for all $z^0 \in T^C$ (and under the assumption that the cone C is regular, cf. Sect. 1.5) the Schwarz representation holds*

$$f(z) = i \int_{\mathbb{R}^m} \mathcal{S}_C(z - t, z^0 - t) \, d\mu(t) + (a, z) + b, \quad z \in T^C \tag{25}$$

where $b = b(z^0) = \text{Re } f(z^0) - (a, x^0)$. Moreover, (a, y) is the best linear minorant of the indicator h (Im $f; y$) in the cone C.

In connection with Theorem 12 we may pose the following question: when is the Poisson transform $P_C[d\mu]$ for $\mu(x) = \text{Im } f(x)$, $f \in H_+(T^C)$ pluriharmonic in T^C? The explicit answer to this question is obtained in the two cases: for $C = \mathbb{R}^m_+$ and for $C = V^+ = V^+(3)$.

In the case of the octant $C = \mathbb{R}^m_+$ the Poisson integral $P_{\mathbb{R}^m_+}[d\mu]$ is always pluriharmonic in $T^{\mathbb{R}^m_+}$ so the other assertions of the Theorem 12 are also true in this case Korányi-Pukansky (1963), Vladimirov (1969b, 1979)). The vector a

from the assertion (iii) of the Theorem is defined in this case by the equations

$$a_j = \lim_{y_j \to \infty} \frac{\operatorname{Im} f(iy)}{y_j}, \quad j = 1, \ldots, m.$$

In Vladimirov (1969b) also the following criterion of pluriharmonicity for an arbitrary nonnegative measure μ is proved: such a measure μ is the imaginary part of the boundary value of a function $f \in H_+(T^{\mathbb{R}^m})$ if and only if the following two conditions are satisfied:

$$\int_{\mathbb{R}^m} \frac{d\mu(x)}{(1 + x_1^2) \cdots (1 + x_m^2)} < \infty,$$

$$\int_{\mathbb{R}^m} \left(\frac{x_1 - i}{x_1 + i}\right)^{\alpha_1} \cdots \left(\frac{x_m - i}{x_m + i}\right)^{\alpha_m} \frac{d\mu(x)}{(1 + x_1^2) \cdots (1 + x_m^2)} = 0$$

for any $\alpha = (\alpha_1, \ldots, \alpha_m) \in \mathbb{Z}^m$ such that $\alpha \notin \mathbb{Z}_+^m \cup \mathbb{Z}_-^m$.

For the future tube $\tau^+ = \tau^+(3)$ the assertions of Theorem 12 are satisfied for functions $f \in H_+(\tau^+)$ whose indicators have the following properties (Vladimirov (1974)).

$$h(\operatorname{Im} f; y) = h_0(y) + (a, y), \quad h_0(y) \geq 0, \quad a \in \bar{V}^+, \quad y \in V^+,$$

$$\lim_{|y'| \to 1 - 0} \int_{|s|=1} h_0(1, s|y'|) \, ds = 0, \quad y' = (y_1, y_2, y_3).$$

In this case the formulas (24), (25) take the form

$$\operatorname{Im} f(z) = \left(\frac{2}{\pi}\right)^3 \int_{\mathbb{R}^4} \frac{(y^2)^2 \, d\mu(t)}{|(z - t)^2|^4} + (a, y), \quad z \in \tau^+,$$

$$f(z) = \frac{i}{\pi^3} [(z + i)^2]^2 \int_{\mathbb{R}^4} \frac{d\mu(t)}{[(z - t)^2(t + i)^2]^2} - i \left(\frac{2}{\pi}\right)^3 \int_{\mathbb{R}^4} \frac{d\mu(t)}{|(t + i)^2|^4} + (a, z) + b$$

where $a \in \bar{V}^+$, $b = \operatorname{Re} f(i)$, $i = (i, 0, 0, 0)$, (a, y) is the best linear minorant of the indicator $h(\operatorname{Im} f; y)$ in the cone V^+.

Using the criterion for pluriharmonicity of the Poisson integrals of distributions in the generalized unit disc B_2 (Vladimirov (op. cit.)), we obtain the following condition for pluriharmonicity of a nonnegative measure on $\mathbb{R}^4$: a measure μ is the imaginary part of the boundary value of a function $f \in H_+(\tau^+)$ if and only if the following conditions are satisfied

$$\int_{\mathbb{R}^4} \frac{d\mu(x)}{|(x + i)^2|^4} < \infty,$$

$$\int_{\mathbb{R}^4} \Delta_{q_1 q_2}^{jl}[X(x)] \frac{d\mu(x)}{|(x + i)^2|^4} = 0, \quad 2j = 2, 3, \ldots, l = -1, -2, \ldots, -2j + 1;$$

$$-j \leq q_1, q_2 \leq j$$

where $\Delta_{q_1 q_2}^{jl}(X)$ for $l = 0, \pm 1, \ldots, 2j = 0, 1, \ldots; -j \leq q_1, q_2 \leq j$ are spherical functions on the group $U(2)$ and $x \to X(x)$ is the mapping $\mathbb{R}^4 \to U(2)$ given by formula (3) from Chap. 1, Sect. 2.1.

The description of holomorphic functions with nonnegative imaginary part in bounded symmetric domains was obtained in Aizenberg-Dautov (1976).

4.3. Tauberian Theorems. Vladimirov (1976) gave a multidimensional generalization of the *Hardy-Littlewood Tauberian Theorem* for measures. These results were then extended to the case of temperate distributions having support in a proper convex cone in Drozhzhinov-Vladimirov-Zavialov (1984), Drozhzhinov (1982), Drozhzhinov-Zavialov (1979, 1985). We shall present here one of the results of Drozhzhinov-Zavialov (1985). Let Γ be a proper convex cone in $\mathbb{R}^m$ and V_k, $k = 0, 1, \ldots,$ a semigroup of linear nondegenerate transformations of $\mathbb{R}^m$ preserving the cone Γ. We can take, for example, for V_k the dilatation of the cone Γ with ratio equal to k. We say that a function g on Γ has an asymptotic with respect to the semigroup $\{V_k\}$ and a nonnegative function $\rho(k)$ depending on k iff

$$\lim_{k \to \infty} \frac{g(V_k \xi)}{\rho(k)} = \beta(\xi) \neq 0 \tag{26}$$

for any $\xi \in \Gamma$. If $g \in \mathscr{S}'(\Gamma)$ and the limit (26) exists in the space $\mathscr{S}'(\Gamma)$ we shall say that the function g has a *quasiasymptotic* with respect to the semigroup $\{V_k\}$ and the function ρ. An equivalent formulation asserts that some primitive function $g^{(-N)}(\xi)$ of g has a "usual" asymptotic with respect to the semigroup $\{V_k\}$ and the function $(\det V_k)^N \cdot \rho(k)$. If a function g has a quasiasymptotic then the functions ρ and β cannot be arbitrary. The function ρ is necessarily *automodel* (or *regularly varying*) which means that for any $a > 0$ the limit of $\rho(ak)/\rho(k)$ exists as $k \to \infty$ (uniformly on compacta in $\mathbb{R}_+$) equal to a^γ. The function β is necessarily homogeneous of degree γ with respect to V_k, i.e. $\beta(V_k \xi) = k^\gamma \beta(\xi)$.

We say that a function $f \in \mathcal{O}(T^C)$ is *slowly growing with respect to a semigroup* $\{U_k\}$ (leaving the cone C invariant) *and a function* ρ if it satisfies the estimate

$$(\det U_k) \frac{|f(U_k z)|}{\rho(k)} \leq M \frac{(1 + |z|)^a}{[\varDelta(y)]^b}, \quad z \in T^C \tag{27}$$

for $k > k_0$ and some positive constants a, b, M (here $\varDelta(y)$ is the distance from y to the boundary of the cone C).

Theorem 13. *Let $f \in H(C)$ be the Laplace transform of a function $g \in \mathscr{S}'(C^*)$. The function g has a quasiasymptotic with respect to a semigroup $\{V_k\}$ preserving the cone $\mathbb{C}^*$ and a function $\rho(k)$ if and only if the function f is slowly growing with respect to the semigroup $\{U_k\}$, $U_k = (V_k^*)^{-1}$ and the function ρ and*

$$(\det U_k) \frac{f(U_k z)}{\rho(k)} \to \alpha(z), \quad z \in T^C \tag{28}$$

as $k \to \infty$ uniformly on compacta in T^C. Moreover, α is the Laplace transform of β.

The sufficient condition can be considerably weakened. Namely, it is sufficient to require that the function $f(z)$ satisfy (27) only along one "imaginary" direction, i.e. for $z = x + i\delta e$ where e is some fixed vector of C, $0 < \delta \le 1$. For the limit (28) it is sufficient to require its existence for points $z = x^0 + iy$ where x^0 is fixed and y belongs to an open subset of C. There are more detailed versions of Theorem 13 for some particular cases (especially, for functions from $H_+(T^C)$), cf. Drozhzhinov-Vladimirov-Zavialov (1984), Drozhzhinov (1982), Drozhzhinov-Zavialov (1979, 1985). Using this theorem, extensions of the Fatou and Lindelöf Theorems for functions in $H^\infty(T^C)$ (cf. Chap. 2, Sect. 3.2) and in $H_+(T^C)$ (cf. Drozhzhinov (op. cit.)) have been proved.

4.4. Linear Passive Systems. The results given in Sect. 4.1–4.3 can be applied to the theory of *linear passive systems*, i.e. systems of linear convolution equations

$$Z * u = f$$

where $Z(\xi)$ is a real $N \times N$-matrix function with components from $\mathscr{D}'(\mathbb{R}^m)$ which is passive with respect to a proper open convex cone Γ. The last assertion by definition means that

$$\mathrm{Re} \int_{-\Gamma} \langle Z * \varphi, \varphi \rangle \, d\xi \ge 0$$

for any vector-function $\varphi \in [\mathscr{D}(\mathbb{R}^m)]^N$.

Theorem 14 (Vladimirov (1969d, 1972)). *A matrix Z is passive with respect to a cone Γ if and only if its Laplace transform $L[Z]$ is a nonnegatively-real holomorphic matrix function in T^C with $C = \mathrm{int}\ \Gamma^*$, i.e. the function $L[Z]$ is holomorphic and has nonnegative real part in T^C with $L[Z](z) = \overline{L[Z](-\bar{z})}$ for any $z \in T^C$.*

Using this theorem, we can apply the results formulated above to passive systems. We note that the equations of Dirac, Maxwell and many other equations of mathematical physics are passive systems.

Bibliographical Notes

The Cauchy-Bochner representation considered in the first section of this chapter was constructed by Bochner (1944) for tube cones and classical domains. For Sect. 1.1, 1.3, 1.5 and related results cf. Vladimirov (1979). The Cauchy-Bochner integral for classical domains was studied in Hua (1958). For a discussion of the properties of the Cauchy-Bochner integral in L^p spaces cf. Stein (1971). Uniform estimates for the Cauchy-Bochner representation in the future tube are obtained in Jöricke (1983), Sergeev (1986). The general properties of the Poisson integral given in Sect. 2.1 and at the beginning of Sect. 2.3 were considered in Vladimirov (1979), Stein-Weiss (1971); the Poisson integral for classical domains was studied in Hua (1958). The second part of Sect. 2.3 which deals with admissible limits is of introductory character; a detailed exposition and review of earlier results is contained in Koranyi (1972). Section 2.4 is based on the results of Rudin (1969). For further results related to Sect. 2.5 cf. Berline-Vergne (1981), Lasalle (1984a, b). The derivation of the Cauchy-Bochner and Jost-Lehmann-Dyson integral representations from the general Cauchy-Fantappiè representation was proposed in Bros-Itzykson-Pham (1966), Lu (1965); we note also that basically the idea of the derivation of the Cauchy-Bochner representation from the Cauchy-Fantappiè representation is contained in formula (1) of Sect. 1.1. For a detailed exposition and proof of the estimates of Sect. 3.2–3.4 cf. Sergeev (1986). For a further development of the results of Sect. 4.1, 4.2 cf. Vladimirov (1979). The assertions of Sect. 4.3 obtained in Drozhzhinov-Vladimirov-Zavialov (1984), Drozhzhinov-Zavialov (1979, 1985) can also be extended and improved. For a detailed exposition of the properties of linear passive systems (Sect. 4.4) cf. Vladimirov (1979, 1969d, 1972).

References*

Aizenberg, L.A., Dautov, Sh.A. (1976): Holomorphic functions of several complex variables with nonnegative real part. Traces of holomorphic and plurisubharmonic functions on the Shilov boundary. Mat. Sb., Nov. Ser. 99, No. 3, 342–355. Engl. transl.: Math. USSR, Sb. 28, 301–313 (1978), Zbl.341.32002

Aizenberg, L.A., Yuzhakov, A.P. (1979): Integral Representations and Residues in Multidimensional Complex Analysis. Novosibirsk: Nauka. 335 pp. Engl. Transl.: Transl. Math. Monogr. Vol. 58, Providence, 283 pp. (1983), Zbl.445.32002

Aleksandrov, A.B. (1983): On the boundary values of functions holomorphic in the ball. Dokl. Akad. Nauk SSSR 271, No. 4, 777–779. Engl. transl.: Sov. Math., Dokl. 28, 134–137 (1983), Zbl.543.32002

Aleksandrov, A.B. (1984): Inner functions on compact spaces. Funkts. Anal. Prilozh. 18, No. 2, 1–13. Engl. transl.: Funct. Anal. Appl. 18, 87–98 (1984), Zbl.574.32006

Araki, H. (1963): A generalization of Borchers' theorem. Helv. Phys. Acta 36, No. 1, 132–139, Zbl.112,432

Atiyah, M.F. (1979): Geometry of Yang-Mills Fields. Pisa: Scuola Normale Superiore. 99 pp., Zbl.435.58001

Bekolle, D. (1984): Le dual de l'espace des fonctions holomorphes intégrables dans des domaines de Siegel. Ann. Inst. Fourier 34, No. 3, 125–154, Zbl.513.32032

Bell, S. (1982): Proper holomorphic mappings between circular domains. Comment. Math. Helv. 57, No. 3, 532–538, Zbl.511.32013

Bell, S. (1985): Proper holomorphic correspondences between circular domains. Math. Ann. 270, No. 3, 393–400, Zbl.554.32019

---

* For the convenience of the reader, references to reviews in Zentralblatt für Mathematik (Zbl.), compiled using the MATH database, have, as far as possible, been included in this bibliography.

Berline, N., Vergne, M. (1981): Equations de Hua et noyau de Poisson. Lect. Notes Math. *880*, Berlin, Heidelberg, New York: Springer-Verlag, 1–51, Zbl.521.32024

Beurling, A. (1972): Analytic continuation across a linear boundary. Acta Math. *128*, No. 3, 153–182, Zbl.235.30003

Bochner, S. (1944): Group invariance of Cauchy's formula in several variables. Ann. Math., II, Ser. *45*, No. 4, 686–707, Zbl.60,243

Bogolubov, N.N., Vladimirov, V.S. (1958): A theorem on analytic continuation of generalized functions. Nauchn. Dokl. Vyssh. Shkoly, Fiz.-Mat. Nauki 1958, No. 3, 26–35 (Russian), Zbl.116,85

Bogolubov, N.N., Vladimirov, V.S. (1971): Representation of n-point functions. Tr. Mat. Inst. Steklova *112*, 5–21. Engl. transl.: Proc. Steklov Inst. Math. *112*, 1–18 (1973), Zbl.254.32015

Bogolubov, N.N., Medvedev, B.V., Polyvanov, M.K. (1958): Problems of the Theory of Dispersion Relations. Moscow: Fizmatgiz. 203 pp. (Russian), Zbl.83,435

Bony, J.M. (1976): Propagation des singularités différentiables pour une classe d'opérateurs différentiels à coefficients analytiques. Astérisque, 34–35, 43–91, Zbl.344.35075

Borchers, H.J. (1961): Über die Vollständigkeit lorentzinvarianter Felder in einer zeitartigen Röhre. Nuovo Cimento *19*, No. 4, 787–793, Zbl.111,432

Bros, J. (1977): Analytic completion and decomposability properties in tuboid domains. Publ. Res. Inst. Math. Sci. *12*, Suppl., 19–37, Zbl.372.32002

Bros, J., Epstein, H., Glaser, V. (1967): On the connection between analyticity and Lorentz covariance of Wightman functions. Commun. Math. Phys. *6*, No. 1, 77–100, Zbl.155,323

Bros, J., Iagolnitzer, D. (1975): Tuboides et structure analytique des distributions. Sémin. Goulaouic-Lions-Schwartz, 1974–1975, Exposé *16*, 19 pp., Zbl.333.46028

Bros, J., Iagolnitzer, D. (1976): Tuboides dans $\mathbb{C}^n$ et géneralisation d'un théorème de Cartan et Grauert. Ann. Inst. Fourier *26*, No. 3, 49–72, Zbl.336.32003

Bros, J., Itzykson, C., Pham, F. (1966): Représentations intégrales de fonctions analytiques et formule de Jost-Lehmann-Dyson. Ann. Inst. Henri Poincaré, New. Ser., Sect. A *5*, No. 1, 1–35, Zbl.163,225

Bros, J., Messiah, A., Stora, R. (1961): A problem of analytic completion related to the Jost-Lehmann-Dyson formula. J. Math. Phys. *2*, No. 4, 639–651, Zbl.131,441

Burns, D., Stout, E.L. (1976): Extending functions from submanifolds of the boundary. Duke Math. J. *43*, No. 5, 391–404, Zbl.328.32013

Cartan, E. (1935): Sur les domaines bornés homogènes de l'espace de n variables complexes. Abh. Math. Semin. Univ. Hamb. *11*, No. 1–2, 116–162, Zbl.11,123

Chern, S.S. (1956): Complex Manifolds. Chicago: Univ. of Chicago, 181 pp., Zbl.88,378

Chirka, E.M. (1973): Theorems of Lindelöf and Fatou in $\mathbb{C}^n$. Mat. Sb., Nov. Ser. *92*, No. 4, 622–644. Engl. transl.: Math. USSR, Sb. *21*, 619–639 (1975), Zbl.297.32001

Chirka, E.M., Khenkin, G.M. (1975): Boundary properties of holomorphic functions of several complex variables. Itogi Nauki Tekh., Ser. Sovrem. Probl. Math. *4*, 13–142. Engl. transl.: J. Sov. Math. *5*, 612–687 (1976), Zbl.375.32005

Dadok, J., Yang, P. (1985): Automorphisms of tube domains and spherical hypersurfaces. Am. J. Math. *107*, No. 4, 999–1013, Zbl.586.32035

Danilov, L.I. (1985): On regularity of proper cones in R^n. Sib. Math. Zh. *26*, No. 2, 198–201, Zbl.581.32002

Drozhzhinov, Yu.N. (1982): Multidimensional Tauberian theorems for holomorphic functions of a bounded argument and quasiasymptotics of passive systems. Mat. Sb., Nov. Ser. *117*, No. 1, 44–59. Engl. transl.: Math. USSR, Sb. *45*, 45–61 (1983), Zbl.497.32001

Drozhzhinov, Yu.N., Vladimirov, V.S., Zavialov, B.I. (1984): Tauberian type theorems for generalized functions. Tr. Mat. Inst. Steklova *163*, 42–48. Engl. transl.: Proc. Steklov Inst. Math. *163*, 53–60 (1985), Zbl.568.46032

Drozhzhinov, Yu.N., Zavialov, B.I. (1979): Tauberian theorems for generalized functions supported in cones. Mat. Sb., Nov. Ser. *108*, No. 1, 78–90. Engl. transl.: Math. USSR, Sb. *36*, 75–86 (1980), Zbl.405.46033

Drozhzhinov, Yu.N., Zavialov, B.I. (1982): On a multidimensional analog of Lindelöf's theorem. Dokl. Akad. Nauk SSSR 262, No. 2, 269–270 (Russian)

Drozhzhinov, Yu.N., Zavialov, B.I. (1985): Multidimensional Tauberian comparison theorems for generalized functions in cones. Mat. Sb., Nov. Ser. 126, No. 4, 515–542. Engl. transl.: Math. USSR, Sb. 54, 499–524 (1986), Zbl.585.46033

Dyson, F.J., (1958): Integral representations of causal commutators. Phys. Rev., II. Ser. 110, No. 6, 1460–1464, Zbl.85,434

Fefferman, C. (1970): Inequalities for strongly singular convolution operators. Acta Math. 124, No. 1–2, 9–36, Zbl.188,426

Freeman, M. (1977): Real submanifolds with degenerate Levi form. Proc. Symp. Pure Math. 30, part 1, 141–147, Zbl.354.53010

Freeman, M. (1977): Local biholomorphic straightening of real submanifolds. Ann. Math., II. Ser. 106, No. 2, 319–352, Zbl.372.32005

Fuks, B.A. (1963): Special Topics from the Theory of Analytic Functions of Several Complex Variables. Moscow: Fizmatgiz. 427 pp. English transl.: Transl. Math. Monogr., Vol. 14, Providence (1965), Zbl.146,308

Furstenberg, H. (1963): A Poisson formula for semi-simple Lie groups. Ann. Math., II. Ser. 77, No. 2, 335–386, Zbl.192,127

Gindikin, S.G. (1964): Analysis on homogeneous domains. Usp. Mat. Nauk 19, No. 4, 3–92. Engl. transl.: Russ. Math. Surv. 19, No. 4, 1–89 (1964), Zbl.144,81

Hahn, K.T. (1972): Properties of holomorphic functions of bounded characteristic on star-shaped circular domains. J. Reine Angew. Math. 254, 33–40, Zbl.246.32002

Heinzner, P., Sergeev, A.G. (1991): The extended matrix disk is a domain of holomorphy. Izv. Akad. Nauk SSSR, Ser. Mat. 55, No. 3, 647–657. Engl. transl.: Math. USSR, Izv. 38, 637–645 (1992)

Helgason, S. (1978): Differential Geometry, Lie Groups and Symmetric Spaces. New York: Academic Press, 628 pp., Zbl.451.53038

Hill, C.D., Kazlow, M. (1977): Function theory on tube manifolds. Proc. Symp. Pure Math. 30, part 1, 153–156, Zbl.383.32002

Hörmander, L. (1971): Fourier integral operators, I. Acta Math. 127, 79–183, Zbl.212,466

Hua, L.-K. (1958): Harmonic Analysis of Functions of Several Complex Variables in the Classical Domains. Peking: Science Press. Engl. transl.: Providence: Am. Math. Soc. 1979. 186 pp., Zbl.90,96

Hua, L.-K., Look, K.H. (1983): Theory of harmonic functions in classical domains. Hua L.K., Selected Papers. Berlin, Heidelberg, New York: Springer-Verlag, 743–806, Zbl.518.01022.

Johnson, K.D. (1978): Differential equations and the Bergman-Shilov boundary on the Siegel upper half-plane. Ark. Mat. 16, No. 1, 95–108, Zbl.395.22013

Johnson, K.D. (1984a): Generalized Hua-operators and parabolic subgroups. The cases of $SL(n, \mathbb{C})$ and $SL(n, \mathbb{R})$. Trans. Am. Math. Soc. 281, No. 1, 417–429, Zbl.531.22010

Johnson, K.D. (1984b): Generalized Hua-operators and parabolic subgroups. Ann. Math., II. Ser. 120, No. 3, 477–496, Zbl.576.22016

Johnson, K.D., Koranyi, A. (1980): The Hua operators on bounded symmetric domains of tube type. Ann. Math., II. Ser. 111, No. 3, 589–608, Zbl.468.32007

Jöricke, B. (1982): The two constants theorem for functions of several complex variables. Math. Nachr. 107, 17–52 (Russian), Zbl.526.32003

Jöricke, B. (1983): Continuity of the Cauchy projection in Hölder norms for classical domains. Math. Nachr. 112, 227–244 (Russian), Zbl.579.32006

Jost, R., Lehmann, H. (1957): Integral Darstellung kausaler Kommutatoren. Nuovo Cimento, X. Ser. 5, No. 7, 1598–1610, Zbl.77,424

Kataoka, K. (1981): On the theory of Radon transformations of hyperfunctions. J. Fac. Sci., Univ. Tokyo, Sect. 1A 28, No. 2, 331–413, Zbl.576.32008

Khenkin, G.M., (= Henkin, G.M.), Leiterer, J. (1984): Theory of Functions on Complex Manifolds. Berlin: Akademie-Verlag, 226 pp., Zbl.573.32001

Khenkin, G.M., (= Henkin, G.M.), Sergeev, A.G. (1980): Uniform estimates of solutions of the $\bar{\partial}$-equation in pseudoconvex polyhedra. Mat. Sb., Nov. Ser. 112, No. 4, 522–567. Engl. transl.: Math. USSR, Sb. 40, 469–507 (1981), Zbl.452.32012

Khenkin, G.M., (= Henkin, G.M.), Tumanov, A.E. (1976): Interpolation submanifolds of pseudoconvex manifolds, Math. Program. Rel. Probl., Cent. Ehkon. Mat. Inst. Akad. Nauk SSSR, Mosk. 1974, 74–86. Engl. transl.: Transl., II. Ser., Am. Math. Soc. *115*, 59–69 (1980), Zbl.455.32009

Khenkin, G.M., (= Henkin, G.M.), Tumanov, A.E. (1983): Local characterization of holomorphic automorphisms of Siegel domains. Funkts. Anal. Prilozh. *17*, No. 4, 49–61. Engl. transl.: Funct. Anal. Appl. *17*, 285–294 (1983), Zbl.572.32018

Khurumov, Yu.V. (1983): Lindelöf's theorem in $\mathbb{C}^n$. Dokl. Akad. Nauk SSSR *273*, No. 6, 1325–1328. Engl. transl.: Sov. Math. Dokl. *28*, 806–809 (1983), Zbl.567.32002

Knapp, A.V., Williamson, R.E. (1971): Poisson integrals and semisimple groups. J. Anal. Math. *24*, 53–76, Zbl.247.31002

Koecher, M. (1957): Positivitätsbereiche im $\mathbb{R}^n$. Am. J. Math. *79*, No. 3, 575–596, Zbl.78,12

Komatsu, H. (1972): A local version of Bochner's tube theorem. J. Fac. Sci., Univ. Tokyo, Sect. IA *19*, No. 2, 201–214, Zbl.239.32012

Koranyi, A. (1965): The Poisson integral for generalized half-planes and bounded symmetric domains. Ann. Math., II. Ser. *82*, No. 2, 332–350, Zbl.138,66

Koranyi, A. (1969): Boundary behavior of Poisson integrals on symmetric spaces. Trans. Am. Math. Soc. *140*, 393–409, Zbl.179,151

Koranyi, A. (1972): Harmonic functions on symmetric spaces. Symmetric Spaces. Pure Appl. Math. *8*, 379–412, Zbl.291.43016

Koranyi, A. (1976): Poisson integrals and boundary components of symmetric spaces. Invent. Math. *34*, No. 1, 19–35, Zbl.328.22017

Koranyi, A. (1979): Compactifications of symmetric spaces and harmonic functions. Lect. Notes Math. *739*, Berlin, Heidelberg, New York: Springer-Verlag, 341–366, Zbl.425.43014

Koranyi, A., Malliavin, P. (1975): Poisson formula and compound diffusion associated to an overdetermined elliptic system on the Siegel half-plane of rank two. Acta Math. *134*, No. 1–2, 185–209, Zbl.318.60066

Koranyi, A., Pukanszky, L. (1963): Holomorphic functions with positive real part on polycylinders. Trans. Am. Math. Soc. *108*, 449–456, Zbl.136,71

Koranyi, A., Vagi, S. (1976): Isometries of H^p spaces of bounded symmetric domains. Can. J. Math. *28*, No. 2, 334–340, Zbl.344.32025

Koranyi, A., Vagi, S. (1979): Rational inner functions on bounded symmetric domains. Trans. Am. Math. Soc. *254*, 179–193, Zbl.439.32006

Koranyi, A., Wolf, J.A. (1965): Realisation of hermitian symmetric spaces as generalized half-planes. Ann. Math., II. Ser. *81*, No. 2, 265–288, Zbl.137,274

Labonde, J.-M. (1985): Ensembles pics pour $A(U^n)$. C.R. Acad. Sci., Paris, Sér. I *301*, No. 13, 671–673, Zbl.584.32031

Lassalle, M. (1984a): Les équations de Hua d'un domaine borné symétrique du type tube. Invent. Math. *77*, No. 1, 129–161, Zbl.582.32042

Lassalle, M. (1984b): Sur la valeur au bord du noyau de Poisson d'un domaine borné symétrique. Math. Ann. *268*, No. 4, 417–423, Zbl.579.32052

Leray, J. (1959): Le calcul différentiel et intégral sur une variété analytique complexe (Problème de Cauchy, III). Bull. Soc. Math. Fr. *87*, 81–180, Zbl.199,412

Lindahl, L.-A. (1972): Fatou's theorem for symmetric spaces. Ark. Mat. *10*, No. 1, 33–47, Zbl.246.22010

Löw, E. (1984): Inner functions and boundary values in $H^\infty(\Omega)$ and $A(\Omega)$ in smoothly bounded pseudoconvex domains. Math. Z. *185*, No. 2, 191–210, Zbl.508.32005

Lu Qui-keng (1965): On the Cauchy-Fantappiè formula. Acta Math. Sin. *16*, No. 3, 344–363, Zbl.173,329

Manin, Yu.I. (1984): Gauge Fields and Complex Geometry. Moscow: Nauka. Engl. transl.: Berlin, Heidelberg, New York: Springer-Verlag, 1988, Zbl.576.53002

Martineau, A. (1964): Distributions et valeurs au bord des fonctions holomorphes. Proc. Intern. Summer Course on the Theory of Distributions. Lisboa, 195–326

Martineau, A. (1970): Le "edge of the wedge theorem" en théorie des hyperfonctions de Sato. Proc. Int. Conf. Funct. Anal Rel. Topics, Tokyo 1969, 95–106, Zbl.193,415

Matsushima, Y. (1972): On tube domains. In: Symmetric Spaces, Pure Appl. Math. 8, 255–270, Zbl.232.32001

Mitchell, J., Sampson, G. (1982): Singular integrals on bounded symmetric domains in $\mathbb{C}^n$. J. Math. Anal Appl. 90, No. 2, 371–380, Zbl.506.32017

Monopoles (1985): (Collection of papers translated into Russian). Ed.: Monastyrski, M.I., Sergeev, A.G.; Moscow: Mir

Morimoto, M. (1973): Edge of the wedge theorem and hyperfunction. Lect. Notes Math. 287. Berlin, Heidelberg, New York: Springer-Verlag, 41–81, Zbl.262.46043

Morimoto, M. (1980): Analytic functionals on the Lie sphere. Tokyo J. Math. 3, No. 1, 1–35, Zbl.454.46032

Murakami, S. (1972): On automorphisms of Siegel domains, Lect. Notes Math. 286. Berlin, Heidelberg, New York: Springer-Verlag, 95 pp., Zbl.245.32001

Nagel, A. (1976): Smooth zero sets and interpolation sets for some algebras of holomorphic functions on strictly pseudoconvex domains. Duke Math. J. 43, No. 2, 323–348, Zbl.343.32016

Penrose, R. (1980): The complex geometry of the natural world. Proc. Int. Congr. Math., Helsinki 1978, Vol. 1, 189–194, Zbl.425.53033

Penrose, R. (1968): The structure of space-time. Battelle Rencontres, 1967, Lect. Math. Phys., 121–235, Zbl.174,559

Pflug, P. (1974): Über polynomiale Funktionen auf Holomorphiegebieten. Math. Z. 139, No. 2, 133–139, Zbl.278.32011

Pinchuk, S.I. (1992): CR transformations of real manifolds in $\mathbb{C}^n$. Indiana Univ. Math. J. 41, No. 1, 1–16

Piatetski-Shapiro, I.I. (1961): Geometry of Classical Domains, and Theory of Automorphic Functions. Moscow: GOSIZDAT. 191 pp. French transl.: Paris: Dunod 1966, Zbl.137,275, Zbl.142,51

Polyakov, P.L. (1985): Solution of the $\bar{\partial}$-equation with estimates in tube domains. Usp. Mat. Nauk 40, No. 1, 213–214. Engl. transl.: Russ. Math. Surv. 40, No. 1, 235–236 (1985), Zbl.593.32013

Rigoli, M., Travaglini, G. (1983): A remark on mappings of bounded symmetric domains into balls. Lect. Notes Math. 992. Berlin, Heidelberg, New York: Springer-Verlag, 387–390, Zbl.552.32020

Rossi, H., Vergne, M. (1976): Equations de Cauchy-Riemann tangentielles associées à un domaine de Siegel. Ann. Sci. Ec. Norm. Supér., IV Sér. 9, No. 1, 31–80, Zbl.398.32018

Rothaus, O.S. (1960): Domains of positivity. Abh. Math. Semin. Univ. Hamb. 24, 189–235, Zbl.96,279

Rudin, W. (1969): Function Theory in Polydiscs. New York: Benjamin, 188 pp. Zbl.177,341

Rudin, W. (1971a): Harmonic analysis in polydiscs. Actes Congr. Int. Math., Nice, 1970, t. 2, 489–493, Zbl.233.32002

Rudin, W. (1971b): Lectures on the Edge-of-the-Wedge theorem. Reg. Conf. Ser. Math. 6. Providence: Am. Math. Soc., 30 pp., Zbl.214,90

Rudin, W. (1978): Peak-interpolation sets of class C^1. Pac. J. Math. 75, No. 1, 267–279, Zbl.383.32007

Rudin, W. (1980): Function Theory in the Unit Ball of $\mathbb{C}^n$. New York, Berlin, Heidelberg: Springer-Verlag, 436 pp., Zbl.495.32001

Saerens, R. (1984): Interpolation manifolds. Ann. Sc. Norm. Super. Pisa, Cl. Sci., IV. Ser. 11, No. 2, 177–211, Zbl.579.32023

Sato, M. (1959–1960): Theory of hyperfunctions, I, II. J. Fac. Sci. Univ. Tokyo, Sect. I 8, No. 1, 139–193, Zbl.87,314. No. 2, 387–437, Zbl.97,314

Sato, M., Kawai, T., Kashiwara, M. (1973): Microfunctions and pseudodifferential equations. Lect. Notes Math. 287. Berlin, Heidelberg, New York: Springer-Verlag, 265–529, Zbl.277.46039

Schapira, P. (1970): Théorie des Hyperfonctions. Lect. Notes Math. 126, Berlin, Heidelberg, New York: Springer-Verlag, 157 pp., Zbl.192,473

Schmid, W. (1969): Die Randwerte holomorpher Funktionen auf Hermiteschen symmetrischen Räumen. Invent. Math. 9, No. 1, 61–80, Zbl.219.32013

Sergeev, A.G. (1975): Multiplicative theory of hyperfunctions. Usp. Mat. Nauk 30, No. 1, 257–258 (Russian), Zbl.379,46034

Sergeev, A.G. (1978): Multidimensional factorization problem. Proc. All-Union Conf. on PDEs. Moscow, 440–441 (Russian)

Sergeev, A.G. (1983): Complex geometry and integral representations in the future tube in $\mathbb{C}^3$. Teor. Mat. Fiz. 54, No. 1, 99–110. Engl. transl.: Theor. Math. Phys. 54, 62–70 (1983), Zbl.529.32001

Sergeev, A.G. (1985): Estimates for the Bergman projector in the future tube. Multidim. Compl. Anal., Krasnojarsk, SOAN SSSR, 161–172 (Russian)

Sergeev, A.G. (1986): Complex geometry and integral representations in the future tube. Izv. Akad. Nauk SSSR, Ser. Mat. 50, No. 6, 1241–1275, 1343–1344. Engl. transl.: Math. USSSR, Izv. 29, 597–628 (1987), Zbl.618.32001

Sergeev, A.G. (1988): On the behavior of solutions of the $\bar{\partial}$-equation on the boundary of the future tube. Dokl. Akad. Nauk SSSR 298, No. 2, 294–298. Engl. transl.: Sov. Math., Dok. 37, No. 1, 83–87 (1988), Zbl.691.32007

Sergeev, A.G. (1989): On complex analysis in the future tube. Compl. Anal. Appl. 87, Sofia, 450–459

Sergeev, A.G. (1991): On complex analysis in tube cones. Proc. Sympos. Pure Math. 52, Part 1, 173–190

Sergeev, A.G., Vladimirov, V.S. (1985): A compactification of Minkowski space and complex analysis in the future tube. Ann. Pol. Math. 46, No. 1, 439–454 (Russian), Zbl.602.32010

Siciak, J. (1969): Separately analytic functions and envelopes of holomorphy of some lower dimensional subsets of $\mathbb{C}^n$. Ann. Pol. Math. 22, No. 2, 145–171, Zbl.185,152

Siegel, C.L. (1949): Analytic Functions of Several Complex Variables. Princeton: Inst. Adv. Stud., 200 pp., Zbl.36,50

Stein, E.M. (1971): Some problems in harmonic analysis suggested by symmetric spaces and semisimple groups. Actes Congr. Int. Math., Nice, 1970, 1, 173–189, Zbl.252.43022

Stein, E.M. (1972): Boundary behaviour of holomorphic functions of several complex variables. Princeton: Princeton Univ. Press, 72 pp., Zbl.242.32005

Stein, E.M. (1983): Boundary behavior of harmonic functions on symmetric spaces: maximal estimates for Poisson integrals. Invent. Math. 74, No. 1, 63–83, Zbl.522.43007

Stein, E.M., Weiss, G. (1971): Introduction to Fourier Analysis on Euclidean Spaces. Princeton: Princeton Univ. Press, 297 pp., Zbl.232.42007

Stein, E.M., Weiss, G., Weiss, M. (1964): H^p classes of holomorphic functions in tube domains. Proc. Natl. Acad. Sci. USA 52, No. 4, 1035–1039, Zbl.126,94

Stein, E.M., Weiss, N.J. (1969): On the convergence of Poisson integrals. Trans. Am. Math. Soc. 140, 34–54, Zbl.182.108

Stoll, M. (1974): Integral formulae for pluriharmonic functions on bounded symmetric domains. Duke Math. J. 41, No. 2, 393–404, Zbl.287.32020

Stoll, M. (1976a): Harmonic majorants for plurisubharmonic functions on bounded symmetric domains with applications to the spaces H_φ and N_*. J. Reine Angew. Math. 282, 80–87, Zbl.318.32014

Stoll, M. (1976b): The space N_* of holomorphic functions on bounded symmetric domains. Ann. Pol. Math. 32, No. 1, 95–110, Zbl.284.32013

Stoll, M. (1985): Mean growth and Fourier coefficients of some classes of holomorphic functions on bounded symmetric domains. Ann. Pol. Math. 45, No. 2, 161–183, Zbl.579.32022

Stout, E.L. (1981): Interpolation manifolds. In: Recent Developments in Several Complex Variables. Ann. Math. Stud. 100, 373–391, Zbl.486.32010

Tillmann, H.G. (1961a): Distributionen als Randverteilungen analytischer Funktionen. Math. Z. 76, No. 1, 5–21, Zbl.97,96

Tillmann, H.G. (1961b): Darstellung der Schwartzschen Distributionen durch analytische Funktionen. Math. Z. 77, No. 2, 106–124, Zbl.99,97

Twistors and Gauge Fields (1983): (Collection of papers translated into Russian). Ed.: Zharinov, V.V.; Moscow: Mir, 364 pp.

Uhlmann, A. (1963): The closure of Minkowski space. Acta Phys. Pol. 24, No. 2, 295–296, Zbl.115,423

Upmeier, H. (1984): Toeplitz C^*-algebras on bounded symmetric domains. Ann. Math., II. Ser. 119, No. 3, 549–576, Zbl.549.46031

Vinberg, E.B. (1963): The theory of convex homogeneous cones. Tr. Mosk. Mat. O.-va *12*, 303–358. Engl. transl.: Trans. Mosc. Math. Soc. 1963, 340–403 (1965), Zbl.138,433

Vladimirov, V.S. (1960): On constructing the envelope of holomorphy for domains of special type. Dokl. Akad. Nauk SSSR *134*, No. 2, 251–254. Engl. transl.: Sov. Math., Dokl. *1*, 1039–1042 (1960), Zbl.118,303

Vladimirov, V.S. (1961): On constructing the envelope of holomorphy for domains of special type and their applications. Tr. Mat. Inst. Steklova *6*, 101–144. Engl. transl.: Am. Math. Soc., Transl., II. Ser. *48*, 107–150 (1965), Zbl.118,303

Vladimirov, V.S. (1964): Methods of the Theory of Functions of Several Complex Variables. Moscow: Nauka, 410 pp. French transl.: Les fonctions de plusieurs variables complexes et leur application. Paris: Dunod 1967, 338 pp., Zbl.125,319

Vladimirov, V.S. (1965): The problem of linear conjugation of holomorphic functions of several complex variables. Izv. Akad. Nauk SSSR, Ser. Mat. *29*, No. 4, 807–834. Engl. transl.: Am. Math. Soc., Transl., II. Ser. *71*, 203–232 (1968), Zbl.166,337

Vladimirov, V.S. (1969a): A generalization of the Cauchy-Bochner integral representation. Izv. Akad. Nauk SSSR, Ser. Mat. *33*, No. 1, 90–108. Engl. transl.: Math. USSR, Izv. *3*, 87–104 (1969), Zbl.183,87

Vladimirov, V.S. (1969b): Holomorphic functions with nonnegative imaginary part in a tube domain over a cone. Mat. Sb., Nov. Ser. *79*, No. 1, 128–152. Engl. transl.: Math. USSR, Sb. *8*, 125–146 (1969), Zbl.183,87

Vladimirov, V.S. (1969c): Bogolubov's "edge-of-the-wedge" theorem, its development and applications. Problems of Theoretical Physics. Moscow: Nauka, 61–67 (Russian)

Vladimirov, V.S. (1969d): Linear passive systems. Theor. Mat. Fiz. *1*, No. 1, 67–94. Engl. transl.: Theor. Math. Phys.

Vladimirov, V.S. (1971): Analytic functions of several complex variables and axiomatic quantum field theory. Actes Congr. Int. Math. Nice, 1970, t. 3. Paris: Gauthier-Villars, 21–26

Vladimirov, V.S. (1972): Multidimensional linear passive systems. Meh. Splosn. Sredy rodstv. Probl. Anal., 121–134 (Russian), Zbl.263.93019

Vladimirov, V.S. (1974, 1974, 1977): Holomorphic functions with positive imaginary part in the future tube. Mat. Sb. *93*, No. 1, 3–17; II, *94*, No. 4, 499–515: IV, *104*, No. 3, 341–370. Engl. transl.: Math. USSR, Sb. *22*, 1–16; II, *23*, 467–482; III, *27*, 263–268; IV, *33*, 301–325 (1975–1977); Zbl.291.32003; Zbl.313.32001; Zbl.319.32004; Zbl.383.32001

Vladimirov, V.S. (1976): Multidimensional generalization of a Tauberian theorem of Hardy-Littlewood. Izv. Akad. Nauk SSSR, Ser. Mat. *40*, No. 5, 1084–1101. Engl. transl.: Math. USSR, Izv. *10*, 1031–1048 (1978), Zbl.359.40001

Vladimirov, V.S. (1978a): Holomorphic functions with nonnegative imaginary part in tube domains over cones. Dokl. Akad. Nauk SSSR *239*, No. 1, 26–29. Engl. transl.: Sov. Math., Dokl. *19*, 254–258 (1978), Zbl.448.32003

Vladimirov, V.S. (1978b): Growth estimates for boundary values of positive pluriharmonic functions in a tube domain over a proper cone. Complex Analysis and its Applications. Moscow: Nauka, 137–148 (Russian), Zbl.447.31006

Vladimirov, V.S. (1979): Generalized Functions in Mathematical Physics, 2nd. ed. Moscow: Nauka, 319 pp. Engl. transl.: Moscow: Mir, 362 pp, Zbl.515.46034

Vladimirov, V.S. (1982): Several complex variables in mathematical physics. Lect. Notes Math. *919*. Berlin, Heidelberg, New York: Springer-Verlag, 358–386, Zbl.493.32014

Vladimirov, V.S. (1983): Functions of several complex variables in mathematical physics. In: Problems of Mathematics and Mechanics. Novosibirsk: Nauka, 15–32. Engl. transl.: Transl., II. Ser., Am. Math. Soc. *136*, 19–33 (1987), Zbl.625.32001

Vladimirov, V.S. (1984): Blaschke products in the generalized unit disc and complete orthonormal systems in the future tube. Tr. Mat. Inst. Steklova *166*, 44–51. Engl. transl.: Proc. Steklov Inst. Math. *166*, 45–52 (1986), Zbl.574.32007

Vladimirov, V.S., Zharinov, V.V. (1970): On a representation of Jost-Lehmann-Dyson type. Teor. Mat. Fiz. *3*, 305–319. Engl. transl.: Theor. Math. Phys. *3*, No. 3, 525–536 (1970), Zbl.201,582

Weiss, N.J. (1972): Fatou's theorem for symmetric spaces. In: Symmetric Spaces, Pure Appl. Math. 8, 413–441, Zbl.242.43011

Wolf, J.A. (1972): Fine structure of Hermitian symmetric spaces. In: Symmetric Spaces, Pure Appl. Math. 8, 271–357, Zbl.257.32014

Yang, P. (1984): Geometry of tube domains. Proc. Symp. Pure Math. 41, 277–283, Zbl.579.32050

Yang, P.C. (1982): Automorphisms of tube domains. Am. J. Math. 104, No. 5, 1005–1024, Zbl.514.32018

Zakharyuta, V.P. (1976): Separately analytic functions, generalization of the Hartogs theorem and envelopes of holomorphy. Mat. Sb., Nov. Ser. 101, No. 1, 57–76. Engl. transl.: Math. USSR, Sb. 30, 51–67 (1978), Zbl.357.32002

Zharinov, V.V. (1980): On an exact squence of modules and Bogolubov's "edge-of-the-wedge" theorem. Dokl. Akad. Nauk SSSR 251, No. 1, 19–22. Engl. transl.: Sov. Math., Dokl. 21, 357–360 (1980), Zbl.478.46046

Zharinov, V.V. (1983): Distributive lattices and their applications in complex analysis. Tr. Mat. Inst. Steklova 162, 3–80. Engl. transl.: Proc. Steklov Inst. Math. 162, Providence, 79 pp. (1985), Zbl.574.32017

Zygmund, A. (1958): Trigonometric Series, Vol. 1, 2. 2nd. ed. Cambridge: Cambridge University Press, Zbl.85,56, Zbl.11,17

Author Index

Subject Index

Encyclopaedia of Mathematical Sciences
Editor-in-Chief: R. V. Gamkrelidze

Springer

Encyclopaedia of Mathematical Sciences
Editor-in-Chief: R. V. Gamkrelidze

Algebra

Volume 11: **A. I. Kostrikin, I. R. Shafarevich** (Eds.)

Algebra I
Basic Notions of Algebra
1990. V, 258 pp. 45 figs.
ISBN 3-540-17006-5

Volume 18: **A. I. Kostrikin, I. R. Shafarevich** (Eds.)

Algebra II
Noncommutative Rings. Identities
1991. VII, 234 pp. 10 figs.
ISBN 3-540-18177-6

Volume 37: **A. I. Kostrikin, I. R. Shafarevich** (Eds.)

Algebra IV
Infinite Groups. Linear Groups
1993. Approx. 220 pp. 9 figs.
ISBN 3-540-53372-9

Volume 38: **A. I. Kostrikin, I. R. Shafarevich** (Eds.)

Algebra V
Homological Algebra
1994. Approx. 230 pp.
ISBN 3-540-53373-7

Volume 57: **A. I. Kostrikin, I. R. Shafarevich** (Eds.)

Algebra VI
Combinatorial and Asymptotic Methods of Algebra
1994. Approx. 260 pp. 4 figs.
ISBN 3-540-54699-5

Volume 58: **A. N. Parshin, I. R. Shafarevich** (Eds.)

Algebra VII
Combinatorial Group Theory. Applications to Geometry
1993. Approx. 260 pp. 38 figs.
ISBN 3-540-54700-2

Volume 73: **A. I. Kostrikin, I. R. Shafarevich** (Eds.)

Algebra VIII
Representations of Finite-Dimensional Algebras
1992. VI, 177 pp. 98 figs.
ISBN 3-540-53732-5

The manufacturer's authorised representative in the EU is Springer
Nature Customer Service Centre GmbH, Europaplatz 3, 69115 Heidelberg,
Germany. If you have any concerns regarding our products, please
contact ProductSafety@springernature.com

Printed and bound by CPI Group (UK) Ltd, Croydon, CR0 4YY
24/04/2026
02096348-0005